AF252141

Foundations of Decision-Making Agents

Logic, Probability and Modality

Foundations of Decision-Making Agents

Logic, Probability and Modality

Subrata Das

World Scientific

Imperial College Press

Published by

World Scientific Publishing Co. Pte. Ltd.

5 Toh Tuck Link, Singapore 596224

USA office: 27 Warren Street, Suite 401-402, Hackensack, NJ 07601

UK office: 57 Shelton Street, Covent Garden, London WC2H 9HE

British Library Cataloguing-in-Publication Data
A catalogue record for this book is available from the British Library.

Cover image by Sebastien Das

FOUNDATIONS OF DECISION-MAKING AGENTS
Logic, Probability and Modality

ISBN-13 978-981-277-983-0
ISBN-10 981-277-983-3

Editor: Tjan Kwang Wei

Printed in Singapore by World Scientific Printers (S) Pte Ltd

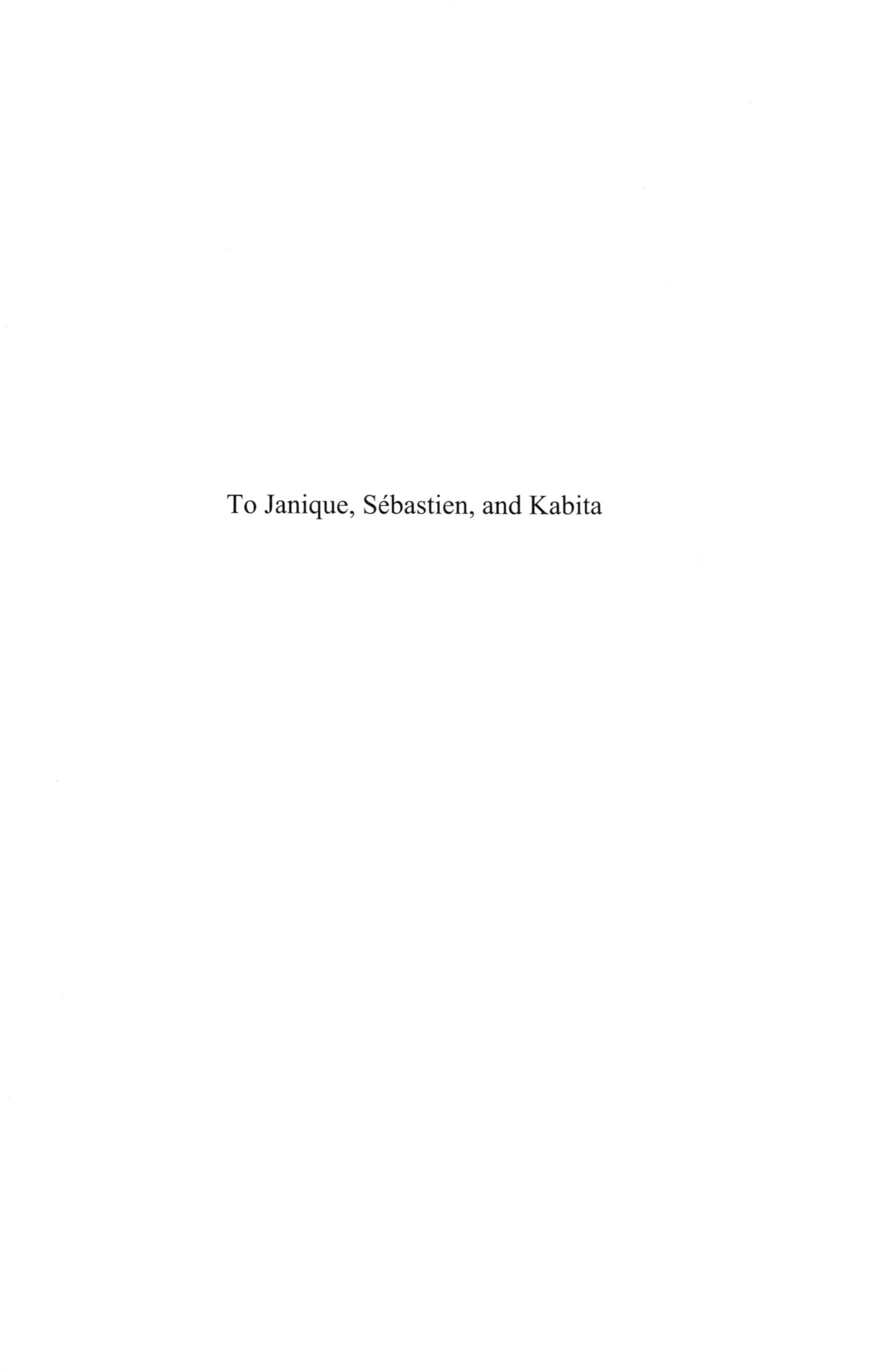

To Janique, Sébastien, and Kabita

Preface

Artificial Intelligence (AI) systems simulating human behavior are often called *intelligent agents*. By definition, these intelligent agents exhibit some form of human-like intelligence. For example, intelligent agents can learn from the environment, make plans and decisions, react to the environment with appropriate actions, express emotions, or revise beliefs. Intelligent agents typically represent human cognitive states using underlying beliefs and knowledge modeled in a knowledge representation language. We use the term *epistemic state* to refer to an actual or possible cognitive state that drives human behavior at a given point in time; the accurate determination (or estimation) of these epistemic states is crucial to an agent's ability to correctly simulate human behavior.

This book provides three fundamental and generic approaches (logical, probabilistic, and modal) to representing and reasoning with agent epistemic states, specifically in the context of decision making. In addition, the book introduces a formal integration of these three approaches into a single unified approach we call $\mathcal{P}3$ (Propositional, Probabilistic, and Possible World), that combines the advantages of the other approaches. Each of these approaches can be applied to create the foundation for intelligent, decision-making software agents. Each approach is symbolic in nature (unlike the sub-symbolic neural network approach, for example), yielding agent "thought processes" that are sufficiently transparent to provide users with understandable explanations of agent "reasoning." The classical logic and probability theory thrust of our approach naturally excludes from our consideration other potential symbolic approaches to decision-making such as fuzzy logics. The "symbolic argumentation" approach ($\mathcal{P}3$) to decision making combines logic and probability, and therefore offers several advantages over the traditional approach to decision-making based on simple rule-based expert systems or expected utility theory.

The generic logical, probabilistic, and modal reasoning techniques that we discuss in the book, such as logical deduction and evidence propagation, are applied to various decision-making problems that an agent will encounter in various circumstances. However, the book does not focus on particular applications of these reasoning techniques to such problems as planning, learning, or belief revision. Neither does the book focus on communication and collaboration within agent societies. Such applications require extensive treatments in themselves and so are beyond the scope of the book; our purpose here is to provide the formal foundation necessary to support decision-making.

This book is divided into four parts by the four reasoning approaches we develop. The first part focuses on logic-based modeling of an epistemic state as a set of propositions expressed by sentences from classical propositional and first-order logics. The second part takes a probabilistic approach to modeling epistemic states. In this approach, states are represented by a probability measure defined over some space of events, which are represented as random variables. The third part focuses on extending logic-based modeling with modalities to yield various systems of modal logics, including epistemic logics. These logics are based on mental constructs that represent information and are suitable for reasoning with possible human cognitive states. Finally, the fourth part describes the combination of these three rather disparate approaches into the single integrated $\mathcal{P}3$ approach within an argumentation framework especially appropriate for simulating human decision making.

The purpose of this book is not to carry out an elaborate philosophical discussion on epistemic state modeling, but rather to provide readers with various practical ways of modeling and reasoning with agent epistemic states used in decision making. Therefore, the resolution-based logic programming paradigm is introduced under logic-based modeling; Bayesian belief networks are introduced as a way of modeling and reasoning with epistemic states where beliefs are represented by probability measures defined over some state space; and modal resolution schemes are introduced under modal logic-based modeling and reasoning with actual and possible worlds.

Frameworks specific to decision making, such as the logic-based production rule or belief-network-based influence diagrams that incorporate the notions of action and utility, are also detailed in the book. The use of logic-based reasoning (especially resolution theorem proving) and probabilistic reasoning (especially Bayesian belief networks) is widespread when implementing agents capable of making decisions, generating plans, revising beliefs, learning, and so on. But the book goes a step further by bridging the gap between these two approaches, augmenting them with informational mental constructs, such as beliefs and knowledge, and then producing a coherent approach that culminates in the penultimate chapter of the book in the form of a symbolic argumentation approach for weighing the pros and cons of decision-making options.

In the book I have tried to balance discussions on theory and practice. As for theory, I have stated results and developed proofs wherever necessary to provide sufficient theoretical foundations for the concepts and procedures introduced. For example, if I have introduced a specific resolution-based theorem-proving procedure, then I also provide the associated soundness and completeness of the procedure with respect to the stated semantics. On the practical side, I have provided detailed algorithms that can be encoded in computer programs in a straightforward manner to incorporate aspects of human behavior into intelligent decision-making agents. For example, an implementation of the evidence propagation algorithms for belief networks and influence diagrams can be incorporated into agents that make decisions within a probabilistic framework.

The decision-making problem is present in every industry; finance, medicine, and defense are a few of the big ones. Computer professionals and researchers in the decision science community within commerce and industry will find the book most useful for building intelligent and practical decision-aiding agents or, more generally, for embedding intelligence in various systems based on sound logical and probabilistic reasoning; the material is especially suited for building safety-critical decision aiding applications based on well-founded theories. This book may also be used as an AI textbook on logic and probability for undergraduate and graduate courses, and as a reference book for researchers in universities.

The chapters in the book and their dependencies are shown in the figure below. Readers are advised not to skip Chapter 1, which includes an informal overview of the rest of the book. A basic understanding of this chapter will prevent the reader from becoming lost later on. The beginner should not be dismayed by the use of unfamiliar terminology in this first chapter, as these terms will be explained in later chapters.

Readers who have been exposed to the concepts of basic mathematics and probability may omit Chapter 2, which provides a background on mathematical preliminaries. Chapters 3 and 4, on classical logics and logic programming, should be read in sequence. Chapter 6, on Bayesian belief networks, can be read independently of Chapters 3 and 4, but Chapter 8, on modal logics, requires an understanding of classical logic as presented in Chapter 3. Chapters 3, 4, 6, and 8 provide the semantic foundations of our three problem modeling approaches, namely, propositional, probabilistic, and possible world. Because production rules are a special form of rules in logic programming with embedded degrees of uncertainty I recommend to reading at least the logic programming section in Chapter 4 and the background material on probability theory before reading

Chapter 5. Chapter 7, on influence diagrams, requires understanding the basics of Bayesian belief networks as it extends this understanding with the concepts of action and utility. The final chapter (Chapter 9) on symbolic argumentation requires an overall understanding of the contents of Chapters 3-8. Ultimately, three different decision-making frameworks are provided, in Chapters 5, 7, and 9.

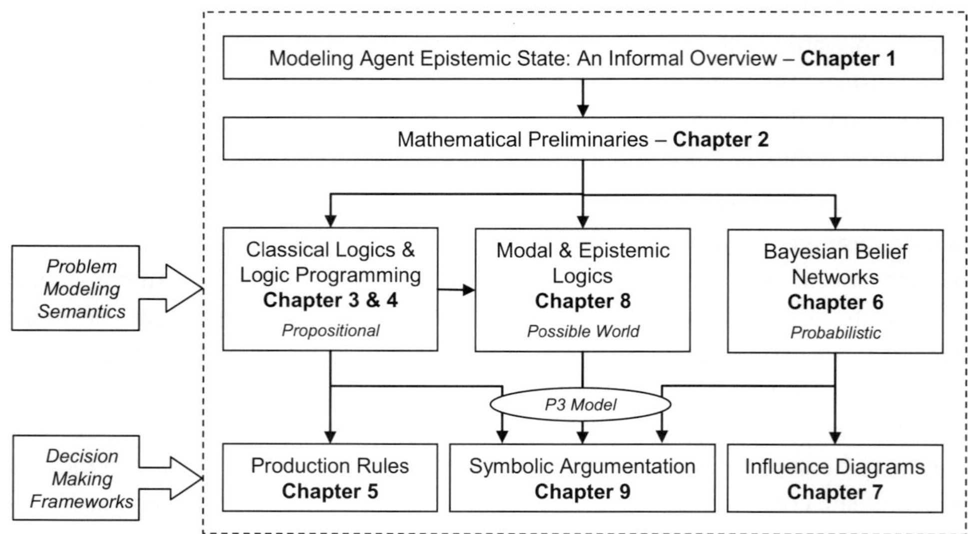

Figure 1-1: Chapters and their dependencies

There is another suggested way to follow this book for those readers who, with some exposure to mathematical logic and the theory of probability, are looking for some well-founded computational decision support technologies to build deployable systems. They can skip detailed theoretical materials as presented in some chapters (e.g. 3, 4, and 8), and focus more on the computational aspects (e.g. chapter 5, 6, 7, and 9).

My sincere thanks go to my wife Janique, my son Sébastien, and my daughter Kabita, for their love, patience and inspiration throughout the preparation of this book (My children's perception of the book are very different from each other. Sébastien's view is very philosophical, which he manifested in his drawing for the title page of a sage from the Museum of Fine Arts in Boston. However, Kabita is more pragmatic, expressed herself through a dog's decision-making dilemma as depicted below.) Special thanks go to Karen DeSimone for her careful proofreading of the manuscript that enormously enhanced readability and enriched the content of the book. Thanks are due to my colleagues, Dave

Lawless and Josh Introne, with whom I have had many technical discussions on various aspects of this book. Thanks also to all of my colleagues at Charles River Analytics, especially Alex Feinman, Paul Gonsalves, Partha Kanjilal, Rich Reposa, and Greg Zacharias, and Avi Pfeffer at Harvard University, for their part in creating a stimulating intellectual environment which inspired me to write this book. Thanks to the World Scientific/Imperial College Press, especially Tjan Kwang Wei, for their help in producing the book from the beginning. Finally, I thank my parents, brothers, sisters and other family members back in one of several thousands of small villages in India for patiently accepting my absence and showing their encouragement and support through their many phone calls.

This book is based on the work of many researchers, and I express my gratitude to those who have directly or indirectly contributed so much to the fields of deductive logic, probability theory, and artificial intelligence. I have made my best effort to make this book informative, readable, and free from mistakes, and I would welcome any criticism or suggestion for improvement.

Subrata Das

Cambridge, MA

March 2007

Dog's dilemma

Contents

Chapter 1

Modeling Agent Epistemic States: An Informal Overview

Recent advances in intelligent agent research (AGENTS, 1997-2001; AAMAS, 2002-2006) have culminated in various agent-based applications that autonomously perform a range of tasks on behalf of human operators. Just to name a few, the kinds of tasks these applications perform include information filtering and retrieval, situation assessment and decision making, and interface personalization. Each of these tasks requires some form of human-like intelligence that must be simulated and embedded within the implemented agent-based application. The concept of *epistemic states* is often used to represent an actual or a possible cognitive state that drives the human-like behavior of an agent. Established traditional artificial intelligence (AI) research in the areas of knowledge representation and inferencing has been transitioned to represent and intelligently reason about the various mental constructs of an agent, including beliefs, desires, goals, intentions, and knowledge (Cohen and Levesque, 1990; Rao and Georgeff, 1991), simulating its human-like cognitive states.

1.1 Models of Agent Epistemic States

The most commonly used models of agent epistemic states are the propositional, probabilistic, and possible world models (Gärdenfors, 1988):

1. An epistemic state in a *propositional model* is represented as a set of propositions that the agent accepts in the epistemic state. These propositions are expressed by sentences in an object language.

2. An epistemic state in a *probabilistic model* is represented by a probability measure defined over the states of some collection of random variables. This probability measure provides the agent's degree of belief about each state of each random variable in the epistemic state.

3.   An epistemic state in a *possible world model* is represented by a set of possible worlds that includes the agent's actual state or world, along with any worlds compatible with the agent's knowledge and beliefs. Each world consists of those propositions that the agent accepts in its epistemological worldview.

Irrespective of the model chosen to represent an agent's epistemic state, the state model must always be built via a substantial knowledge engineering effort; this effort may occasionally be aided by machine learning techniques (Mitchell, 1997) to automatically extract knowledge from data. Part of the knowledge engineering effort is the knowledge acquisition process that a scientist or engineer goes through when extracting knowledge from a domain or subject matter expert (such as a medical doctor), to be used by the agent for problem solving in a particular domain (for example, diagnosing diseases). The knowledge gained from this process can then be represented and stored in a computer, using the syntax of logic, frames, or graphical representations, such as semantic networks and probabilistic belief networks. The focus in this text is on logical syntaxes and graphical belief networks (and their amalgamations) that encode an agent's epistemic state that represents knowledge about an uncertain environment. The agent then uses this encoded knowledge for making decisions under uncertainty, that is, to choose between two or more different options. For example, the agent may need to choose to believe in one possibility among many (for example, a patient has cancer or an ulcer), and thus an agent is revising its own belief, or may need to adopt a plan (or take some action), so that the agent is choosing among many possible options (e.g. surgery, chemotherapy, medication) for action.

A note of caution: When we refer to an agent *making a decision*, we do not really mean that agents *always* make decisions autonomously. In some situations, when not enough evidence is available to choose one option over the others, or in cases when some dilemma occurs, agents may very well present the human decision maker with the viable options along with their accumulated supports, backed by explanations. In these situations, agents autonomously provide "decision aids" to human decision makers. Regardless of their role as a decision maker or aid provider, labeling them with the term "agent" is always justified at least due to their autonomous role of routinely searching through huge volumes of data (e.g. Internet web pages) for relevant evidence (a task for which human performance is usually poor) to be used for the decision making process.

In the process of introducing representation and reasoning within various epistemic models, we will often rely on the example of a game (for example, a sporting event) which is scheduled to occur sometime on the current day. In our example, the decision that an agent will usually be trying to make is to determine the status of the game, (that is, whether the game is "on," "cancelled," or "delayed") while preparing to go to town based on accumulated evidence.

1.2   Propositional Epistemic Model

The language in the propositional epistemic model of an agent is usually governed by classical propositional logic; it is uniquely identified by its particular syntax, set of axioms, and inference rules. In general, logic is a systematic study of valid arguments. Each argument consists of certain propositions, called *premises*, from which another proposition, called the *conclusion*, follows. Consider the following argument by an agent to determine the status of a game:

If the field is wet or there is no transportation,

then the game is cancelled. (Premise)

The field is wet. (Premise)

Therefore, the game is cancelled. (Conclusion)

In the above argument, the first two statements are premises. The first is a *conditional* statement, and the second is *assertional.* The third statement is the conclusion, or *argument.* (The term "therefore" is a sign of argument.) The above argument is valid. (Technically, a valid argument is one in which the conclusion must be true whenever the premises are true.)

Valid arguments are studied independently of the premises from which arguments are drawn. This is achieved by expressing valid arguments in their *logical* or *symbolized form.* The valid argument above is symbolized at an abstract level as:

$$P \vee Q \rightarrow R, P \vDash R$$

where the symbols P, Q, and R stand for the proposition "the field is wet," "no transportation," and "the game is cancelled," respectively. So the first premise of the argument, which is a conditional premise, is symbolized as $P \vee Q \rightarrow R$, where '$\rightarrow$' reads "implies". The second premise is symbolized by just the proposition P. The symbol $\vDash$ is the consequence relationship and R is a consequence of the premises. R is arrived at by the use of another argument

$P \vDash P \vee Q$, and applying an *inference rule* of the logic, called *modus ponens*, on $P \vee Q \rightarrow R$ and $P \vee Q$. The inference rule states that given $X \rightarrow Y$ and X is true then it follows that Y is true, where X and Y are arbitrary sentences of the logic.

Although the above argument is valid, it is not known if the outfield is wet in reality; it has just been assumed for the purpose of argument. The validity of an argument is neither concerned with the actual subject matter nor with the truth or falsehood of the premises and conclusion in reality. Mathematical logic does not study the truth or falsehood of the particular statements in the premises and conclusion, but rather focuses on the process of reasoning, i.e. whether or not the assumed truth of the premises implies the truth of the conclusion via permissible symbol manipulation within the context. This phenomenon, to some extent, stimulates interesting philosophical argument on whether an agent, whose epistemic state is modeled using mathematical logic, is really capable of human-like thinking or is just "blindly" manipulating symbols (Searle, 1984).

Often arguments are advanced without stating all the premises, as is evident from the following argument constructed out of the elements of the previous argument:

The field is wet. (Premise)

Therefore, the game is cancelled. (Conclusion)

Due to the lack of an appropriate premise, this example is not a valid argument in the context of classical propositional logic unless the agent intelligently assumes that "If the field is wet then the game is cancelled" is by default a premise. Consider the following example, which has an incomplete premise:

If the field is wet, then the game is cancelled. (Premise)

The field is wet, or the field sensing equipment is inaccurate. (Premise)

Therefore, the game is cancelled. (Conclusion)

The premise of the above argument does not definitely imply that the game is cancelled, since the field may be dry and the field sensing equipment incorrectly signaled a wet field. Hence it is an invalid argument.

There are various kinds of arguments that cannot be stated by propositional logic without an appropriate combinatorial enumeration over the underlying objects. Consider the following example of a valid argument that requires the syntax of the so-called *first-order* logic:

Every wet field is unsuitable for playing a game.

The field at Eden Garden is wet.

Therefore, Eden Garden is unsuitable for playing a game.

Clearly, the assertional second premise does not occur in the conditional first premise, so that propositional logic fails to produce the desired conclusion. One solution to this shortcoming is to produce a specific instantiation of the first premise for the specific Eden Garden field (a sporting venue) and then apply Modus Ponens. But first the conditional premise of the above argument is symbolized as a general sentence:

$$\forall x(Field(x, Wet) \rightarrow Unsuitable(x))$$

where x is a variable or placeholder for terms representing things such as field names, *Wet* is a constant symbol representing one possible type of field condition, *Field* is a binary relation or predicate symbol that relates a field to its condition, and *Unsuitable* is a unary relation representing a property of a field. The symbols $\forall$ and $\rightarrow$ are read as "for every" (or "for all") and "implies" respectively. The expression $Field(x, Wet)$, which appears on the left , is treated as the *antecedent* of the premise. The expression following the symbol is the *consequent*. Note that there are other ways that the argument can be symbolized depending on the context and inference need. For example, one can define a unary predicate or property *Wet Field* of a field replacing the binary predicate *Field*. (This of course blocks any sort of inference that takes into account the dryness property of the field.)

The second premise of the argument, which is an assertional premise, is symbolized based on the predicate symbol *Field* as follows

$$Field(Eden\,Garden, Wet)$$

A first-order symbolization of the argument now looks like this:

$$\forall x(Field(x, Wet) \rightarrow Unsuitable(x)), Field(Eden\,Garden, Wet) \vdash$$
$$Unsuitable(Eden\,Garden)$$

This symbolization is within the framework of *first-order logic*. An axiomatic deduction or inferencing of the conclusion of the argument appears as follows:

Step 1: $\forall x(\textit{Field}(x,\textit{Wet}) \rightarrow \textit{Unsuitable}(x))$ Given Premise

Step 2: $\textit{Field}(\textit{Eden Garden},\textit{Wet})$ Given Premise

Step 3: $\textit{Field}(\textit{Eden Garden},\textit{Wet}) \rightarrow \textit{Unsuitable}(\textit{Eden Garden})$
                                                    Axiom on Step 1

Step 4: $\textit{Unsuitable}(\textit{Eden Garden})$ Modus Pones on Steps 2 & 3

The premises in steps 1 and 2 are considered *proper axioms*. The above deduction is a *proof* of the conclusion *Unsuitable(Eden Garden)* and therefore is a *theorem* that follows from the first-order system with the two proper axioms. Step 3 is derived from step 1 by particularization of the first premise. If the first premise is true for all fields, then it is also true for the field Eden Garden. Step 4 is arrived at by the application of Modus Ponens. These steps demonstrate the basic *axiomatic* theorem-proving approach.

In contrast to the axiomatic theorem proving approach, *resolution* theorem proving is based on the refutation principle. This principle states that if a set of premises is inconsistent then there must be a refutation. Based on this principle, the negation of what is to be proved via a query is assumed as a premise and a refutation is established if the query is to be derived (this very general technique is formally known as *reduction ad absurdum*, or proof by contradiction) . A query could be as general as "Is there any unsuitable playing field?", or could be more specific as in the case of our example, i.e. "Is Eden Garden unsuitable?" The negation of the conclusion in our example arguments context is "It is not the case that Eden Garden is unsuitable" is considered as a premise. The symbolized form of this premise is:

$$\neg\textit{Unsuitable}(\textit{Eden Garden})$$

where $\neg$ is read as "It is not the case." As this assumption is inconsistent with the other two premises, a refutation will result. To do this, first the symbolized premises are converted to their equivalent *clausal form*. In this example the assertional premises are already in clausal form, while the conditional premise can be converted to clausal form to give:

$$\forall x(\neg\textit{Field}(x,\textit{Wet}) \vee \textit{Unsuitable}(x))$$

where $\vee$ is read as "or." By convention, a sentence in clausal form is normally written by removing $\forall x$-type signs from its front and these are assumed by default. In clausal form, the first premise of the argument can now be read as:

Every field is either not wet or is unsuitable for a game

which has the same meaning (semantically) as the original form of the first premise. Using this equivalent clausal form, the steps of the refutation of the argument are as follows:

Step 1: $\neg Field(x, Wet) \vee Unsuitable(x)$ Premise

Step 2: $Field(Eden\,Garden, Wet)$ Premise

Step 3: $\neg Unsuitable(Eden\,Garden)$ Negation of query –
                                    assumed as premise

Step 4: $Unsuitable(Eden\,Garden)$ Resolution of clause in Steps 1 and 2

Step 5: $\square$ (empty clause)   Resolution of clause in Steps 3 and 4

The complement of the atomic premise $Field(Eden\,Garden, Wet)$ in step 2 and the subexpression $\neg Field(x, Wet)$ of the premise in step 1 can be made equal by a *most general unification* $\{x\,/\,Eden\,Garden\}$, that is, by substituting *Eden Garden* in place of x. The clause $Unsuitable(Eden\,Garden)$ is then obtained by applying the *resolution principle* (Robinson, 1965) to steps 1 and 2. This means applying most general unifiers to both expressions and then merging and canceling the complementary expressions. To complete above derivation, the resolution principle can be applied again to steps 3 and 4. After canceling the complementary atomic expressions, the resultant expression is empty and therefore a refutation of the argument has been arrived at ($\square$ denotes a refutation). This shows that the assumption in step 3 is wrong and therefore its complement is true, that is, Eden Garden is unsuitable for a game.

If-then types of sentences are clauses of a special type that many believe are natural representations of human knowledge. So-called production rules in expert systems are based on the if-then syntax. Many subject matter experts are comfortable expressing their knowledge in this form. In the logic programming community, a set of these rules constitute a program that is guaranteed to be consistent, and most logic programming systems can directly handle such programs without needing any meta-interpreters. Therefore, the use of these rules in building an agent's knowledge is both practical and consistent with established theories.

In building an agent's propositional epistemic state, *logical omniscience* is assumed which means that the agent knows all tautologies and that its knowledge is closed under Modus Ponens. However, omniscience is not realizable in practice since real agents are resource-bounded. Attempts to define knowledge in the presence of bounds include restricting what an agent knows to a set of formulae which is not closed under inference or under all instances of a given axiom.

1.3   Probabilistic Epistemic Model

The knowledge that constitutes an epistemic state of an intelligent software agent will often be imprecise, vague, and uncertain. One major drawback of the propositional approach to modeling epistemic state is that it does not even provide the granularity needed to represent the uncertainty in sentences that we use in everyday life. The truth value of a logical sentence in the propositional approach is either true or false, and there is nothing in between. Therefore no special treatment can be given to a sentence like "If the field is wet then the game is 90% likely to be cancelled." The probabilistic model provides this granularity with probabilities representing uncertainty in sentences.

A *probabilistic model* of epistemic states is represented by a probability measure defined over some space of events represented as random variables. This probability measure provides the agent's degree of belief in each possible world state. A *random variable* is a real-valued function defined over a sample space (that is, the domain of the variable) and its value is determined by the outcome of an experiment, known as an *event*. A *discrete random variable* is a random variable whose domain is finite or denumerable. Probabilistic modeling of an agent's epistemic state involves first identifying the set of random variables in the domain and then developing a joint probability distribution of the identified variables to determine the likelihood of the epistemic state that the agent is in. To illustrate this, consider an example domain to determine the status of a game depending on the field condition and the transport. The random variables in this context are *Game*, *Field*, and *Transport*; they are defined over the sample spaces or states {*on, cancelled, delayed*}, {*dry, wet*}, and {*present, absent*} respectively.

The *probability distribution* of a random variable is a function whose domain contains the values that the random variable can assume, and whose range is a set of values associated with the probabilities of the elements of the domain. The *joint probability distribution* of the three discrete random variables *Game*, *Field*, and *Transport* is a function whose domain is the set of triplets (x, y, z), where x, y, and z are possible values for *Game*, *Field*, and *Transport*, respectively, and whose range is the set of probability values corresponding to the ordered pairs in its domain. Therefore, the following probability measure

$$p(Game = cancelled, Field = wet, Transport = present) = 0.8$$

indicates that 0.8 is the probability that the game is cancelled *and* the field is wet *and* transport is present. The probabilities in a joint probability distribution have to be computed from previous data based on the frequency probability approaches. Given a distribution, you can then compute *conditional probabilities* of interest by applying multiplication and marginalization rules. In the frequency approach, probability is derived from observed or imagined frequency distributions. For example, the conditional probability that the game is on given that the field is wet *and* transport is absent is computed as follows:

$$p(Game = on \mid Field = wet, Transport = present) =$$

$$\frac{p(Game = on, Field = wet, Transport = present)}{\sum_{\substack{x \in \{on, cancelled, \\ delayed\}}} p(Game = x, Field = wet, Transport = present)}$$

In the Bayesian approach, the probability on a particular statement regarding random variable states is not based on any precise computation, but describes an individual's personal judgment (degree of belief) about how likely a particular event is to occur based on experience. The Bayesian approach is more general and expected to provide better results in practice than frequency probabilities alone because it incorporates subjective probability. The Bayesian approach can therefore obtain different probabilities for any particular statement by incorporating prior information from experts. The Bayesian approach is especially useful in situations where no historic data exists to compute prior probabilities. Bayes' rule allows one to manipulate conditional probabilities as follows:

$$p(A \mid B) = \frac{p(B \mid A)p(A)}{p(B)}$$

This rule estimates the probability of A in light of the observation B. Even if B did not happen, one can estimate the probability $p(A \mid B)$ that takes into account the prior probability $p(A)$.

Computing the joint probability distribution for a domain with even a small set of variables can be a daunting task. Our example domain with the variables *Game*, *Field*, and *Transport*, and state space sizes 3, 2, and 2 respectively, requires $3 \times 2 \times 2 - 1$ probabilities for the joint distribution. A domain with just 10 binary variables will require $2^{10} - 1$ probabilities. This problem can be mitigated by building *Bayesian belief networks* based on qualitative information in the domain, such as the following:

The field condition and the transport situation together determine the status of the game.

This is simply qualitative information about some dependency between the variables *Field*, *Transport*, and *Game*. We can add even more qualitative information relating the game status and radio commentary as follows:

There is radio commentary when the game is on.

Note that we have no explicit information stated on any dependency between the newly introduced variable *Commentary* and the two variables *Field* and *Transport*.

A Bayesian belief network is a graphical, probabilistic knowledge representation of a collection of random variables (e.g. *Game*, *Field*, *Transport*, and *Commentary*) describing some domain. Nodes of a belief network denote random variables; links are added between the nodes to denote causal relationships between the variables. The topology encodes the qualitative knowledge about the domain. (These descriptions are usually in the form of "causal relationships" among variables.) Conditional probability tables (CPTs) encode the quantitative details (strengths) of the causal relationships, such as the following:

If the field is dry and transport is present there is a 90% chance that the game is on.

The CPT of a variable without any parents consists of just its *a priori* (or prior) probabilities. The belief network of Figure 1-1 encodes the relationships over the domain consisting of the variables, *Game*, *Field*, *Transport*, and *Commentary*; its topology captures the commonsense about the variable dependencies discussed above. Each variable has a mutually exclusive and exhaustive set of possible *states*. For example, *dry* and *wet* are the states of the variable *Field*.

As shown in the figure, the CPT specifies the probability of each possible value of the child variable conditioned on each possible combination of parent variable values. The probability that the game is on given the field is dry and the transport is present is 0.90, whereas the probability that the game is cancelled given the same two conditions is 0.09. Similarly, the probability of radio commentary given the game is on is 0.95. The prior probability that the field is dry is 0.8. Each CPT column must sum up to 1.0 as the states of a variable are mutually exclusive and exhaustive. Therefore, the probability that the game is delayed given that the field is dry and transport is present is $1.0 - 0.9 - 0.09$, or 0.01.

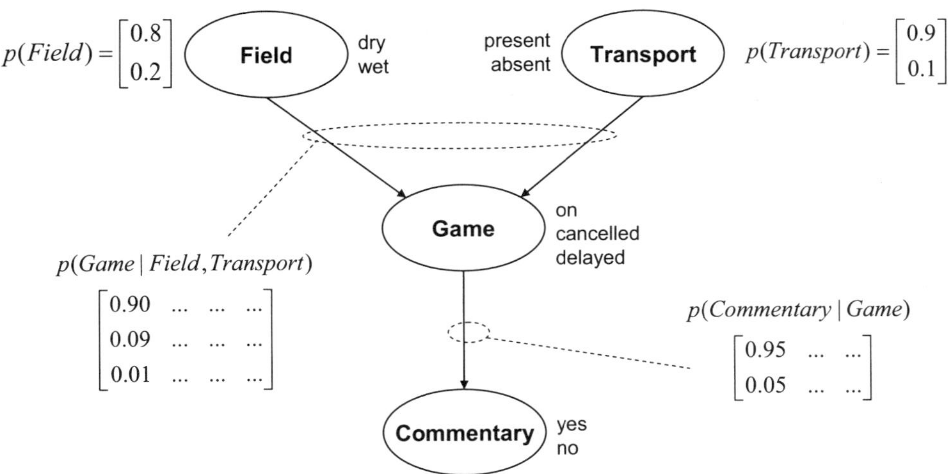

Figure 1-1: Example Bayesian Belief Network

As mentioned above, the structure of a belief network encodes other information as well. Specifically, the lack of links between certain variables represents a lack of direct causal influence, that is, they indicate conditional independence relations. This belief network encodes many independence relations, for example:

$$Commentary \perp \{Field, Transport\} \mid Game$$

which states that *Commentary* is independent of *Field* and *Transport* given *Game*. In other words, once the value of *Game* is known, the values of *Field* and *Transport* add no further information about *Commentary*. Therefore, independence between two nodes is represented by the absence or blocking of links between the two nodes. Whether any dependence between a pair of nodes exists or not is determined by a property called *d-separation*. Conditional independence among variables in a belief network allows factorization of a joint probability distribution, thus eliminating the need for acquiring all the probabilities in a distribution. For example, the joint distribution of the variables in the belief network in Figure 1-1 is factorized as follows:

$$p(Field, Transport, Game, Commentary) =$$

$$p(Field)p(Transport)p(Game \mid Field, Transport)p(Commentary \mid Game)$$

Only 13 ($=1+1+8+3$) probabilities are required to compute the joint probability distribution in this case, as opposed to 23 ($=2 \times 2 \times 3 \times 2 - 1$) when no causal relationship information is given.

When new evidence is posted to a variable (i.e. its state is determined) in a belief network, that variable updates its own state probabilities that constitute its *belief vector* and then sends out messages indicating updated predictive and diagnostic support vectors to its children and parent nodes respectively. The messages are used by the other nodes, which update their own belief vectors, and also propagate their own updated support vectors. The probability distribution of a variable in a belief network at any stage is the *posterior probability* of that variable given all the evidence posted so far. In the case of polytrees, the separation of evidence yields a propagation algorithm (Pearl, 1988) in which update messages need only be passed in one direction between any two nodes after the posting of evidence. See (Jensen, 1996) for a propagation algorithm for the more general case of directed acyclic graphs (DAGs).

It is fairly straightforward to model a decision-making problem in terms of a belief network, where a set of mutually exclusive decision options can be encoded as the states of a random variable representing the decision node within the network. Other attributes from the problem domain are also encoded as random variables connected to the decision node as per their causal relationships. Moreover, belief networks are suitable for synthesizing knowledge at higher levels of abstraction by performing aggregations on low-level evidence. Our experience with various subject matter experts suggests that their thought process during information aggregation is very similar to what the graphical representation of belief networks depicts via their dependencies among random variables.

A belief network is built on the concept of random variables, but a decision making process often involves reasoning explicitly with actions and utilities. *Influence diagrams* are belief networks augmented with decision variables and a utility function to solve decision problems. There are three types of nodes in an influence diagram: *chance nodes* (i.e. belief network nodes), *decision nodes*, and value or *utility nodes*. Using an influence diagram, agents will therefore be able to directly reason about, for example, what loss might be incurred if a decision for proceeding with the game under rainy conditions is made.

1.4 Possible World Epistemic Model

If an agent receives information from a dynamic and uncertain environment, then the agent's understanding about the environment will be based on its own "beliefs," and potential realizations of each of the decision making options will naturally yield a possible world that the agent can transition to. Reasoning within an uncertain environment therefore requires an appropriate underlying knowledge representation language incorporating the concepts of belief and possibility in its semantics, along with a sound inference mechanism. Logical formalisms are especially appealing from the knowledge representation point of view. However, fragments of standard first-order logic in the form of Horn clauses used in most of these formalisms are often inadequate. Researchers have therefore moved to higher-order and non-classical logics. The notion of possibility in classical modal logics and the notion of belief in their epistemic interpretations together provide a useful means for representing an agent's uncertain knowledge. The possibility concept naturally blends with the concept of decision options triggering agents to consider different possible worlds corresponding to those options. The language in the possible world epistemic model is governed by epistemic interpretations of the language of modal logics.

The original purpose behind the revival of modern modal logic by Lewis in his book *A Survey of Symbolic Logic* (Lewis, 1918) was to address issues related to *material implication* of Principia Mathematica (Whitehead and Russell, 1925–1927), by developing the concept of *strict implication*. The language of modal propositional logics extends the language of classical propositional logic with two modal operators $\Box$ (necessary) and $\Diamond$ (possibility). For example, when the necessary operator is applied to the sentence "If the field is wet then the game is cancelled" (symbolized as $\Box(P \rightarrow Q)$, where P stand for "the field is wet" and Q stands for "the game is cancelled"), the resultant sentence is read as "It is necessary that if the field is wet then the game is cancelled." The interpretation of $\Box(P \rightarrow Q)$ in terms of possible worlds is that in every world that the agent considers possible, if the field is wet then the game is cancelled. Similarly, the interpretation of $\Diamond(P \rightarrow Q)$ is that in *at least one* of worlds that the agent considers possible, if the field is wet then the game is cancelled.

Modal epistemic logics (Hintikka, 1962) are instances of modal logics constructed by interpreting necessity and possibility in a suitable manner for handling mentalistic constructs for knowledge and belief. If an agent's epistemic state contains $\langle bel \rangle(P \rightarrow Q)$, where the $\langle bel \rangle$ represents the modal operator for belief analogous to the operator $\Box$, then in every world that the agent considers possible (or *accessible* from the current world), the agent believes that if the field is wet, then the game is cancelled. It may so happen that in one of the worlds that the agent considers possible the game is played despite the field being wet. In this case, the agent does not believe in $P \rightarrow Q$.

Various modal systems are built by adding axioms to the base modal system $\mathcal{K}$. For example, the system $\mathcal{T}$ is obtained from $\mathcal{K}$ by adding the axiom $\Box F \to F$, which states that whatever is necessary is also true in the current world. Various other modal systems, including the well-known $\mathcal{S4}$, $\mathcal{B}$, and $\mathcal{S5}$ systems are also obtained by adding appropriate axioms to $\mathcal{K}$. The possible world interpretation of the modal operators as explained above is exactly what Kripke (1963) proposed as semantics for various modal systems. The semantics of each modal system imposes some restrictions on the accessibility among possible worlds. For example, the accessibility relation is reflexive in the case of system $\mathcal{T}$. Axiomatic deductions in modal propositional logics are carried out in a manner similar to standard propositional logic, i.e. using a set of inference rules and axioms.

The syntax of modal first-order logics can be obtained simply from the classical first-order logic by the introduction of modalities into the language. Then we can produce first-order modal logics analogous to various propositional modal systems such as $\mathcal{K}$, $\mathcal{D}$, $\mathcal{T}$, $\mathcal{S4}$, $\mathcal{B}$, and $\mathcal{S5}$. In principle, the semantics of these first-order modal systems could simply be obtained by extending the possible world semantics for modal propositional systems. However, lifting up modal propositional systems to their counterparts in modal first-order logics raises some interesting issues regarding the existence of objects in various possible worlds and interactions between the modal operators and quantifiers. For example, in a possible world there may be more or fewer objects than in the current world. Adding the Barcan Formula (BF) (Barcan, 1946) can usually axiomatize first-order modal logics that are adequate for applications with a fixed or non-growing domain:

$$\text{BF:} \quad \forall x \Box F[x] \to \Box \forall x F[x]$$

where $F[x]$ means F has zero or more free occurrences of the variable x. The above axiom means that if everything that exists necessarily possesses a certain property F then it is necessarily the case that everything possesses F. But in the cumulative domain case there might be some additional objects in accessible worlds that may not possess the property F.

In general, there is no equivalent to the standard, first-order logic, clausal representation of an arbitrary modal formula where each component in the clause is one of the atomic modal formulae P, $\neg P$, $\Diamond P$, and $\Box P$. But a modal formula can be transformed into an equivalent formula in first-order like world-path syntax, which can then be transformed into a set of clauses in the same way as in the first-order logic. For example, the formula $\Diamond \forall x P(x)$ is transformed into

$\forall x P([0a], x)$, meaning that from the initial world 0 there is a world a accessible from 0 such that for all x $P(x)$ holds, where P is interpreted in the world a. On the other hand, $\Box \forall x P(x)$ is transformed into $\forall u \forall x P([0u], x)$, meaning that from the initial world 0 and for every world w that is accessible from a for all x, $P(x)$ holds, where P is interpreted in the world u. Therefore, the symbolization of "It is necessary that every wet field is unsuitable for playing game":

$$\Box \forall x(\, Field(x, Wet) \rightarrow Unsuitable(x))$$

can be transformed into the following clause in world-path syntax:

$$\forall u \forall x(\neg Field([0u], x, Wet) \vee Unsuitable([0u], x))$$

Resolution theorem proving in this setting (Ohlbach, 1988) is carried out on a set of clauses of the above form in a way similar to standard first-order resolution theorem proving except for the unification of terms like $[0u]$, which is carried out by taking into account the type of accessibility relation among possible worlds.

The notion of possibility in classical modal logics is useful for representing and reasoning about the uncertain knowledge of an agent. However, this coarse grain representation of the knowledge of possibilities about assertions through this logic is quite inadequate for practical applications. That is to say, if two assertions are possible in the current world, they are indistinguishable in the modal formalism even if an agent knows that one of them is true in twice as many possible worlds as compared to the other one. Therefore, any epistemic interpretation of the modal operators for necessity and possibility would fail to incorporate the notion of an agent's "degree of belief" in something into its epistemic states.

Therefore, we developed an extended formalism of modal epistemic logic to allow an agent to represent its *degrees of support* about an assertion. The degrees are drawn from qualitative and quantitative dictionaries that are accumulated from the agent's *a priori* knowledge about the application domain. We extend the syntax of traditional modal epistemic logic to include an indexed modal operator $\langle sup_d \rangle$ to represent an agent's degrees of support about an assertion. In this way, the proposed Logic of Agents Beliefs ($\mathcal{LAB}$) can model an argument that merely supports an assertion (to some extent d), but does not necessarily warrant an agent committing to believe in that assertion. The extended modal logic is given a modified form of possible world semantics by introducing the concept of an accessibility *hyperelation*.

1.5   Comparisons of Models

Table 1-1 presents a comparison of various features of the three epistemic models presented in the last three sections. Detailed pros and cons of individual decision making frameworks under these models (e.g. logical rules and belief networks) and their suitability for different application types are discussed in their corresponding chapters.

	Underlying formalism	Qualitative information	Uncertainty handling	Inferencing mechanism	Efficiency enhancement
Proposi-tional	Classical logics	Logical sentences	None	Theorem proving	Horn clauses, resolution
Probabi-listic	Probability theory	Causal networks	Fine-grained	Evidence propagation	Conditional independence
Possible worlds	Modal logics	Modal sentences	Coarse-grained	Theorem proving	Modal resolution

Table 1-1: Comparison of features of epistemic models

The underlying formalism for the propositional epistemic model is the classical propositional and first-order logics, whereas various systems of modal logics formalize the possible world epistemic model. The probabilistic epistemic model, as its name suggests, is formalized by the theory of probability, especially Bayesian probability.

Qualitative domain information in the propositional epistemic model is expressed by user-defined relations among domain objects and by logical connectives among various assertions, producing sentences. Qualitative domain information in the probabilistic epistemic model is in the form of causal relationships among various domain concepts represented as random variables, producing belief networks. Qualitative domain information in the possible world epistemic model is also expressed by user-defined relations and logical connectives. Additionally, modalities are allowed to apply to sentences to produce qualitative domain information.

The truth-value of an assertion in the classical logics is Boolean (either true or false), and therefore the propositional epistemic model is unable to handle uncertainties in assertions. Modal logics provide a very coarse-grained uncertainty representation through their modal operator for possibility. An agent's belief in an assertion does not imply that the assertion is true in the world the agent is in. The belief rather describes the agent's uncertain assumption of the true nature of the assertion. Uncertainty handling in the probabilistic epistemic model is fine-grained in the sense that an agent can choose any number from the dense real-valued dictionary [0,1] of probability as its degree of belief in a random variable state.

The underlying inferencing mechanism for deriving implicit assumptions in the propositional and possible world epistemic models is based on axiomatic theorem proving. The inferencing mechanism for deriving prior and posterior probability distributions in the probabilistic epistemic model is based on the application of Bayes' rule, or takes the form of evidence propagation when models are available as causal belief networks.

To enhance inferencing efficiency in the propositional epistemic model, the resolution theorem proving technique is employed. The possible world epistemic model employs the same concept in the context of modal logics. To further enhance inferencing efficiency, sentences in the propositional epistemic model are often expressed in the Horn clause syntax, which is less general than full first-order language but expressive enough to deal with most common applications. Expert system rules are just positive Horn clauses. Various conditional independences among random variables are assumed in the probabilistic epistemic model.

1.6 $\mathcal{P}3$ Model for Decision-Making Agents

So far we have described three models of epistemic states, namely propositional, possible world, and probabilistic, and compared their features. Propositional and possible world models are ideal for representing agent knowledge at a higher level natural language-like syntax and for thinking semantics in terms of possible worlds, and the probabilistic model provides the required granularity with probabilities representing uncertainty in sentences that we use in everyday life, but not one of these models by itself can sufficiently capture an agent's epistemic states. Consequently, there is a need for suitably combining these three models into one. The focus of the final part of this book is to develop such an integrated model of agent epistemic states, called $\mathcal{P}3$ (<u>P</u>ropositional, <u>P</u>robabilistic, and <u>P</u>ossible World). The $\mathcal{P}3$ model is based on a logical language with embedded modalities and probabilities and is particularly suitable for making decisions in the context of choosing between possible courses of action or between possible assertions for belief revision. An intelligent agent continuously performs this kind of cognitive task when making decisions.

The human decision-making process can be regarded as a complex information processing activity. According to (Rasmussen, 1983), the process is divided into three broad categories that correspond to activities at three different levels of complexity. At the lowest level is skill-based sensorimotor behavior, representing the most automated, largely unconscious level of skill-based performance such as deciding to brake upon suddenly seeing a car ahead. At the next level is rule-based behavior exemplified by simple procedural skills for well-practiced, simple tasks, such as inferring the condition of a game-playing field based on the current rainy weather. Knowledge-based behavior represents the most complex cognitive processing. It is used to solve difficult and sometimes unfamiliar problems, and for making decisions that require dealing with various factors and uncertain data. Examples of this type of processing include determining the status of a game given the observation of transport disruption.

The proposed $\mathcal{P}3$ model, grounded in the logic $\mathcal{LAB}$ introduced earlier, supports embedding the human decision-making process within an agent at the knowledge base level by providing suggestions on alternative courses of action, and helping to determine the most suitable one. Human decision-makers often weigh the available alternatives and select the most promising one based on the associated pros and cons. The $\mathcal{P}3$ model will represent these pros and cons as logical sentences with embedded probabilities as follows:

$$\langle bel \rangle Heavy\, Rain \rightarrow \langle sup_{0.7} \rangle Cancelled$$

The above sentence can be interpreted as follows: if the agent believes that it rained heavily then it asserts that there is a 70% chance (equivalently, generates an amount of support 0.7) that the game will be cancelled. An agent may obtain evidence from different sources as support for the cancellation, as well as support against the cancellation, as follows:

$$\langle bel \rangle Club\, Financial\, Crisis \rightarrow \langle sup_{0.6} \rangle \neg Cancelled$$

The above sentence states that if the club is in financial crisis then there is a 60% chance that the cancellation will be avoided. These types of $\mathcal{P}3$ sentences that provide support both for and against various decision options, constitute *arguments* used by the agent to solve a decision problem. Such an argumentation-based decision-making framework has been developed in (Das et al. 1997; Fox and Das, 2000).

The set of evidence for and against a certain decision option F must be aggregated to come up with the overall support for F. This aggregation process can be implemented using Dempster's combination rule within the framework of the Dempster-Shafer theory of belief functions (Shafer, 1976; Yager et al., 1994), but this requires supports to be interpreted as a mass distribution, as opposed to a probability distribution, along with an appropriate evidence independence assumption. If supports are drawn from the dictionary of probability, any aggregation process must be consistent with the theory of probability, which is a special case of Dempster-Shafer theory. The belief network technology can perform such an aggregation process on arguments "for" options, but not on arguments "against" an option. Its propagation mechanism is consistent with the theory of probability. To illustrate this, let us explain the use of belief network technology for aggregation by considering the following two sentences supporting the option *Cancelled*:

$$\langle bel\rangle Heavy\,Rain \rightarrow \langle sup_{0.7}\rangle Cancelled$$

$$\langle bel\rangle Terrorist\,Threat \rightarrow \langle sup_{0.9}\rangle Cancelled$$

The agent will now go through an aggregation process to decide whether or not to believe in game cancellation. One approach is based on the assumption that the heavy rain and terrorist threat conditions are a set of independent conditions that are likely to cause the cancellation of game and that this likelihood does not diminish or increase when several of these conditions prevail simultaneously. This is the effect of the *noisy-or* technique in belief networks, which is usually applied to generate conditional probabilities in large tables for nodes with multiple parents. In many applications this approach makes sense, as experts will often enumerate a list of factors causally influencing a particular event along with their associated uncertainty. The belief network in Figure 1-2 is constructed out of the two sentences mentioned above for aggregating all evidence for the cancellation of the game.

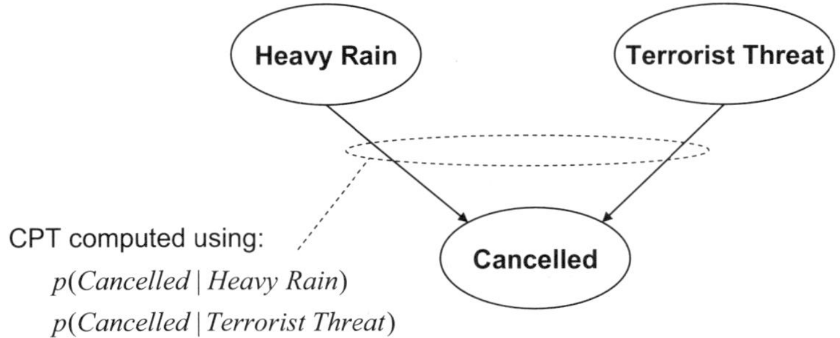

Figure 1-2: Aggregation with belief networks

As shown in the figure, there are two parents of the node *Cancelled*, corresponding to the two sentences. The sentences provide the following probabilities:

$$p(C\,|\,HR) = 0.7$$

$$p(C\,|\,TT) = 0.9$$

where $C \equiv Cancelled$, $HR \equiv Heavy\,Rain$, $TT \equiv Terrorist\,Threat$

By applying the noisy-or technique, the following probabilities for the table are computed:

$$p(C\,|\,HR,\neg TT) = 0.7$$

$$p(C\,|\,HR,TT) = p(C\,|\,HR) + p(C\,|\,TT) - p(C\,|\,HR) \times p(C\,|\,TT) = 0.97$$

$$p(C\,|\,\neg HR,TT) = 0.9$$

$$p(C\,|\,\neg HR,\neg TT) = 0$$

As evidence is posted into the network, the posterior probability (or the agent's degree of belief in the cancellation node) changes. An agent may decide to accept the cancellation of the game as its belief and $\langle bel \rangle Cancelled$ will be added to its database.

To summarize the $\mathcal{P}3$ model, we use the syntax of modal propositional logics for representing arguments, and include probabilities to represent their strengths. The use of modal logics allows agents to consider decision options in terms of the intuitive possible world concept, each of which is the result of committing to one decision option. When aggregating a set of arguments to choose a decision option, we can either apply the Dempster-Shafer theory of belief functions with an appropriate evidence independence assumption, or apply the belief network evidence propagation technique on a restricted set of arguments.

Before we delve into the details of various agent epistemic models, it is worth pointing out the role of belief revision in the context of our proposed agent decision-making paradigm. As an agent perceives an environment, it revises its own belief about the environment, which amounts to updating some of its own knowledge representing its own epistemic state. The monotonic nature of logic-based modeling of epistemic states does not automatically allow such a belief revision process. The agent can potentially continue accumulating knowledge as long as the knowledge base is not inconsistent (i.e. deriving a formula F and its negation). A special inference rule or some kind of meta-level reasoning is required for belief revision in case an inconsistency arises between the observation and what is already in the epistemic state.

The belief network-based probabilistic epistemic model is more flexible in accommodating observations from uncertain environments because a consistent probability assignment can always be obtained to yield a consistent epistemic state unless it is a hard inconsistency (for example, the agent already believes in F with degree of support 1.0, but subsequently observed its negation with degree of support 1.0) or unless a soft probabilistic consistency concept is incorporated as in (Das and Lawless, 2005). Various propagation algorithms for computing posterior probabilities of random variables based on their priors and observations do these assignments. Our approach to decision making here deals with a "snapshot" of the agent epistemic state rather than its evolution over time. Moreover, the argumentation based $\mathcal{P}3$ model does not require an epistemic state to be consistent because a hard inconsistency will at worst yield a dilemma amongst decision options.

Chapter 2

Mathematical Preliminaries

This chapter provides the background mathematical notations and concepts needed to understand the rest of the book. The topics discussed in this chapter include sets, relations, and functions (Section 2.2), graphs and trees (Section 2.3), basic probability theory (Section 2.4), and a concise introduction to the theory of algorithmic complexity (Section 2.5). (We will use this theory to analyze expected run time performance of logical inference schemes and evidence propagation algorithms.) But first, we explain our conventions for symbol usage.

2.1　Usage of Symbols

In general, the following conventions are used for symbols representing terms, random variables, clauses, sets, and so on:

Item	Convention	Example
<ul><li>Variables in a logical language</li><li>Variables representing random variable states</li></ul>	<ul><li>Italicized, lower case letters</li></ul>	<ul><li>$x, y, z, ..., x_1, y_1, z_1, ...$</li></ul>
<ul><li>Probabilistic random variables</li><li>Graph nodes</li></ul>	<ul><li>Italicized, upper case letters</li><li>Italicized string starting with an upper case letter</li></ul>	<ul><li>$X, Y, Z, N, U, V, W, ...,$ $X_1, Y_1, Z_1, N_1, U_1, V_1, W_1, ...$</li><li>*Rain, Game*</li></ul>
<ul><li>Predicate symbols</li></ul>	<ul><li>Italicized upper case letters</li><li>Italicized string starting with an upper case letter</li></ul>	<ul><li>$P, Q, R, ..., P_1, Q_1, R_1, ...$</li><li>*Even, Employee*</li></ul>
<ul><li>Constant symbols in a logical language</li><li>Random variable states</li></ul>	<ul><li>Italicized, lower case letters</li><li>Italicized string starting with a lower case letter</li></ul>	<ul><li>$a, b, c, ..., a_1, b_1, c_1, ...$</li><li>*medium, yes, john, cancelled*</li></ul>

23

	Italicized, lower case letters	$f,g,h,...,f_1,g_1,h_1,...$
• Function symbols	• Italicized string starting with a lower case letter	• *succ*
• Atoms • Literals	• Italicized, upper case letters	$A,B,L,M,...,$ $A_1,B_1,L_1,M_1,...$
• Formulae • Goals • Clauses	• Italicized, upper case letters	• $F,G,H,F_1,G_1,H_1,...$
• Sets	• Bold, Times-Roman typeface, upper case letters	• $\mathbf{A},\mathbf{B},\mathbf{S},...,\mathbf{A_1},\mathbf{B_1},\mathbf{S_1},...$
• Special logical systems • Special logical sets	• *Monotype Corsiva* typeface string starting with an upper case letter	• *Pos, PL, S5, LAB*
• Prolog program code	• `Courier` typeface	• `append([], L, L)`

New terminology is italicized on its first use. Propositions, theorems, figures, and tables are numbered by chapter and section.

2.2   Sets, Relations, and Functions

Given a set $\mathbf{S}$, $a \in \mathbf{S}$ means "a is an element of $\mathbf{S}$," "a belongs to $\mathbf{S}$," or "a is a member of $\mathbf{S}$." Similarly, $a \notin \mathbf{S}$ means "a is not an element of $\mathbf{S}$." A shorthand notation for a set $\mathbf{S}$ is

$$\mathbf{S} = \{x \mid P(x)\}$$

which is read as "$\mathbf{S}$ is the set of all elements x such that the property P holds." When it is possible to list all the elements, or at least to list a sufficient number to clearly define the elements of set $\mathbf{S}$, then $\mathbf{S}$ is written as $\{a,b,c,...\}$. The *empty set* has no elements and is denoted as Φ or $\{\}$.

A set $\mathbf{A}$ is a *subset* of a set $\mathbf{S}$ (or $\mathbf{A}$ is included in $\mathbf{S}$, $\mathbf{A}$ is contained in $\mathbf{S}$, or $\mathbf{S}$ contains $\mathbf{A}$), denoted as $\mathbf{A} \subseteq \mathbf{S}$, if every element of $\mathbf{A}$ is an element of $\mathbf{S}$. That is, $a \in \mathbf{A}$ implies $a \in \mathbf{S}$. So, for any set $\mathbf{A}$, we always have $\mathbf{A} \subseteq \mathbf{A}$ and $\Phi \subseteq \mathbf{A}$. If two sets $\mathbf{A}$ and $\mathbf{B}$ are the same (or *equal*), then that is denoted as $\mathbf{A} = \mathbf{B}$; set equality is equivalent to having both $\mathbf{A} \subseteq \mathbf{B}$ and $\mathbf{B} \subseteq \mathbf{A}$. If $\mathbf{A} \subseteq \mathbf{S}$ and $\mathbf{A} \neq \mathbf{S}$, then $\mathbf{A}$ is a *proper subset* of $\mathbf{S}$ and is denoted as $\mathbf{A} \subset \mathbf{S}$. In other words, if $\mathbf{A} \subset \mathbf{S}$ then there exists at least one $a \in \mathbf{S}$ such that $a \notin \mathbf{A}$.

The *union* (or *join*) of two sets $\mathbf{A}$ and $\mathbf{B}$ is written as $\mathbf{A} \cup \mathbf{B}$, and is the set $\{x \mid x \in \mathbf{A} \text{ or } x \in \mathbf{B}\}$. That is, it is the set of all elements that are members of either $\mathbf{A}$ or $\mathbf{B}$ (or both). The union of an arbitrary collection of sets is the set of all elements that are members of at least one of the sets in the collection. The *intersection* of two sets $\mathbf{A}$ and $\mathbf{B}$, denoted $\mathbf{A} \cap \mathbf{B}$, is the set $\{x \mid x \in \mathbf{A} \text{ and } x \in \mathbf{B}\}$, that is, the set of all elements that are members of both $\mathbf{A}$ and $\mathbf{B}$. The intersection of an arbitrary collection of sets is the set of all elements that are members of every set in the collection. The *difference* of two sets $\mathbf{A}$ and $\mathbf{B}$, denoted $\mathbf{A} - \mathbf{B}$, is the set $\{x \mid x \in \mathbf{A} \text{ and } x \notin \mathbf{B}\}$, that is, the set of all elements that are members of $\mathbf{A}$ but not members of $\mathbf{B}$.

Example

$\{1,4,7,9\}$, $\{\{a,b\},c,d\}$, $\{0,1,2,3,...\}$, $\{on,cancelled,delayed\}$ are all examples of sets. Let $\mathbf{N}$ be the set $\{1,2,3,...\}$ of all natural numbers. Then $\{x \in \mathbf{N} \mid x \text{ is odd}\}$ is the set $\{1,3,5,...\}$. The set $\{4,7\}$ is a proper subset of $\{1,4,7,9\}$. Suppose $\mathbf{A} = \{a,b,c\}$ and $\mathbf{B} = \{a,c,d\}$. Then $\mathbf{A} \cup \mathbf{B} = \{a,b,c,d\}$, $\mathbf{A} \cap \mathbf{B} = \{a,c\}$, $\mathbf{A} - \mathbf{B} = \{b\}$.

Suppose $\mathbf{I}$ is an arbitrary set and for each $i \in \mathbf{I}$ we have a corresponding set $\mathbf{A}_i$. Then the set of sets $\{\mathbf{A}_i \mid i \in \mathbf{I}\}$ is denoted as $\{\mathbf{A}_i\}_{i \in \mathbf{I}}$, and $\mathbf{I}$ is called an *index set*. The join of all sets in the collection $\{\mathbf{A}_i\}_{i \in \mathbf{I}}$ is denoted as $\bigcup_{i \in \mathbf{I}} \mathbf{A}_i$ and the intersection is denoted as $\bigcap_{i \in \mathbf{I}} \mathbf{A}_i$.

Given a set $\mathbf{S}$, the *power set* of $\mathbf{S}$, denoted as $\mathcal{P}(\mathbf{S})$ or $2^{\mathbf{S}}$, is the set $\{\mathbf{A} \mid \mathbf{A} \subseteq \mathbf{S}\}$, that is, the set of all possible subsets of $\mathbf{S}$. If a set $\mathbf{S}$ contains n elements then $\mathcal{P}(\mathbf{S})$ contains 2^n elements.

Example

Suppose $\mathbf{S} = \{a,b,c\}$. Then $\mathcal{P}(\mathbf{S}) = \{\varnothing,\{a\},\{b\},\{c\},\{a,b\},\{a,c\},\{b,c\},\{a,b,c\}\}$.

Given a collection of sets $S_1, S_2, ..., S_n$, the *cartesian product* of these n sets, denoted by $S_1 \times S_2 \times ... \times S_n$ or more concisely as $\prod_{i=1}^{n} S_i$, is the set of all possible ordered n-tuples $\langle a_1, a_2, ..., a_n \rangle$ such that $a_i \in S_i$, for $i = 1, 2, ..., n$. When each S_i is equal to S then the product is written as S^n.

Example

Suppose $S_1 = \{a_1, a_2\}$, $S_2 = \{b\}$, $S_3 = \{c_1, c_2, c_3\}$. Then $S_1 \times S_2 \times S_3 = \{ \langle a_1, b, c_1 \rangle$, $\langle a_1, b, c_2 \rangle$, $\langle a_1, b, c_3 \rangle$, $\langle a_2, b, c_1 \rangle$, $\langle a_2, b, c_2 \rangle$, $\langle a_2, b, c_3 \rangle$.

A *relation* is formally described by a statement involving elements from a collection of sets. An n-ary relation R on the sets $S_1, S_2, ..., S_n$ is a subset of $S_1 \times S_2 \times ... \times S_n$. An n-ary relation R on a set S is a subset of S^n. If R is a *binary relation* on a set S, then the notation xRy implies that $\langle x, y \rangle \in R$, where $x, y \in S$. A binary relation R on a set S is

- *serial* if for all $x \in S$ there is a $y \in S$ such that xRy
- *reflexive* if xRx, for all $x \in S$
- *symmetric* if xRy implies yRx, for all $x, y \in S$
- *transitive* if xRy and yRz implies xRz, for all $x, y, z \in S$
- *euclidean* if xRy and xRz implies yRz, for all $x, y, z \in S$
- *antisymmetric* if xRy and yRx implies $x = y$, for all $x, y \in S$
- *equivalence* if it is reflexive, symmetric, and transitive

Example

As relations on the real numbers, $x < y$ is serial and transitive; $x \leq y$ is serial, reflexive, transitive, and antisymmetric; and $x = y$ is a simple equivalence relation.

Given an equivalence relation $\mathbf{R}$ on a set $\mathbf{S}$, the equivalence class of $a \in \mathbf{S}$ is the set $\{x \in \mathbf{S} \mid a\mathbf{R}x\}$ and is denoted by $Eq_\mathbf{S}[a,\mathbf{R}]$. The following theorem provides an association between an equivalence relation and equivalence classes.

Theorem 2-1: Let $\mathbf{S}$ be a set and $\mathbf{R}$ be an equivalence relation on $\mathbf{S}$. Then $\mathbf{R}$ induces a decomposition of $\mathbf{S}$ as a union of mutually disjoint subsets of $\mathbf{S}$. (Informally, we say that $\mathbf{R}$ *partitions* the set $\mathbf{S}$.) Conversely, given a decomposition on $\mathbf{S}$ as a union of mutually disjoint subsets of $\mathbf{S}$, an equivalence relation can be defined on $\mathbf{S}$ such that each of these subsets are the distinct equivalence classes.

Proof: Since $x\mathbf{R}x$ holds, $x \in Eq_\mathbf{S}[x,\mathbf{R}]$. Suppose $Eq_\mathbf{S}[a,\mathbf{R}]$ and $Eq_\mathbf{S}[b,\mathbf{R}]$ are two equivalence classes and let $x \in Eq_\mathbf{S}[a,\mathbf{R}] \cap Eq_\mathbf{S}[b,\mathbf{R}]$. Then since $x \in Eq_\mathbf{S}[a,\mathbf{R}]$, $a\mathbf{R}x$. Since $x \in Eq_\mathbf{S}[b,\mathbf{R}]$, $b\mathbf{R}x$, i.e., $x\mathbf{R}b$, since $\mathbf{R}$ is symmetric. Now, $a\mathbf{R}x$ and $x\mathbf{R}b$, and therefore by transitivity of $\mathbf{R}$, $a\mathbf{R}b$.

Suppose $x \in Eq_\mathbf{S}[b,\mathbf{R}]$ and thus $b\mathbf{R}x$. $a\mathbf{R}b$ is already proved. Therefore, by transitivity of $\mathbf{R}$, $a\mathbf{R}x$ and hence $x \in Eq_\mathbf{S}[a,\mathbf{R}]$. Thus, $Eq_\mathbf{S}[a,\mathbf{R}] \subseteq Eq_\mathbf{S}[b,\mathbf{R}]$. Similarly, we can prove that $Eq_\mathbf{S}[b,\mathbf{R}] \subseteq Eq_\mathbf{S}[a,\mathbf{R}]$. Thus, $Eq_\mathbf{S}[b,\mathbf{R}] = Eq_\mathbf{S}[a,\mathbf{R}]$ and hence equivalence classes are either identical or mutually disjoint.

Now, every $s \in \mathbf{S}$ is in an equivalence class, namely $Eq_\mathbf{S}[s,\mathbf{R}]$, we have our decomposition of $\mathbf{S}$ into a union of mutually disjoint subsets, i.e. the equivalence classes under $\mathbf{R}$.

To prove the converse, suppose $\mathbf{S} = \underset{\alpha \in \mathbf{I}}{\cup} \mathbf{S}_\alpha$, where $\mathbf{I}$ is an index set and $\mathbf{S}_\alpha \cap \mathbf{S}_\beta = \varnothing$, when $\alpha \neq \beta$ and $\alpha, \beta \in \mathbf{I}$. Define an equivalence relation $\mathbf{R}$ on $\mathbf{S}$ as $x\mathbf{R}y$ if and only if x and y belong to the same $\mathbf{S}_\alpha$. It is then straightforward to verify that $\mathbf{R}$ is an equivalence relation on $\mathbf{S}$.

Example

Let $\mathbf{S}$ be the set of all integers. Define a relation Mod_5 on $\mathbf{S}$ as $xMod_5y$ if and only if $x - y$ is divisible by 5, for $x, y \in \mathbf{S}$. (We say x is *congruent* to y modulo 5.) Then:

- $x - x = 0$ and 0 is divisible by 5. Therefore, $xMod_5x$, so Mod_5 is reflexive.

- If $x - y$ is divisible by 5 ($x Mod_5 y$) then $y - x$ is also divisible by 5 ($y Mod_5 x$). Therefore, Mod_5 is symmetric.

- If $x - y$ is divisible by 5 ($x Mod_5 y$) and $y - z$ is divisible by 5 ($y Mod_5 z$), then $(x - y) + (y - z) = x - z$ is also divisible by 5 ($x Mod_5 z$). Therefore, Mod_5 is transitive.

Hence, Mod_5 is an equivalence relation on $\mathbf{S}$. (In fact, such an equivalence relation Mod_i exists on $\mathbf{S}$, for every positive integer i.) The equivalence class of 3 is $Eq_\mathbf{S}[3, Mod_5] = \{...,-7,-2,3,8,13,...\}$. The mutually disjoint equivalence classes corresponding to Mod_5 are $\{...,-10,-5,0,5,10,...\}$, $\{...,-9-4,1,6,11,...\}$, $\{...,-8,-3,2,7,12,...\}$, $\{...,-7,-2,3,8,13,...\}$, and $\{-6,-1,4,9,14,...\}$.

Functions provide a way of describing associations between elements of different sets.

Let $\mathbf{A}$ and $\mathbf{B}$ be two non-empty sets. Then a *function f* (or *mapping*) from $\mathbf{A}$ to $\mathbf{B}$, denoted as $f : \mathbf{A} \to \mathbf{B}$, is a subset $\mathbf{C}$ of $\mathbf{A} \times \mathbf{B}$ such that for every $x \in \mathbf{A}$ there exists a unique $y \in \mathbf{B}$ for which $\langle x, y \rangle \in \mathbf{A} \times \mathbf{B}$.

Clearly, any function $f : \mathbf{A} \to \mathbf{B}$ is a particular kind of relation on $\mathbf{A}$ and $\mathbf{B}$. The element x is said to have been "mapped into y" (or "y is the image of x" or "y is the value of f at x") and y is denoted as $f(x)$; we usually write $y = f(x)$. Also, $f(\mathbf{A})$ denotes the set $\{y \in \mathbf{B} : \text{for some } x \in \mathbf{A}, \langle x, y \rangle \in \mathbf{C}\}$ and is called the *image* of f. The set $\mathbf{A}$ is called the *domain* of f and $\mathbf{B}$ is called the *range* of f.

A mapping $f : \mathbf{A} \to \mathbf{B}$ is said to be an *identity mapping* if $f(x) = x$, for all $x \in \mathbf{A}$ (so that implicitly $\mathbf{A} \subseteq \mathbf{B}$). The mapping is said to be *onto* (or *surjective*) if given $y \in \mathbf{B}$, there exists an element $x \in \mathbf{A}$ such that $f(x) = y$. The mapping is said to be *one-to-one* (or *injective*) if for all $x, y \in \mathbf{A}$, $x \neq y$ implies that $f(x) \neq f(y)$. The mapping is a *one-to-one correspondence* between the two sets $\mathbf{A}$ and $\mathbf{B}$ (or *bijective*) if it is a one-to-one and onto mapping from $\mathbf{A}$ to $\mathbf{B}$.

Two mappings $f : \mathbf{A} \to \mathbf{B}$ and $g : \mathbf{A} \to \mathbf{B}$ are *equal* if for all $x \in \mathbf{A}$, $f(x) = g(x)$ (we usually write $f \equiv g$). The *composition* of the two mappings $f : \mathbf{A} \to \mathbf{B}$ and $g : \mathbf{B} \to \mathbf{C}$, denoted as $g \circ f$ (or just gf in some cases), is a mapping $g \circ f : \mathbf{A} \to \mathbf{C}$ defined by $(g \circ f)(x) = g(f(x))$, for all $x \in \mathbf{A}$.

Now, $((f \circ g) \circ h)(x) = (f \circ g)(h(x)) = f(g(h(x))) = f((g \circ h)(x)) = (f \circ (g \circ h))(x)$, that is, $(f \circ g) \circ h \equiv f \circ (g \circ h)$, where f, g and h are mappings with appropriate domains and ranges. Thus, the mapping composition operation follows the associative property, making the positions of parentheses irrelevant when mappings are composed.

Example

Suppose $\mathbf{I}$, $\mathbf{E}$, $\mathbf{O}$, and $\mathbf{N}$ are, respectively, the sets of all integers, even integers, odd integers and natural numbers (i.e. positive integers). Let $\mathbf{N_0}$ be $\mathbf{N} \cup \{0\}$. Then the mapping $f : \mathbf{E} \to \mathbf{O}$ defined by $f(x) = x + 1$ is a one-to-one correspondence between $\mathbf{E}$ and $\mathbf{O}$. The mapping $f : \mathbf{I} \to \mathbf{N_0}$ defined by $f(x) = |x|$ is onto but not one-to-one. The mapping $f : \mathbf{I} \to \mathbf{N_0}$ defined by $f(x) = x^2$ is neither one-to-one nor onto.

2.3   Graphs and Trees

The fundamental modeling tools known as graphs and trees are introduced in this section. Graphs and trees are most often presented graphically (hence their names) as in the examples below; we give their formal non-graphical definitions here, as this background is useful for issues such as theoretical discussions and algorithm development.

A *simple graph* $\mathcal{G}$ is a pair $\langle \mathbf{V}, \mathbf{E} \rangle$, where $\mathbf{V}$ is a non-empty set of elements called *vertices* (or *nodes*), and $\mathbf{E}$ is a set of unordered pairs of distinct elements of $\mathbf{V}$ called *edges*. Edges are denoted as $N_i N_j$, where $N_i, N_j \in \mathbf{V}$. The definition of a *directed graph* (or *digraph*) is given in the same way as a graph except that the set $\mathbf{E}$ is a set of ordered pairs ($N_i N_j \neq N_j N_i$) of elements of $\mathbf{V}$ called *directed edges*. A simple graph is said to have been *obtained* from a directed graph by removing the direction of each of its edges.

A *path* (of length m) in a simple graph $\langle \mathbf{V}, \mathbf{E} \rangle$ is a finite sequence of edges of the form

$$N_0 N_1, N_1 N_2, ..., N_{m-1} N_m$$

where $N_i N_{i+1} \in \mathbf{E}$ for each term in the sequence, and where the $N_i \in \mathbf{V}$ are distinct vertices (except possibly $N_0 = N_m$). Informally, a path consists of a sequence of "hops" along edges of the graph to distinct vertices. The sequence above can be written in abbreviated form as

$$N_0 - N_1 - N_2 - ... - N_{m-1} - N_m$$

Therefore, if $N_0 - N_1 - ... - N_m$ is a path in a simple graph then $N_m - N_{m-1} - ... - N_0$ is also a path in the graph. Similarly, a *path* (of length m) in a directed graph $\langle \mathbf{V}, \mathbf{E} \rangle$ is a finite sequence of directed edges of the form

$$N_0 \rightarrow N_1 \rightarrow N_2 \rightarrow ... \rightarrow N_{m-1} \rightarrow N_m$$

where each ordered pair $N_i N_{i+1} \in \mathbf{E}$ in the sequence is understood to be a directed edge of the digraph.

Two nodes are *connected* if there is path between them. A *cycle* (of length $m+1$) in a simple graph $\langle \mathbf{V}, \mathbf{E} \rangle$ is a finite sequence of edges of the form $N_0 - N_1 - N_2 - ... - N_m - N_0$, where $m \geq 2$. A *cycle* (of length $m+1$) in a directed graph $\langle \mathbf{V}, \mathbf{E} \rangle$ is a finite sequence of edges of the form $N_0 \rightarrow N_1 \rightarrow N_2 \rightarrow ... \rightarrow N_m \rightarrow N_0$. Thus, if $N_0 = N_m$ in a path then we return to our starting point, and the path is called a *cycle* (or directed cycle in the case of a digraph).

Example

Figure 2-1a represents a simple graph $\langle \mathbf{V}, \mathbf{E} \rangle$, where $\mathbf{V} = \{N_1, N_2, N_3, N_4, N_5\}$ and $\mathbf{E} = \{N_1 N_2, N_1 N_3, N_2 N_4, N_3 N_4, N_3 N_5\}$. Since each edge is an unordered pair of elements, $N_i N_j = N_j N_i$, for all i, j. An example path of length 3 in this simple graph is $N_1 - N_2 - N_4 - N_3$. An example path of length 4 in this simple graph is $N_1 - N_2 - N_4 - N_3 - N_1$.

Figure 2-1b represents a directed version of the graph, where $\mathbf{E} = \{N_2 N_1, N_4 N_2, N_3 N_4, N_1 N_3, N_3 N_5\}$. An example path of length 3 in this directed graph is $N_1 \rightarrow N_3 \rightarrow N_4 \rightarrow N_2$. An example path of length 4 in this directed graph is $N_1 \rightarrow N_3 \rightarrow N_4 \rightarrow N_2 \rightarrow N_1$.

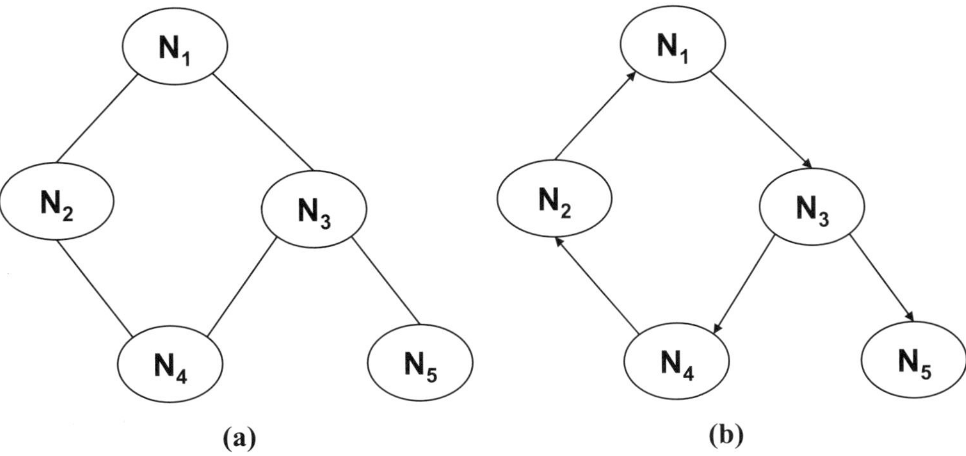

Figure 2-1: Simple and directed graphs

A simple graph is said to be *acyclic* if it has no cycles. A directed graph is said to be *acyclic* (or a *directed acyclic graph* or simply a DAG) if it has no cycles. Neither of the simple and directed graphs in Figure 2-1 is acyclic. Examples of simple and directed acyclic graphs are shown in Figure 2-2.

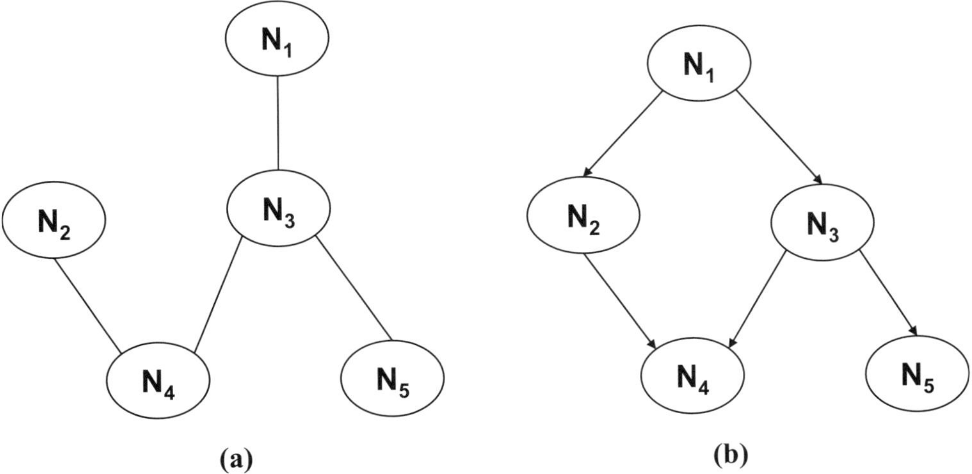

Figure 2-2: Simple and directed acyclic graphs

A simple graph is a *polytree* if and only if any two vertices of the graph are connected by exactly one path. A directed graph is a *polytree* if and only if its underlying simple graph is a polytree. Example polytrees are shown in Figure 2-3.

Suppose $N_0 \rightarrow N_1 \rightarrow N_2 \rightarrow ... \rightarrow N_{m-1} \rightarrow N_m$ is a path of a directed graph. The vertices occurring in this path are described in genealogical terms as follows:

- N_{i+1} is a *child* of N_i, for $0 \leq i < m$

- N_{i-1} is a *parent* of N_i, for $0 < i \leq m$

- N_i is an *ancestor* of N_j, for $0 \leq i < j \leq m$

- N_j is a *descendant* of N_i, for $0 \leq i < j \leq m$

A *leaf* of a directed polytree is a node without any children in any path of the tree containing the node. A *root* of a directed tree is a node without any parent in any path of the tree containing the node.

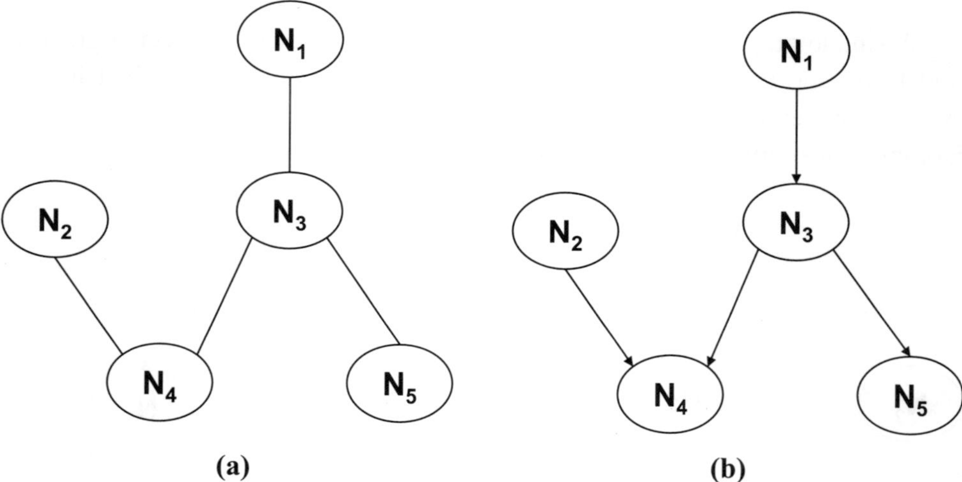

(a)   **(b)**

Figure 2-3: Simple and directed polytrees

<u>Example</u>

Figure 2-3a is a simple polytree and Figure 2-3b is a directed polytree. Consider the directed polytree in the figure: N_3 is the only child of N_1; N_4 and N_5 are the children of N_3; N_2 and N_3 are the parents of N_4; and N_3 is a parent of N_4 and N_5. N_3, N_4, N_5 are the descendants of N_1; N_1 and N_3 are the ancestors of N_5; N_1 and N_2 are the root nodes; and N_4 and N_5 are the leaf nodes.

A directed graph is a *tree* if it is a polytree with only one root node. The *level* of a vertex in such a tree is the number of edges in the path between the vertex and the root. The *depth* of such tree is the maximum level of the vertices in the tree. The level of the root in a tree is 0. Example trees are shown in Figure 2-4.

The root and leaf nodes of a simple polytree are not well-defined as its edges are undirected. For example, the node N_4 in the simple polytree in Figure 2-3a can be taken as a root node as well as a leaf node. To resolve this kind of ambiguity, we can designate a set of such vertices as root nodes and convert a simple polytree into a *rooted tree*. For example, the nodes N_1 and N_2 in Figure 2-3a can be designated as roots. Unless otherwise stated, each reference in this book to the term "tree" will be implicitly regarded as "rooted tree."

<u>Example</u>

Figure 2-4a is a simple tree and Figure 2-4b is a directed tree. Vertex N_1 is the root node of the directed tree, whereas the vertex N_1 has been designated as the root of the simple tree. The level of the vertices N_1, N_2, N_3, N_4, and N_5 are 0, 1, 1, 2, and 2 respectively. Therefore the depth of each of these two trees is 2.

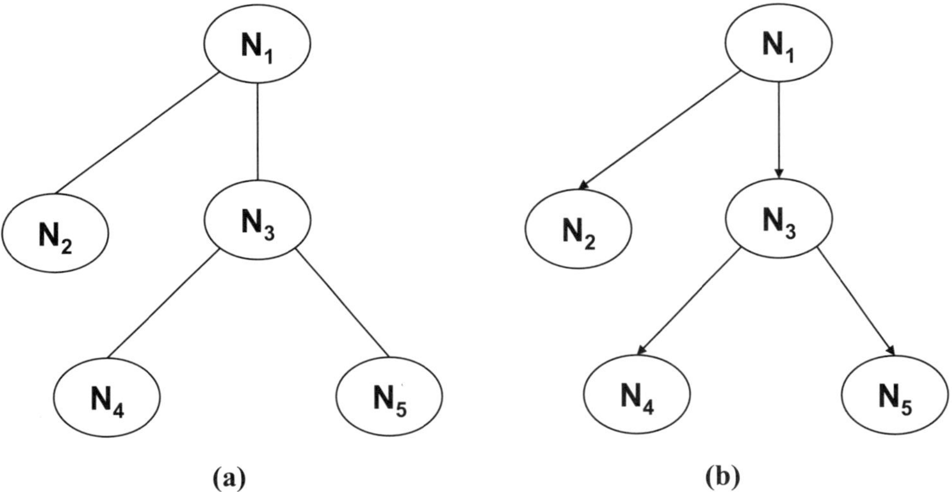

Figure 2-4: Simple and directed trees

2.4   Probability

Probabilities are defined in terms of likely outcomes of random experiments. A repetitive process, observation, or operation that determines the results of any one of a number of possible outcomes is called a *random experiment*. An *event* is an outcome of a random experiment. The set of all possible outcomes of an experiment is called the *sample space* or *event space.*

Example

Random experiments and outcomes include: tossing a coin hundred times to determine the number of heads, rolling a pair of dice a couple of hundred times to determine the number of times the sum of the upturned faces is 7, observing the weather throughout the month of March to determine the number of sunny mornings, and sensing day temperatures over a month to determine the number of hot days. Therefore, tossing a head, rolling a six and a three, and a sunny morning are example events. The sets $\{head, tail\}$, $\{(1,1),(1,2),...,(6,6)\}$, $\{sunny, rain, snow\}$, and $\{t : t \in [0^o C, 100^o C]\}$ are, respectively, examples of sample spaces for these experiments.

A *probability* provides a quantitative description of the likely occurrence of a particular event. The probability of an event x, denoted as $p(x)$, is conventionally expressed on a scale from 0 to 1, inclusive.

Example

In the single die experiment, the probability of rolling a six is 1/6. There are 36 possible combinations of numbers when two dice are rolled. The sample points for the two events x and y consisting of sums of 7 and 10 are respectively $x = \{(1, 6), (2, 5), (3, 4), (4, 3), (5, 2), (6, 1)\}$ and $y = \{(4, 6), (5, 5), (6, 4)\}$. Hence, we have $p(x) = 6/36$, $p(y) = 3/36$.

As defined above, an event consists of a single outcome in the sample space. Let us generalize this definition by calling it an *elementary* (or *simple event* or *atomic event*), and by defining a *compound event* as an event that consists of multiple simple events. In general, an event is either a simple event or a compound event. Set theory can be used to represent various relationships among events. In general, if x and y are two events (which may be either simple or compound) in the sample space S then:

- $x \cup y$ means either x or y occurs (or both occur)

- $x \cap y$ or xy means both x and y occur

- $x \subseteq y$ means if x occurs then so does y

- $\bar{x}$ means event x does not occur (or equivalently, the complement of x occurs)

- Φ represents an impossible event

- S is an event that is *certain* to occur

Two events x and y are said to be *mutually exclusive* if $x \cap y = \Phi$. (The occurrence of both x and y is impossible, and therefore the two events are mutually exclusive.) On the other hand, two events x and y are said to be *independent* if $p(x \cap y) = p(x) \times p(y)$. As a result, when dealing with independent events x and y in an event space, the sets x and y must have a point (event) in common if both x and y have nonzero probabilities. Mutually exclusive, non-impossible events x and y cannot be independent as $x \cap y = \Phi$, so that $p(x \cap y) = 0$, but $p(x) \times p(y) \neq 0$.

Example

Suppose in the two-dice experiment we want to find the probability that the first die shows even and the second die shows odd. We consider the event x as the set of all sample points with the first element even and event y as the set of all sample points with the second element odd. Therefore, x is $\{(2,1),(2,2),...,(6,6)\}$ and y is $\{(1,1),(2,1),...,(6,5)\}$. Each of these two events has 18 points and the two sets have nine points in common. Hence, $p(x) = 18/36$, $p(y) = 18/36$, and $p(x \cap y) = 9/36$. Therefore, $p(x \cap y) = p(x) \times p(y)$ holds. So by definition, x and y are independent.

There are three approaches that provide guidelines on how to assign probability values:

- The classical approach

- The relative frequency approach

- The axiomatic approach

In the *classical approach,* the probability of an event x in a finite sample space S is defined as follows:

$$p(x) = \frac{n(x)}{n(S)}$$

where $n(X)$ is the cardinality of the (finite) set X. Since $x \subseteq S$, $0 \le p(x) \le 1$ and $p(S) = 1$.

In the *relative frequency* approach, the probability of an event x is defined as the ratio of the number (say, n) of outcomes or occurrences of x to the total number (say, N) of trials in a random experiment. The choice of N depends on the particular experiment, but if an experiment is repeated at least N times without changing the experimental conditions, then the relative frequency of any particular event will (in theory) eventually settle down to some value. The probability of the event can then be defined as the limiting value of the relative frequency:

$$p(x) = \lim_{n \to \infty} \frac{n}{N}$$

where n is the number of occurrences of x and N is total number of trials. For example, if a die is rolled many times then the relative frequency of the event "six" will settle down to a value of approximately 1/6.

In the *axiomatic* approach, the concept of probability is axiomatized as follows:

- $p(x) \ge 0$, where x is an arbitrary event

- $p(S) = 1$, where S is a certain event (i.e. the whole event space)

- $p(x \cup y) = p(x) + p(y)$, where x and y are mutually exclusive events.

Note that while the axiomatic approach merely provides guidance on how to assign values to probabilities, the classical and relative frequency approaches specify what values to assign.

A *subjective probability* describes an individual's personal judgment about how likely a particular event is to occur. It is not based on any precise computation, but is an assessment by a subject matter expert based on his or her experience (that is, it's a "guesstimate").

Now we turn to formally defining random variables and probability distributions, the concepts central to the development of probabilistic models for decision-making. A *random variable* is a function defined over an event space (that is, the domain of a random variable consists of random events from the sample space) and its value is determined by the outcome of an event. A *discrete random variable* is a random variable whose range is finite or denumerable. The elements in the range (i.e. possible values) of a random variable are called its *states*.

Example

Consider the process of rolling a pair of dice, whose sample space is $\{(1,1),(1,2),...,(6,6)\}$. Consider the random variable *Dice* defined over this sample space, where its values are determined by the sum of the upturned faces, that is, $Dice(i,j) = i + j$, for each sample point (i,j). For example, $Dice(2,3)$ is equal to 5. Therefore, *Dice* is discrete, with a range of $\{2,3,4,...,12\}$. Consider another random variable *Weather* defined over the sample space of the morning weather conditions in a particular month, where the current weather determines its value on a particular morning. The possible values of the discrete random variable *Weather* might be $\{sunny, rain, snow\}$. The domain $\{t : t \in [0^{o}C, 100^{o}C]\}$ of the random variable *Temperature* is continuous, and the range could be kept the same as the domain. If the range is considered as, for example, $\{hot, warm, normal, cold, freezing\}$ then it becomes a discrete random variable.

The *probability distribution* of a random variable is a function whose domain is the range of the random variable, and whose range is a set of values associated with the probabilities of the elements of the domain. The probability distribution of a discrete random variable is called a *discrete probability distribution*. The probability distribution of a continuous random variable is called a *continuous probability distribution*. In this book, we are mainly concerned about discrete probability distributions.

The probability distribution of a discrete random variable is represented by its *probability mass function* (or *probability density function* or *pdf*). A probability density function f of a random variable X with states $\{x_1,...,x_n\}$ is defined as follows: $f(x_i)$ is the probability that X will assume the value x_i.

__Examples__

Consider the random variable *Dice* defined as the sum of the upturned faces of two dice, and therefore having the range $\{2,3,4,...,12\}$. Now, $\{(i,j) \in \{(1,1),(1,2),...,(6,6)\} : i+j=5\}$ is equal to $\{(1,4),(2,3),(3,2),(4,1)\}$. Therefore, $p(Dice=5)=4/36$. Similarly, we have the following: $p(Dice=2)=1/36$, $p(Dice=3)=2/36$, $p(Dice=4)=3/36$, $p(Dice=5)=4/36$, $p(Dice=6)=5/36$, $p(Dice=7)=6/36$, $p(Dice=8)=5/36$, $p(Dice=9)=4/36$, $p(Dice=10)=3/36$, $p(Dice=11)=2/36$, $p(Dice=12)=1/36$. Also, the sum of all the probabilities is equal to 1, that is, $\sum_{x \in \{2,3,...,12\}} p(Dice=x)=1$.

Consider the random variable *Weather* with range $\{sunny, rain, snow\}$. Define $p(Weather=sunny)=0.55$, $p(Weather=rain)=0.15$, and $p(Weather=snow)=0.30$. Figure 2-5 represents the graphs of the two probability density functions associated with the random variables *Dice* and *Weather*. Such a graphical depiction of a probability mass function is called a *probability histogram*.

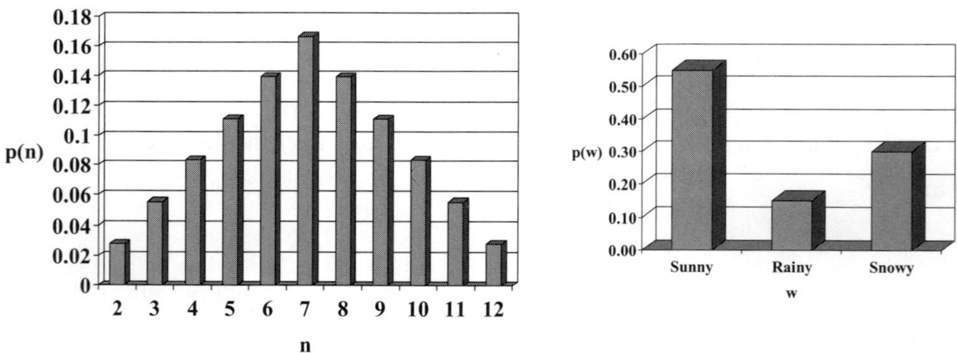

Figure 2-5: Probability density functions for the random variables *Dice* and *Weather*

The *joint probability distribution* of two discrete random variables X and Y, denoted as $p(X,Y)$ or $p(XY)$, is a function whose domain is the set of ordered pairs (x,y) of events, where x and y are possible values for X and Y, respectively, and whose range is the set of probability values corresponding to the ordered pairs in its domain. Such a probability is denoted by $p(X = x, Y = y)$ (or simply $p(x, y)$ when X and Y are clear from the context) and is defined as

$$p(x, y) = p(X = x, Y = y) = p(X = x \,\&\, Y = y)$$

The definition of the joint probability distribution can be extended to three or more random variables. In general, the joint probability distribution of the set of discrete random variables $X_1, ..., X_n$, denoted as $p(X_1, ..., X_n)$ or $p(X_1...X_n)$, is given by

$$p(x_1, ..., x_n) = p(X_1 = x_1, ..., X_n = x_n) = p(X_1 = x_1 \,\&\, ... \,\&\, X_n = x_n)$$

The notion of conditional probability distribution arises when you want to know the probability of an event, given the occurrence of another event. For example, the probability of snowy weather later today given that the current temperature is freezing. Formally, the *conditional probability distribution* of the two random variables X and Y, denoted as $p(X \mid Y)$, is a function whose domain is the set of ordered pairs (x, y), where x and y are possible values for X and Y, respectively, and is a function whose range is the set of probability values corresponding to the ordered pairs. The conditional probability distribution is defined as follows:

$$p(X \mid Y) = \frac{p(XY)}{p(Y)}, \text{ if } p(Y) > 0$$

Following are some important results for conditional probabilities that follow from this definition:

Multiplication Rule

$$p(X_0, X_1, ..., X_n) = p(X_0)p(X_1 \mid X_0)p(X_2 \mid X_0 X_1)...p(X_n \mid X_0 X_1...X_{n-1}),$$

$$\text{if } p(X_0, X_1, ..., X_n) > 0$$

Total Probability Rule

$$p(X) = \sum_{i=1}^{n} p(X \mid Y_i)p(Y_i), \text{ given } p(Y_i) > 0, \text{ for every } i,$$

$$\text{and given } \sum_{i=1}^{n} p(Y_i) = 1$$

Special Case: $p(X) = p(X \mid Y)p(Y) + p(X \mid \overline{Y})p(\overline{Y})$, if $0 < p(Y) < 1$

Marginalization Rule

$$p(X) = \sum_{i=1}^{n} p(X, Y_i), \text{ given } p(Y_i) > 0, \text{ for every } i, \text{ and given } \sum_{i=1}^{n} p(Y_i) = 1$$

$$\text{Special Case: } p(X) = p(X, Y) + p(X, \overline{Y}), \text{ if } 0 < p(Y) < 1$$

Bayes' Rule

$$p(Y_j \mid X) = \frac{p(X \mid Y_j) p(Y_j)}{\sum_{i=1}^{n} p(X \mid Y_i) p(Y_i)}, \text{ if } p(X) > 0 \text{ and } p(Y_i) > 0, \text{ for every } i, \text{ and}$$

$$\sum_{i=1}^{n} p(Y_i) = 1$$

$$\text{Special Case: } p(X \mid Y) = \frac{p(Y \mid X) p(X)}{p(Y)}, \text{ if } p(X) > 0 \text{ and } p(Y) > 0$$

2.5   Algorithmic Complexity

An algorithm is a program that is guaranteed to give a correct answer to a problem within a certain time. We say an algorithm runs in *polynomial time* when there is a polynomial

$$p(x) = a_0 + a_1 x + a_2 x^2 + \ldots + a_k x^k$$

such that the time taken to run the algorithm is less than or equal to $p(x)$, where x is the "input length" (essentially, the amount of data needed to describe an instance of the problem), and $a_0, a_1, a_2, \ldots, a_k$ are non-negative integers. Formally, if an algorithm runs in polynomial time, then we say that the algorithm complexity (or simply, the algorithm) is *"of order x^k"* or $O(x^k)$, where k is the highest power in $p(x)$ above. This theory of computational complexity started with Cook's paper (Cook, 1971), and its close relationship to combinatorial optimization can be found in (Karp, 1972).

A problem for which an algorithm of order x^k exists is said to be in the *polynomial class* or P *class*. Unfortunately, time-complexity functions are not always bounded this way; for example, some algorithms belong to the *exponential* class or *EXP* class. These algorithms have "exploding" time functions which contain exponential factors, like 2^n or $n!$ (where n is again the input length of the problem). They grow extremely quickly, much faster than any polynomial function. A problem with time complexity bounded by a polynomial function is considered *tractable*; otherwise, it is *intractable*.

A decision problem (or recognition problem) is one that takes the form of a question with a "yes" or "no" answer. Consider, for example, the Traveling Salesman Problem (TSP), Problem which decides if there is some tour or circuit in a complete weighted graph which visits every node exactly once, with total path weight less than some given value. This differs from the corresponding optimization problem of finding the shortest tour of all the cities, which requires exponential running time. We say that a decision problem belongs to the NP (*nondeterministic polynomial*) complexity class if every "yes" instance has a certificate whose validity can be checked in polynomial time. For example, if the TSP decision problem has the answer "yes" then one certificate is a list of the orders in which the nodes should be visited. It takes only time $O(x)$ to add up the length of this tour and check it is less than the given value. By reversing the roles played by "yes" and "no" we obtain a complexity class known as *Co-NP*. In particular, for every decision problem in NP there is an associated decision problem in Co-NP obtained by framing the NP question in the negative, for example, the question "Do all traveling salesman tours have length greater than certain value?" For Co-NP problems, every "no" instance has a certificate whose validity can be checked in polynomial time.

The complexity class NP-complete is the set of decision problems that are the hardest problems in NP in the sense that they are the ones most likely not to be in P. Formally, A problem is *NP-complete* if

- It is in NP
- Every NP problem can be reduced to it in polynomial time

An optimization problem for which the related decision problem is NP-complete is termed *NP-hard*. Such a problem is at least as hard or harder than any problem in NP. Thus an NP-hard problem is any problem such that every NP problem can be converted to it (reduced to it) in polynomial time. Therefore, if we are given an algorithm that solves an NP-hard problem, then the algorithm can also be used to solve any problem in NP with no more than $O(x^k)$ extra time. The class of NP-complete problems is the intersection of the classes of NP-hard and NP problems. The running time of a *P problem* on a deterministic Turing machine has a polynomial-length input, and NP problems are the polynomial-time problems on nondeterministic Turing machines.

     Decision-Making Agents

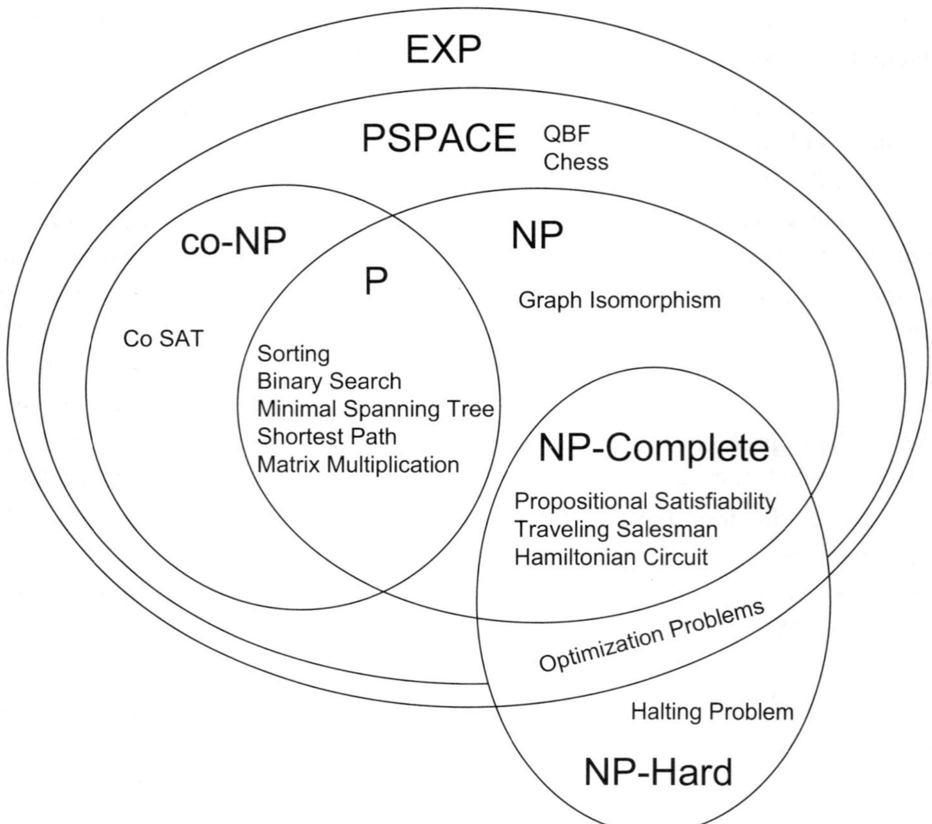

Figure 2-6: Relationships among complexity classes

The set of all P problems is contained in the set of all NP problems, which, in turn, is contained in the set of all EXP problems. Some examples of P problems are sorting, binary search, minimal spanning tree, shortest path, and matrix multiplication. The most well-known NP-complete problem is the propositional satisfiability (SAT) problem which determines whether a given propositional formula is satisfiable (this will be discussed in detail in the next chapter on classical logic). SAT has 2^N possible solutions if there are N propositional variables, and so has exponential time complexity (assuming we must check them all to determine if there is a solution). But a possible solution is easy to check in polynomial time, and therefore it is an NP-complete problem. Another NP-complete problem is the Hamiltonian Circuit Problem which finds a circuit in a graph that passes through each vertex exactly once. The most famous outstanding question in complexity theory is whether P = NP, that is, whether the NP problems actually do have polynomial-time solutions that haven't yet been discovered. Although there is strong suspicion that this is not the case, no one has been able to prove it.

As mentioned previously, decision problems are associated with optimization problems. For example, for the TSP the associated optimization question is "What is the length of the shortest tour?" If an optimization problem asks for a certain type of structure with the minimum "cost" among such structures, we can associate with that problem a decision problem that includes a numerical bound B as an additional parameter and that asks whether there exists a structure of the required type having a cost no more than B. The problem of finding the best clique tree (which will be defined in the context of junction tree algorithms in the chapter on belief networks) is such an optimization problem which is NP-hard.

The *graph isomorphism problem* (Are two graphs isomorphic?) is suspected to be neither in P nor NP-complete, though it is obviously in NP. There exists no known P algorithm for graph isomorphism testing, although the problem has also not been shown to be NP-complete. The *subgraph isomorphism problem* (Is a graph isomorphic to a subgraph of another graph?) is NP-complete. The *halting problem* (Given an algorithm and an input, will the algorithm ever stop?) is a decision problem, but is not NP-complete. It is an NP-hard problem.

The following types of algorithms are used in practice to deal with intractable problems and problems not admitting reasonably efficient algorithms:

- Approximation algorithms that settle for less than optimum solutions
- Heuristic algorithms that are probably efficient for most cases of the problem
- Randomized algorithms that are probably correct in typical problem instances

For *undecidable* problems, there is no algorithm that always solves them, no matter how much time or space is allowed. The halting problem is an undecidable problem. In general, it cannot be solved in finite time. First-order logic (to be discussed in the next chapter) is also undecidable in the sense that, in general, there is no effective procedure to determine whether a formula is a theorem or not.

The *space complexity* of an algorithm is concerned with the amount of memory used but not time. The difference between space complexity and time complexity is that space can be reused. Space complexity is not affected by determinism or nondeterminism, as deterministic machines can simulate nondeterministic machines using a small amount of space (Savitch's theorem). A problem is said to be in the class PSPACE if it can be solved in space polynomial in the size of its input. Clearly, a Turing machine that uses polynomial time also uses at most polynomial space. However, a Turing machine that uses polynomial space may use an exceedingly large amount of time before halting. Therefore, NP is a subset of PSPACE. A problem is said to be PSPACE-complete if it is in PSPACE and if every other PSPACE problem is polynomial-time reducible to it. It is widely believed that PSPACE-complete problems are strictly harder than the NP-complete problems. Proving the validity of Quantified Boolean Formulae (QBF) and perfect chess playing are examples of PSPACE-complete problems.

2.6   Further Readings

There are plenty of well-written text books in each of the areas covered in this chapter. Here I provide only a few popular ones. Stoll's book (1963) is a very well-written and easy to understand book on set theory (and logic). The graph theory book by Deo (1974) is one of the first comprehensive texts on graph theory with applications to computer science and engineering. Wilson's book (1996) is another good introductory book on graph theory. Two good books on probability are (Feller, 1968) and (Chung, 2000). Finally, (Papadimitriou, 1993) is a good text book on complexity theory.

Chapter 3

Classical Logics for the
Propositional Epistemic Model

This chapter presents the classical logics to help provide a foundation for building decision-making agents based on the propositional epistemic model. We start with classical propositional logic and then move directly to full first-order logic. The model theoretic semantics of these logics are developed, and the soundness and completeness theorems are established. We introduce resolution theorem proving as a way to efficiently carry out deductions within the classical logics. A subsequent chapter then introduces special subsets of the classical logics that constitute the basis of logic programming, a declarative (as opposed to procedural) programming paradigm. Later, we present the logic programming language Prolog, which can be effectively used to build the complex reasoning processes involved when implementing the logical and modal epistemic models, via "meta-interpreters."

The presentation style in this chapter is rather informal, and is directed towards providing an agent with a practical and sound approach to reasoning with the propositional epistemic model. In many places supporting propositions and theorems have been stated without proofs based on rather subjective selection criteria to enhance reading flow and make room for more relevant material. In some cases, proofs follow immediately from results preceding them. For the rest, readers can consult standard textbooks on logic. Plenty of examples have been given throughout the chapter. Finally, please note that the relevance of all this foundational material to building decision-making agents was touched on in the introduction and will become clearer in subsequent chapters. Overall, the soundness and completeness results of this chapter validate resolution theorem proving, which is the fundamental technique for reasoning with logic-based agent epistemic models.

3.1   Propositional Logic

A *proposition* is a declarative sentence that is either *true* or *false*, but not both. Examples of propositions are:

 Field is wet

 It is raining at the field

Propositions will be symbolized as, for example, $P, Q, R, \ldots$, and are called *atoms* or *atomic formulae*. Compound propositions are formed by modifying with the word "not" or by connecting sentences via the words/phrases "and," "or," "if … then," and "if and only if." These five words/phrases are called *logical connectives* and are usually symbolized as shown in Table 3-1.

Connective	*Symbol*
not	$\neg$
and	$\wedge$
or	$\vee$
if … then	$\rightarrow$
if and only if	$\leftrightarrow$

Table 3-1: Standard logical connectives

<u>Example</u>

Examples of compound propositions are as follows:

 If sprinkler is on then the field is wet

 Sprinkler is not on

 It is raining or sprinkler is on

The above composite propositions are symbolized respectively in the *propositional language* as:

 $P \rightarrow Q$

 $\neg P$

 $R \vee P$

where:

P stands for "Sprinkler is on"

Q stands for "Field is wet"

R stands for "It is raining"

Each of these compound propositions is called a *well-formed formula (wff)* or *formula* as part of the propositional language, and the symbols occurring in the formulae are part of the propositional alphabet. Formal definitions of these terms are given below.

A *propositional alphabet* consists of the following:

- Two parentheses (and)

- A set of propositional variables $P, Q, R, ...$ as atoms

- A set of logical connectives $\neg, \wedge, \vee, \rightarrow,$ and $\leftrightarrow$

Well-formed formulae or *formulae* in propositional logic are defined as follows:

- An atomic formula is a formula, for example, $P, Q, R, ...$, as mentioned previously.

- If F is a formula, then $(\neg F)$ is a formula.

- If F and G are formulae, then $(F \wedge G)$, $(F \vee G)$, $(F \rightarrow G)$, $(F \leftrightarrow G)$ are formulae.

Given a propositional alphabet, the *propositional language* comprises the set of all formulae constructed from the symbols of the alphabet. An expression is a formula only if it can be shown to be a formula by the above three conditions. A formula of the form $(\neg F)$ is called the *negation* of the formula F. Formulae of the forms $(F \wedge G)$ and $(F \vee G)$ are called the *conjunction* and *disjunction*, respectively, of the formulae F and G. A formula of the form $(F \rightarrow G)$ is called a *conditional* formula; F is called the *antecedent* and G is called the *consequent*. A formula of the form $(F \leftrightarrow G)$ is called a *biconditional* formula. We will see later that disjunctions are one way that uncertain knowledge can be represented within the propositional epistemic model of an agent.

The following conventions are used to avoid using parentheses in a formula. The connective $\neg$ is applied to the smallest formula following it, then $\wedge$ is to connect the smallest formulae surrounding it, and so on for the rest of the connectives $\vee$, $\rightarrow$ and $\leftrightarrow$ in that order.

Example

If parentheses are restored in the formula

$$\neg P \wedge Q \rightarrow R \leftrightarrow Q \vee R \wedge P$$

then the resulting formula would be

$$((((\neg P) \wedge Q) \rightarrow R) \leftrightarrow (Q \vee (R \wedge P)))$$

We are especially interested in assessing the "truth" of a formula as a function of the "truth" of its atoms. To do this, we first assign a *truth value* to each atom; that is, we assign "true" (denoted as $\top$) or "false" (denoted as $\bot$) to the symbolized atomic propositions that occur in a formula. Then we compute the truth value of the formula using special rules to handle the logical connectives. So for every assignment of truth values $\top$ or $\bot$ to the symbolized atomic propositions that occur in a formula, there corresponds a truth value for the formula. This can be determined using the *truth table* of the formula.

Example

Table 3-2 is the combined truth table for the fundamental formulae $\neg P$, $P \wedge Q$, $P \vee Q$, $P \rightarrow Q$, and $P \leftrightarrow Q$; it shows how to compute truth values for a formula containing these logical connectives. Using this basic truth table, the truth table for any formula can then be constructed; for example, the truth table for $(\neg P \leftrightarrow Q) \rightarrow (P \wedge (Q \vee R))$ is displayed in Table 3-3.

P	Q	$\neg P$	$P \wedge Q$	$P \vee Q$	$P \rightarrow Q$	$P \leftrightarrow Q$
$\top$	$\top$	$\bot$	$\top$	$\top$	$\top$	$\top$
$\top$	$\bot$	$\bot$	$\bot$	$\top$	$\bot$	$\bot$
$\bot$	$\top$	$\top$	$\bot$	$\top$	$\top$	$\bot$
$\bot$	$\bot$	$\top$	$\bot$	$\bot$	$\top$	$\top$

Table 3-2: Truth table in propositional logic

P	Q	R	$\neg P$	$\neg P \leftrightarrow Q$	$Q \vee R$	$P \wedge (Q \vee R)$	$(\neg P \leftrightarrow Q) \rightarrow (P \wedge (Q \vee R))$
$\top$	$\top$	$\top$	$\bot$	$\bot$	$\top$	$\top$	$\top$
$\top$	$\top$	$\bot$	$\bot$	$\bot$	$\top$	$\top$	$\top$
$\top$	$\bot$	$\top$	$\bot$	$\top$	$\top$	$\top$	$\top$
$\top$	$\bot$	$\bot$	$\bot$	$\top$	$\bot$	$\bot$	$\bot$
$\bot$	$\top$	$\top$	$\top$	$\top$	$\top$	$\bot$	$\bot$
$\bot$	$\top$	$\bot$	$\top$	$\top$	$\top$	$\bot$	$\bot$
$\bot$	$\bot$	$\top$	$\top$	$\bot$	$\top$	$\bot$	$\top$
$\bot$	$\bot$	$\bot$	$\top$	$\bot$	$\bot$	$\bot$	$\top$

Table 3-3: Truth table for $(\neg P \leftrightarrow Q) \rightarrow (P \wedge (Q \vee R))$

Given a formula F, suppose $P_1, P_2, ..., P_n$ are all atomic formulae occurring in F. Then an *interpretation* of F is an assignment of truth values to $P_1, P_2, ..., P_n$, where no P_i is assigned both $\top$ and $\bot$. Hence every row in a truth table for a formula F is an interpretation of F.

A formula F is a *tautology* or is *valid* (denoted as $\vDash F$) if its value is $\top$ under all possible interpretations of F.

Example

The formula $P \rightarrow (P \vee Q)$ is a tautology according to Table 3-4.

P	Q	$P \vee Q$	$P \rightarrow (P \vee Q)$
$\top$	$\top$	$\top$	$\top$
$\top$	$\bot$	$\top$	$\top$
$\bot$	$\top$	$\top$	$\top$
$\bot$	$\bot$	$\bot$	$\top$

Table 3-4: A tautology

A formula F is *false* (or *inconsistent* or a *contradiction*) if and only if its value is $\bot$ under all possible interpretations.

Example

The formula $P \vee Q \leftrightarrow \neg P \wedge \neg Q$ is false by Table 3-5.

P	Q	$\neg P$	$\neg Q$	$P \vee Q$	$\neg P \wedge \neg Q$	$P \vee Q \leftrightarrow \neg P \wedge \neg Q$
$\top$	$\top$	$\bot$	$\bot$	$\top$	$\bot$	$\bot$
$\top$	$\bot$	$\bot$	$\top$	$\top$	$\bot$	$\bot$
$\bot$	$\top$	$\top$	$\bot$	$\top$	$\bot$	$\bot$
$\bot$	$\bot$	$\top$	$\top$	$\bot$	$\top$	$\bot$

Table 3-5: A contradiction

If a formula F is true under an interpretation I, then I *satisfies* F or F is *satisfied* by I and in such cases, I is a *model* of F.

Two formulae F and G are said to be *equivalent* (or F is *equivalent* to G), denoted as $F \equiv G$, if and only if the truth values of F and G are the same under every interpretation. In other words if $F \leftrightarrow G$ is a tautology then F and G are equivalent.

Examples

$$F \vee G \equiv \neg F \rightarrow G$$
$$F \wedge G \equiv \neg (F \rightarrow \neg G)$$
$$F \leftrightarrow G \equiv (F \rightarrow G) \wedge (G \rightarrow F)$$

In view of the above equivalences, it can be seen that a formula F can be transformed to an equivalent formula G ($F \equiv G$) containing only the connectives $\neg$ and $\rightarrow$. In addition, the following two equivalences show the associative properties of the two connectives $\wedge$ and $\vee$:

$$(F \wedge G) \wedge H \equiv F \wedge (G \wedge H)$$
$$(F \vee G) \vee H \equiv F \vee (G \vee H)$$

Hence the positions of parentheses or their order are immaterial between subformulae $F_1, F_2, ..., F_n$ in a formula $F_1 \vee F_2 \vee ... \vee F_n$ or $F_1 \wedge F_2 \wedge ... \wedge F_n$.

A formula of the form $F_1 \vee F_2 \vee ... \vee F_n$ is called a *disjunction* of $F_1, F_2, ..., F_n$ and a formula of the form $F_1 \wedge F_2 \wedge ... \wedge F_n$ is called a *conjunction* of $F_1, F_2, ..., F_n$.

A *literal* is either an atom or the negation of an atom. The *complement of a literal L,* denoted as L^c, is defined as follows: if L is an atom then L^c is the negation of the atom, but if L is the negation of an atom then L^c is the atom itself. A formula F is in *conjunctive normal form* (or CNF) if F has the form $F_1 \wedge F_2 \wedge ... \wedge F_n$ and each F_i is a disjunction of literals. A formula F is in *disjunctive normal form* (or DNF) if F has the form $F_1 \vee F_2 \vee ... \vee F_n$ and each F_i is a conjunction of literals.

Example

The formula $(P \vee Q) \wedge (Q \vee \neg R) \wedge (P \vee R \vee S)$ is in cnf, and $(P \wedge Q \wedge R) \vee (Q \wedge \neg S)$ is in dnf.

The following laws can easily be verified (by constructing their truth tables):

$$F \vee (G \wedge H) \equiv (F \vee G) \wedge (F \vee H)$$
$$F \wedge (G \vee H) \equiv (F \wedge G) \vee (F \wedge H)$$
(Distributive laws)

$$\neg(F \vee G) \equiv \neg F \wedge \neg G$$
$$\neg(F \wedge G) \equiv \neg F \vee \neg G$$
(*De Morgan's* laws)

Using the above laws and the negation elimination $\neg(\neg F) \equiv F$, any formula can be transformed to its equivalent disjunctive or conjunctive normal form and the procedure is informally described in the following. Its counterpart for first-order logic will be described more formally later.

Step 1: Eliminate $\rightarrow$ and $\leftrightarrow$ using the equivalences $F \rightarrow G \equiv \neg F \vee G$ and $F \leftrightarrow G \equiv (\neg F \vee G) \wedge (\neg G \vee F)$.

Step 2: Repeatedly use De Morgan's laws and the equivalence $\neg(F \vee G) \equiv \neg F \wedge \neg G$, $\neg(F \wedge G) \equiv \neg F \vee \neg G$, and $\neg(\neg F) \equiv F$ to bring the negation sign immediately before atoms.

Step 3: Repeatedly use the two distributive laws to obtain the required normal form.

The following example illustrates the above algorithm.

Example

$(\neg P \leftrightarrow Q) \rightarrow (P \wedge (Q \vee R))$	Given formula
$(\neg\neg P \vee Q) \wedge (\neg Q \vee \neg P) \rightarrow (P \wedge (Q \vee R))$	By Step 1
$\neg((\neg\neg P \vee Q) \wedge (\neg Q \vee \neg P)) \vee (P \wedge (Q \vee R))$	By Step 1
$\neg(P \vee Q) \vee \neg(\neg Q \vee \neg P) \vee (P \wedge (Q \vee R))$	By Step 2
$(\neg P \wedge \neg Q) \vee (Q \wedge P) \vee (P \wedge (Q \vee R))$	By Step 2
$(\neg P \wedge \neg Q) \vee (Q \wedge P) \vee (P \wedge Q) \vee (P \wedge R)$	By Step 3

This formula is in disjunctive normal form.

A formula G is said to be a *logical consequence* of a set of formulae $F_1, F_2, ..., F_n$, denoted as $F_1, F_2, ..., F_n \vDash G$, if and only if for any interpretation I in which $F_1 \wedge F_2 \wedge ... \wedge F_n$ is true G is also true.

Example

The formula $P \vee R$ is a logical consequence of $P \vee Q$ and $Q \rightarrow R$.

This completes our brief introduction to propositional logic. Now we give an example to demonstrate how propositional logic establishes the validity of an argument.

Example

Consider the following valid argument:

If it is raining at the field then the field is wet.

The field is not wet.

Therefore, it is not raining at the field.

A symbolization of the two *premises* of the above argument in propositional logic is $P \rightarrow Q$ and $\neg Q$, where P stands for "It is raining at the field" and Q stands for "The field is wet." To prove the validity of the argument, it is necessary to show that $\neg P$ (the symbolization of the *conclusion* of the argument) is a logical consequence of $P \rightarrow Q$ and $\neg Q$. Table 3-6 establishes $P \rightarrow Q, \neg Q \models \neg P$.

P	Q	$P \rightarrow Q$	$\neg Q$	$\neg P$
T	T	T	$\perp$	$\perp$
T	$\perp$	$\perp$	T	$\perp$
$\perp$	T	T	$\perp$	T
$\perp$	$\perp$	T	T	T

Table 3-6: Truth table to establish $P \rightarrow Q, \neg Q \models \neg P$

An alternative approach to truth tables for dealing with more complex problems concerning logical connectives is that of *formal axiomatic theories*. Such a theory $\mathcal{T}$ consists of the following:

- A countable set of *symbols* (also called an *alphabet*) . An *expression* is a finite sequence of such symbols.

- A set of *formulae* (also called the *language* over the alphabet), which is a subset of the set of all expressions of $\mathcal{T}$, and an effective procedure to determine whether an expression is a valid formula or not.

- A set of *axioms* which is a subset of the set of all formulae of $\mathcal{T}$, and an effective procedure to determine whether a formula is an axiom or not.

- A finite set of *rules of inference*, each of which is a relation among formulae. Given an $(n+1)$-ary relation $\mathbf{R}$ as a rule of inference, and given an arbitrary set of formulae $F_1, F_2, ..., F_n, F$, there is an effective procedure to determine whether $F_1, F_2, ..., F_n$ are in relation $\mathbf{R}$ to F (that is, whether $(F_1, F_2, ..., F_n, F) \in \mathbf{R}$). If so, the formula F is called a *direct consequence* of $F_1, F_2, ..., F_n$ by the application of the rule of inference $\mathbf{R}$.

A *proof* (or *axiomatic deduction*) in the axiomatic theory $\mathcal{T}$ is a finite sequence $F_1, F_2, ..., F_n$ of formulae such that each F_i is either an axiom or a direct consequence of some of $F_1, ..., F_{i-1}$ by the application of a rule of inference. A proof $F_1, F_2, ..., F_n$ is called the proof of F_n. A formula F is a *theorem* in the axiomatic theory $\mathcal{T}$, denoted as $\vdash_{\mathcal{T}} F$ (or simply $\vdash F$ when the theory $\mathcal{T}$ is clear from the context) if there is a proof of F in $\mathcal{T}$.

Suppose **S** is a set of formulae in the axiomatic theory $\mathcal{T}$. A formula F is said to be a *theorem* of **S** in $\mathcal{T}$, denoted as $\mathbf{S} \vdash_{\mathcal{T}} F$ (or again simply $\mathbf{S} \vdash F$ when the theory $\mathcal{T}$ is clear from the context), if and only if there is a sequence $F_1, F_2, ..., F_n$ of formulae such that $F_n = F$ and each F_i is either in **S** or a direct consequence of some of $F_1, F_2, ..., F_{i-1}$ by the application of a rule of inference. Such a sequence is called a *proof* of F from **S**. The members of **S** are called *hypotheses* or *premises*. If **S** is $\{F_1, F_2, ..., F_n\}$ then $\{F_1, F_2, ..., F_n\} \vdash F$ would simply be written as $F_1, F_2, ..., F_n \vdash F$. If **S** is empty then F is a theorem, that is $\vdash F$.

An axiomatic theory $\mathcal{T}$ is *decidable* if there exists an effective procedure to determine whether an arbitrary formula is a theorem or not. If there is no such procedure, then $\mathcal{T}$ is called *undecidable*.

3.1.1   *Axiomatic Theory for Propositional Logic*

A formal axiomatic theory $\mathcal{PL}$ for the propositional logic is defined as follows:

- The symbols of $\mathcal{PL}$ are a propositional alphabet:

    - *Parentheses* $(,)$

    - *Logical connectives* $\neg, \rightarrow$ (elimination of the other connectives from the alphabet is deliberate as they can be defined in terms of these two, and hence are considered as abbreviations rather than part of the alphabet).

    - *Propositional variables* $P, Q, R, P_1, Q_1, R_1, ...$

- Formulae of $\mathcal{PL}$ (the *propositional language* over the propositional alphabet) are inductively defined as follows:

    - All propositional variables are formulae.

    - If F is a formula then $(\neg F)$ is a formula.

    - If F and G are formulae then $(F \rightarrow G)$ is a formula.

    - An expression is a formula if and only if it can be shown to be a formula on the basis of the above conditions.

For a particular theory, only those symbols that occur in the theory are used to construct formulae. The normal convention is adopted concerning the omission of parentheses in a formula.

- If F, G, H are any formulae of $\mathcal{PL}$, then the following are axioms of $\mathcal{PL}$:

    - A1: $F \rightarrow (G \rightarrow F)$

    - A2: $(F \rightarrow (G \rightarrow H)) \rightarrow ((F \rightarrow G) \rightarrow (F \rightarrow H))$

    - A3: $(\neg G \rightarrow \neg F) \rightarrow (F \rightarrow G)$

- The rule of inference of $\mathcal{PL}$ is as follows:

    - *Modus Ponens* (MP) : G is a direct consequence of F and $F \rightarrow G$

The study of axiomatic theory for a propositional language is completed by introducing the other logic connectives $\wedge$, $\vee$ and $\leftrightarrow$ through the following syntactic equivalents:

$$F \wedge G \equiv \neg(F \rightarrow \neg G)$$

$$F \vee G \equiv \neg F \rightarrow G$$

$$F \leftrightarrow G \equiv (F \rightarrow G) \wedge (G \rightarrow F)$$

3.1.2 Soundness and Completeness Theorem

The soundness and completeness theorem for propositional logic provides equivalence between the two approaches, namely the truth-table approach and the formal axiomatic approach. These two approaches are also called the *model-theoretic approach* and the *proof-theoretic approach*. This section states the theorem along with other important results in propositional logic. Their proofs have not been provided here because more generalized versions exist in first-order logic and they will be established in a subsequent section.

Proposition 3-1: If F and G are formulae in propositional logic, then $\vDash F$ and $\vDash F \rightarrow G$ implies $\vDash G$.

Theorem 3-1: (Deduction Theorem) If Γ is a set of formulae and F and G are formulae in propositional logic, then $\Gamma, F \vdash G$ implies $\Gamma \vdash F \rightarrow G$.

Theorem 3-2: (Soundness and Completeness Theorem) A formula F in propositional logic is a theorem if and only if it is a tautology, that is, $\vdash F$ if and only if $\vDash F$.

Theorem 3-3: (Strong Soundness and Completeness Theorem) If $F_1, F_2, ..., F_n$ and F are formulae in propositional logic then $F_1, F_2, ..., F_n \vdash F$ if and only if $F_1, F_2, ..., F_n \vDash F$.

Theorem 3-4: (Consistency Theorem) The theory $\mathcal{PL}$ is consistent. In other words, for any formula F, $\vdash F$ and $\vdash \neg F$ cannot hold simulataneously.

Proposition 3-2: $\vdash F \leftrightarrow \neg\neg F$, for any formula F.

Proposition 3-3: $F_1 \leftrightarrow F_2 \vdash G_1 \leftrightarrow G_2$, where G_2 is obtained from G_1 by simultaneously replacing each occurrence of F_1 in G_1 by F_2 .

Example

In contrast to the justification of the valid argument $P \rightarrow Q, \neg Q \vdash \neg P$ by the truth-table method, the following is an axiomatic deduction of the conclusion of the argument:

Step 1: $P \rightarrow Q$ Hypothesis

Step 2: $\neg Q$ Hypothesis

Step 3: $(\neg\neg P \rightarrow \neg\neg Q) \rightarrow (\neg Q \rightarrow \neg P)$ Axiom A3

Step 4: $(P \rightarrow Q) \rightarrow (\neg Q \rightarrow \neg P)$

                From Step 3, Proposition 3-2, and Proposition 3-3

Step 5: $\neg Q \rightarrow \neg P$ Modus Ponens on Steps 1 and 4

Step 6: $\neg P$ Modus Ponens on Steps 2 and 5

3.2 First-Order Logic

There are various kinds of arguments that cannot be conveniently stated in the language of propositional logic. Consider the following argument as an example:

It is raining at the field Eden Garden

Rain makes every field wet

Therefore, the field Eden Garden is wet

The above argument is a valid argument. However, if the three premises are symbolized as P, Q and R respectively, it is not possible to prove R from P and Q within the framework of propositional logic, unless the second premise "Rain makes every field wet" is instantiated for the specific field Eden Garden. The correctness of the above argument relies upon the meaning of the expression "every," which has not been considered in the propositional logic. However, first-order logic handles this kind of argument, and also extends the propositional logic by incorporating more logical notations, such as terms, predicates and quantifiers. The set of symbols (the *first-order alphabet*) in the case of first-order logic is defined as follows:

- Delimiter: , (comma)
- Parentheses: (,)
- Primitive connectives: $\neg$ (negation), $\rightarrow$ (implication)
- Universal quantifier: $\forall$ (for all)
- Individual variables: $x, y, z, x_1, y_1, z_1, \ldots$
- Individual constants: $a, b, c, a_1, b_1, c_1, \ldots$
- For each natural number n, n-ary predicate symbols: $P, Q, R, P_1, Q_1, R_1, \ldots$
1. For each natural number n, n-ary function symbols: $f, g, h, f_1, g_1, h_1, \ldots$

Terms are expressions, which are defined recursively as follows:

- A variable or an individual constant is a term.
- If f is an n-ary function symbol and $t_1, t_2, \ldots, t_n$ are terms then $f(t_1, t_2, \ldots, t_n)$ is a term.
- An expression is a term if it can be shown to be so only on the basis of the above two conditions.

A *predicate* is a function which evaluates to either true or false, or a statement about a relation that may be true or false. If P is an n-ary predicate symbol and

$t_1, t_2, ..., t_n$ are terms, then $P(t_1, t_2, ..., t_n)$ is an *atomic formula* (or *atom* or *positive literal*). A *negative literal* is a formula of the form $\neg A$, where A is an atom. A *literal* is either positive or negative. Based on these primitive notions, the *well-formed formulae* (*wffs*) or *formulae* of first-order logic are recursively defined as follows:

- Every atomic formula is a formula.

- If F is a formula then $\neg F$ is a formula.

- If F is a formula and x is a variable then $\forall x(F)$ is a formula.

- If F and G are formulae then $F \rightarrow G$ is a formula.

- An expression is a formula only if it can be generated by the above four conditions.

For convenience and improved readability of formulae, the other logical connectives, $\wedge$, $\vee$, and $\leftrightarrow$, are also introduced and defined in terms of $\neg$ and $\rightarrow$ just as in the case of propositional logic. Additionally, an *existential quantifier*, denoted as $\exists$, is introduced and defined as follows:

$$\exists x(F) \equiv \neg(\forall x(\neg F))$$

In the formulae $\exists x(F)$ and $\forall x(G)$, F and G are called the *scope* of the quantifiers $\exists x$ and $\forall x$ respectively. As in the case of propositional calculus, the same convention is made about the omission of parentheses in a formula. A formula in propositional logic can be considered as a formula in first-order logic (where the atoms are 0-ary predicates, and there are no variables, functions, or quantifiers). Hence all the results established so far in connection with propositional logic are also applicable to the set of all quantifier and variable-free formulae in first-order logic. Each ground atomic formula (no occurrence of variables) occurring in this set is considered as a propositional symbol.

Given a first-order alphabet, the *first-order language* $\mathcal{L}$ comprises the set of all formulae constructed from the symbols of the alphabet. Using the first-order language, a symbolization of the first two premises of the argument presented in the beginning of this subsection is as follows:

$Rain(Eden\,Garden)$

$\forall x(Rain(x) \rightarrow Field(x, Wet))$

$Field(Eden\,Garden, Wet)$

where *Rain* and *Field* are unary and binary predicate symbols respectively, and *Eden Garden* and *Wet* are constants.

An occurrence of a variable x in a formula F is *bound* (or x *is bound in F*) if x lies within the scope of a quantifier $\forall x$ in F. If the occurrence of x in F is not bound, its occurrence is *free* in F (or x *is free in F*). A variable occurrence in a formula may be both free and bound. This case may be avoided by simultaneously renaming the variables in the quantifier and its associated bound occurrences by new variables. A formula without any free variable is called a *closed formula* (or a *sentence* or *statement*). If $x_1,...,x_n$ are all free variables of F, then formula $\forall x_1...\forall x_n F$ is called the *closure* of F and is abbreviated as $\forall F$.

Example

Suppose $F = \forall x P(x,y)$ and $G = \forall x (P(x,y) \rightarrow \forall y Q(y))$. The variable x is bound in F and G. The variable y is free in F and both free (the first occurrence) and bound (the second occurrence) in G.

Use of the notation $F[x_1,x_2,...,x_n]$ emphasizes that $x_1,x_2,...,x_n$ are some or all of the free variables of F. The notation $F[x_1/t_1,x_2/t_2,...,x_n/t_n]$ denotes substitution of the terms $t_1,t_2,...,t_n$ for all free occurrences of $x_1,x_2,...,x_n$ respectively in F.

A term t is said to be *free for a variable* x in a formula F if no free occurrence of x lies within the scope of any quantifier $\forall y$, where y is a variable occurring in t (which could become problematic if we perform the substitution $F[x/y]$, as x would become y and so become bound).

Example

Consider the formula $\forall x(P(x,y)) \rightarrow Q(z)$ and the term $f(a,x)$. The term is free for z in the formula but not free for y in the same formula.

An interpretation $\Im$ of a first-order language $\mathcal{L}$ consists of the following:

- A non-empty set **D**, called the *domain of interpretation* of $\mathcal{L}$.

- An assignment to each n-ary predicate symbol of an n-ary relation in **D**.

- An assignment to each n-ary function symbol of an n-ary function with domain $\mathbf{D}^n$ and co-domain **D**.

- An assignment to each individual constant of a fixed element of **D**.

In a given interpretation $\Im$ of $\mathfrak{L}$, the logical connectives and quantifiers are given their usual meanings, and the variables are thought of as ranging over $\mathbf{D}$. If t is a closed term (or function or predicate symbol), then $\Im(t)$ denotes the corresponding assignment by $\Im$. If t is a closed term, and has the form $f(t_1, t_2, ..., t_n)$ (where the t_i are all terms), then the corresponding assignment by $\Im$ is $\Im(f)(\Im(t_1), \Im(t_2), ..., \Im(t_n))$ (if t_i is a variable, then $\Im(t_i) = t_i$ and ranges over $\mathbf{D}$).

Suppose $\Im$ is an interpretation with domain $\mathbf{D}$ and $t \in \mathbf{D}$. Then $\Im_{(x_i / t)}$ is an interpretation that is exactly the same as $\Im$ except that the i-th variable x_i always takes the value t rather than ranging over the whole domain.

Let $\Im$ be an interpretation with domain $\mathbf{D}$. Let Σ be the set of all sequences of elements of $\mathbf{D}$. For a given sequence $S = \langle s_1, s_2, ..., s_n \rangle \in \Sigma$ and for a term t consider the following *term assignment* of t with respect to $\Im$ and S, denoted as $S^*(t)$:

- If t is a variable x_j then its assignment is s_j. (We may assume some fixed indexing of the variables of $\mathfrak{L}$.)

- If t is a constant then its assignment is according to $\Im$.

- If $r_1, r_2, ..., r_n$ are the term assignments of $t_1, t_2, ..., t_n$ respectively and f' is the assignment of the n-ary function symbol f, then $f'(r_1, r_2, ..., r_n) \in \mathbf{D}$ is the term assignment of $f(t_1, t_2, ..., t_n)$.

The definition of *satisfaction of a formula* with respect to a sequence and an interpretation can be inductively defined as follows:

- If F is an atomic wff $P(t_1, t_2, ..., t_n)$, then the sequence $S = \langle s_1, s_2, ..., s_n \rangle$ satisfies F if and only if $P'(r_1, r_2, ..., r_n)$. That is, the n-tuple $\langle r_1, r_2, ..., r_n \rangle$ is in the relation P', where P' is the corresponding n-place relation of the interpretation of P.

- S satisfies $\neg F$ if and only if S does not satisfy F.

- S satisfies $F \wedge G$ if and only if S satisfies F and S satisfies G.

- S satisfies $F \vee G$ if and only if S satisfies F or S satisfies G.

- S satisfies $F \rightarrow G$ if and only if either S does not satisfy F or S satisfies G.

- S satisfies $F \leftrightarrow G$ if and only if S satisfies both F and G or S satisfies neither F nor G.

- S satisfies $\exists x_i(F)$ if and only if there is a sequence S_1 that differs from S in at most the i-th component such that S_1 satisfies F.

- S satisfies $\forall x_i(F)$ if and only if every sequence that differs from S in at most the i-th component satisfies F.

A wff F is *true for the interpretation* (alternatively, F can be given the *truth value* $\top$) $\Im$ (written $\vDash_\Im F$ or $\Im(F) = \top$) if and only if every sequence in Σ satisfies F; F is said to be *false for the interpretation* $\Im$ if and only if no sequence of Σ satisfies F.

If a formula is not closed, then some sequences may satisfy the formula, while the rest of the sequences may not. Fortunately, the truth value of a closed formula does not depend on any particular sequence S of interpretation; in this case the satisfaction of the formula depends only on the interpretation $\Im$. The rest of this text mainly deals with such closed formulae.

Let $\Im$ be an interpretation of a first-order language $\mathcal{L}$. Then $\Im$ is said to be a *model* for a closed wff F if F is true with respect to $\Im$. The interpretation is said to be a *model* for a set Γ of closed wffs of $\mathcal{L}$ if and only if every wff in Γ is true with respect to $\Im$. Let Γ be a set of closed wffs of a first-order language $\mathcal{L}$. Then Γ is *satisfiable* if and only if $\mathcal{L}$ has at least one interpretation that is a model for Γ. Γ is *valid* if and only if every interpretation of Γ is a model for Γ. Γ is *unsatisfiable* if and only if no interpretation of $\mathcal{L}$ is a model for Γ.

Let F be a closed wff of a first-order language $\mathcal{L}$. A closed wff G is said to be *implied by* F (or, equivalently, F *implies* G) if and only if for every interpretation $\Im$ of $\mathcal{L}$, $\Im$ is a model for F implies $\Im$ is a model for G. Two closed wffs F and G are said to be *equivalent* if and only if they imply each other. Let Γ be a set of closed wffs of $\mathcal{L}$. A closed wff F is said to be a *logical consequence* of Γ (written $\Gamma \vDash F$) if and only if for every interpretation $\Im$ of $\mathcal{L}$, $\Im$ is a model for Γ implies that $\Im$ is a model for F. Therefore, $\Gamma \vDash F$ means the formulae in Γ collectively imply F.

Example

Consider the following two formulae in a first-order language $\mathcal{L}$:

$$P(a)$$

$$\forall x(P(x) \rightarrow Q(f(x)))$$

Consider an interpretation of $\mathcal{L}$ as follows (concentrating only on the symbols occurring in the above two clauses):

- The domain of interpretation is the set of all natural numbers.
- Assign a to 0.
- Assign f to the successor function f', i.e. $f'(n) = n+1$.
- Suppose P and Q are assigned to P' and Q' respectively under the interpretation. A natural number x is in relation P' if and only if x is even. A natural number x is in relation Q' if and only if x is odd.

The two formulae are obviously true under the given interpretation. Note that if the function f were instead to be interpreted as function f' where $f'(x) = x+2$, then the second formula would have been false.

3.2.1   Axiomatic Theory for First-order Logic

A formal axiomatic theory $\mathcal{T}$ for first-order logic, known as a *first-order theory*, is defined as follows:

- The set of symbols of a first-order theory is a first-order alphabet.
- The language of a first-order theory is a first-order language over the alphabet. For a particular first-order theory $\mathcal{T}$, only those symbols that occur in $\mathcal{T}$ are used to construct the language of $\mathcal{T}$.
- The axioms of $\mathcal{T}$ are divided into the following two classes:
  - *Logical axioms*: If F, G, H are formulae, then the following are logical axioms of the theory:

    A1: $F \rightarrow (G \rightarrow F)$

    A2: $(F \rightarrow (G \rightarrow H)) \rightarrow ((F \rightarrow G) \rightarrow (F \rightarrow H))$

    A3: $(\neg G \rightarrow \neg F) \rightarrow (F \rightarrow G)$

A4: $(\forall xF) \to G$, if G is a wff obtained from F by substituting all free occurrences of x by a term t, where t is free for x in F.

A5: $\forall x(F \to G) \to (F \to \forall xG)$, if F is a wff that contains no free occurrences of x.

- *Proper axioms*: Proper axioms vary from theory to theory. A first-order theory in which there are no proper axioms is called a *first-order predicate logic*.

- The rules of inference of $\mathcal{T}$ are the following:

  - *Modus ponens* (MP): G follows from F and $F \to G$.

  - *Generalization*: $\forall x(F)$ follows from F.

Since the alphabet of a first-order language is considered denumerable and formulae of the language are strings of primitive symbols, the set of all formulae of the language can be proved to be denumerable.

3.2.2 Soundness and Completeness Theorem

Similar to the case of propositional logic, this section provides the equivalence between the model-theoretic and proof-theoretic approaches of first-order logic. To prove this equivalence, some definitions and results must be established.

A set of formulae Γ is *inconsistent* if and only if $\Gamma \vdash F$ and $\Gamma \vdash \neg F$, for some formula F. Γ is *consistent* if it is not inconsistent.

Proposition 3-4: If Γ is inconsistent, then $\Gamma \vdash F \wedge \neg F$, for some formula F.

A set of statements $\mathbf{M}$ is *maximally consistent* if $\mathbf{M}$ is consistent, and for any statement F, if $\mathbf{M} \cup \{F\}$ is consistent then $F \in \mathbf{M}$.

Proposition 3-5: If $\mathbf{M}$ is a maximally consistent set then the following two properties hold in $\mathbf{M}$:

- $F \in \mathbf{M}$ if and only if $\mathbf{M} \vdash F$.

- Exactly one of F and $\neg F$ is a member of $\mathbf{M}$.

Proof:

- Suppose $F \in \mathbf{M}$. Then obviously $\mathbf{M} \vdash F$. Conversely, assume $\mathbf{M} \vdash F$. This means there exists a finite subset $\mathbf{M}_1$ of $\mathbf{M}$ such that $\mathbf{M}_1 \vdash F$. Then $\mathbf{M}_1 \cup \{F\}$ is consistent. If not, there exists a finite subset $\mathbf{M}_2$ of $\mathbf{M}$ and a formula G such that $\mathbf{M}_2, F \vdash G \wedge \neg G$. Since $\mathbf{M}_1 \vdash F$, therefore $\mathbf{M}_2, \mathbf{M}_1 \vdash G \wedge \neg G$, which contradicts the consistency of $\mathbf{M}$. Thus $\mathbf{M}_1 \cup \{F\}$ is consistent. Since $\mathbf{M}$ is a maximally consistent set, $F \in \mathbf{M}$.

- Both F and $\neg F$ cannot be members of $\mathbf{M}$; otherwise the consistency of $\mathbf{M}$ would be contradicted. If neither F nor $\neg F$ is a member of $\mathbf{M}$, then by maximality of $\mathbf{M}$ each of $\mathbf{M} \cup \{F\}$ and $\mathbf{M} \cup \{\neg F\}$ is inconsistent. Since $\mathbf{M}$ is consistent and $\mathbf{M} \cup \{F\}$ is inconsistent, $\mathbf{M} \vdash \neg F$. Similarly, $\mathbf{M} \vdash F$ and hence $\mathbf{M} \vdash F \wedge \neg F$, which again contradicts the consistency of $\mathbf{M}$.

Proposition 3-6: If $\vDash F$ then $\vDash \forall F$, where $\forall F$ is the closure of F.

Proposition 3-7: Suppose $F[x]$ is a formula in which x is free for y. Then the following hold:

- $\vDash \forall x F[x] \rightarrow F[x/y]$

- $\vDash F[x/y] \rightarrow \exists x F[x]$

Proposition 3-8: Suppose F is a formula without any free occurrences of the variable x and $G[x]$ is any formula. Then the following hold:

- $\vDash F \rightarrow \forall x G[x]$ if $\vDash F \rightarrow G[x]$

- $\vDash \exists x G[x] \rightarrow F$ if $\vDash G[x] \rightarrow F$

Proposition 3-9: If F and G are formulae in first-order logic, then $\vDash F$ and $\vDash F \rightarrow G$ imply $\vDash G$.

Theorem 3-5: (Soundness Theorem) If F is a formula in first-order logic then $\vdash F$ implies $\vDash F$.

Proof: Using the soundness theorem for propositional calculus (Theorem 3-2) and Proposition 3-7(1), it can be proved that the theorem is valid for each instance of each axiom schema. Again, using Proposition 3-8(1) and Proposition 3-9, it can be proved that any theorem F is valid which has been obtained by the application of a rule of inference on the previous two theorems. Hence $\vdash F$ implies $\vDash F$.

Theorem 3-6: (Weak Deduction Theorem) If Γ is a set of formulae and F and G are formulae of first-order logic such that F is closed, then $\Gamma, F \vdash G$ implies $\Gamma \vdash F \rightarrow G$.

Proposition 3-10: If S is a set of formulae and F and G are two formulae then $S \vdash F \rightarrow G[x]$ implies $S \vdash F \rightarrow \exists x G[x]$.

Proposition 3-11: If S is a set of formulae and F a formula such that $S \vdash F[x/a]$, where a does not occur in any member of $S \cup F\{[x]\}$, then $S \vdash F[x/y]$, for some variable y which does not occur in a proof $\mathcal{P}$ of $F[x/a]$.

Proof: Since the constant a does not occur in any member of $S \cup F\{[x]\}$, another proof of $F[x/y]$ can be constructed from $\mathcal{P}$ by replacing each occurrence of a in $\mathcal{P}$ by y.

Proposition 3-12: If S is a set of formulae and $F[x]$ a formula such that $S \vdash F[x]$, then $S \vdash F[x/y]$, where y does not occur in F.

In practice, when we deal with consistent agent knowledge bases, the following theorem ensures that such a knowledge base is satisfiable. This means the agent has at least one interpretation which is a model, that is, every sentence in the knowledge base is true in the interpretation.

Theorem 3-7: Let $\mathbf{S}_c$ be a consistent set of statements of a first-order predicate logic $\mathcal{P}$. Then $\mathbf{S}_c$ is satisfiable in a countable domain D. Moreover, the domain D is a one-to-one correspondent to the cardinality of the set of primitive symbols of $\mathcal{P}$.

Proof: Suppose the domain of individual constants of $\mathcal{P}$ is extended by adding $a_1, a_2, \ldots$. The resulting domain is also denumerable. Let $\mathcal{P}'$ be the extended logic and $\exists x F_1[x], \exists x F_2[x], \ldots$ are all the statements of $\mathcal{P}'$ of the form $\exists x F[x]$. Suppose $\mathbf{S}_0 = \mathbf{S}_c$. Let a_{i_1} be the first constant in $a_1, a_2, \ldots$ that does not occur in $\exists x F_1[x]$, where $i_1 = j$, for some $j = 1, 2, \ldots$. Consider the following:

$$\mathbf{S}_1 = \mathbf{S}_0 \cup \{\exists x F_1[x] \rightarrow F_1[x/a_{i_1}]\}$$

Let $a_{i_{j+1}}$ be the first constant in $a_1, a_2, \ldots$ that does not occur in $F_1[x/a_{i_1}], \ldots, F_j[x/a_{i_j}]$. From the set $\mathbf{S}_j$, for $j = 1, 2, \ldots$, the set $\mathbf{S}_{j+1}$ is defined as follows:

$$\mathbf{S}_{j+1} = \mathbf{S}_j \cup \{\exists x F_{j+1}[x] \rightarrow F_{j+1}[x/a_{i_{j+1}}]\}$$

Now, if $\mathbf{S}_1$ is inconsistent, by Proposition 3-4, for some formula H:

$$\mathbf{S}_c, \exists x F_1[x] \rightarrow F_1[x/a_{i_1}] \vdash H \wedge \neg H$$

i.e, $\mathbf{S}_c \vdash (\exists x F_1[x] \rightarrow F_1[x/a_{i_1}]) \rightarrow H \wedge \neg H$ [Theorem 3-6]

i.e, $\mathbf{S}_c \vdash (\exists x F_1[x] \rightarrow F_1[x/y]) \rightarrow H \wedge \neg H$, for some y   [Proposition 3-11]

i.e., $\mathbf{S}_c \vdash (\exists x F_1[x] \rightarrow \exists y F_1[x/y]) \rightarrow H \wedge \neg H$, for some y [Proposition 3-8(2)]

i.e., $\mathbf{S}_c \vdash (\exists x F_1[x] \rightarrow \exists x F_1[x]) \rightarrow H \wedge \neg H$ [Proposition 3-12]

But, $\mathbf{S}_c \vdash \exists x F_1[x] \rightarrow \exists x F_1[x]$,   [Since $F \rightarrow F$ is valid]

Hence, $\mathbf{S}_c \vdash H \wedge \neg H$ [By applying Modus Ponens]

Hence $\mathbf{S}_c$ is inconsistent, which violates the initial assumption that $\mathbf{S}_c$ is consistent. By applying the induction hypothesis, it can easily be proved that each $\mathbf{S}_j$ is consistent for $j = 1, 2, \ldots$.

Let $\mathbf{S} = \bigcup_{j \in \mathbf{N}} \mathbf{S}_j$, where $\mathbf{N}$ is the set of all natural numbers. Construct a set $\mathbf{M}$ as follows:

$$\mathbf{M}_0 = \mathbf{S}$$

$$\mathbf{M}_{j+1} = \mathbf{M}_j \cup \{A\}, \text{ if } \mathbf{M}_j \cup \{A\} \text{ is consistent.}$$

$$\mathbf{M} = \bigcup_{j \in \mathbf{N}} \mathbf{M}_j$$

Thus $\mathbf{S}_c \subseteq \mathbf{S} \subseteq \mathbf{M}$. We claim $\mathbf{M}$ is a maximally consistent set of $\mathcal{P}'$: Let F be any formula such that $\mathbf{M} \cup \{F\}$ is consistent. Suppose that F is the $(n+1)$-th formula in the chosen enumeration of the formulae of $\mathcal{P}'$. Since $\mathbf{M} \cup \{F\}$ is consistent, $\mathbf{M}_n \cup \{F\}$ is consistent. Hence by definition of $\mathbf{M}_{n+1}$, F is a member of $\mathbf{M}_{n+1}$ and therefore a member of $\mathbf{M}$, which shows that $\mathbf{M}$ is maximally consistent. Since $\mathbf{M}$ is a maximally consistent set, the two properties in Proposition 3-5 hold. Suppose a formula of the form $\exists x F[x]$ is in $\mathbf{M}$. From the construction of $\mathbf{S} \subseteq \mathbf{M}$, $\exists x F[x] \to F[x / a_j]$ is in $\mathbf{S}_j$, for some constant a_j. Now

(A) $\qquad F[x / a_j] \in \mathbf{M}$, for some j.

Otherwise, if $F[x / a_j] \notin \mathbf{M}$, for all j, then $\exists x F[x] \to F[x / a_j] \notin \mathbf{M}$, for all j (since $\exists x F[x]$ is in $\mathbf{M}$, and $\mathbf{M}$ is consistent). Hence $\exists x F[x] \to F[x / a_j] \notin \mathbf{S}$, for all j, which violates the fact that $\mathbf{S}_0 \subseteq \mathbf{S}$.

Next, we construct a model for $\mathbf{M}$. Define an interpretation $\mathfrak{I}$ of the language of $\mathcal{P}'$ as follows:

- The domain $\mathbf{D}$ of $\mathfrak{I}$ is the set of all individual constants of $\mathcal{P}'$.

- If P is an n-place predicate symbol in the language of $\mathcal{P}'$, then the corresponding n-ary relation R is defined as $\langle t_1, t_2, ..., t_n \rangle \in R$ if and only if $\mathbf{M} \vdash P(t_1, t_2, ..., t_n)$, for all closed terms $t_1, t_2, ..., t_n$.

- If f is an n-place function symbol in the language of $\mathcal{P}'$, then the corresponding n-place function f' is defined as $f'(t_1, t_2, ..., t_n) = f(t_1, t_2, ..., t_n)$, for all closed terms $t_1, t_2, ..., t_n$.

To prove that the interpretation $\mathfrak{I}$ is a model of $\mathbf{M}$, it suffices to show that

(B) $\qquad F$ is true in $\mathfrak{I}$ if and only if $\mathbf{M} \vdash F$

for each statement F of $\mathcal{P}'$, that is, $\vDash F$ if and only if $\mathbf{M} \vdash F$. Suppose this is established and G is any formula such that $\vdash G$. Then, by generalization, $\mathbf{M} \vdash \forall G$ and thus $\vDash \forall G$ (by (B)). Therefore, $\vDash G$ (since by Proposition 3-6, for any formula F, $\vDash \forall F$ if and only if $\vDash F$). Hence $\mathfrak{I}$ is a model of $\mathbf{M}$.

The proof of (B) is carried out by induction on the number n of connectives and quantifiers of F. Suppose (B) holds for any closed formulae with fewer than k connectives and quantifiers. Consider the following cases:

- Suppose F is a closed atomic formula of the form $P(t_1, t_2, ..., t_n)$. Then (B) follows directly from the definition of P.

- Suppose F is of the form $\neg G$. Then the number of connectives in G is one less than that of F. Now assume that F is true in $\mathfrak{I}$. Then, by induction hypothesis, it is not the case that $\mathbf{M} \vdash G$. Hence by Proposition 3-5(1), G is not a member of $\mathbf{M}$. Again, by Proposition 3-5(2), $\neg G$ is a member of $\mathbf{M}$. Hence by Proposition 3-5(1), $\mathbf{M} \vdash \neg G$, that is, $\mathbf{M} \vdash F$. To prove the converse, assume that $\mathbf{M} \vdash F$, that is, $\mathbf{M} \vdash \neg G$. Since $\mathbf{M}$ is consistent, it is not the case that $\mathbf{M} \vdash G$. By the induction hypothesis, G is not true and hence F is true in $\mathfrak{I}$.

- Suppose F is $G \rightarrow H$. Since F is closed, so are G and H. Let $\mathbf{M} \vdash F$ and assume F is false for $\mathfrak{I}$. Then G is true and H is false for $\mathfrak{I}$. Each of G and H contains fewer connectives than in F. Hence by induction hypothesis, $\mathbf{M} \vdash G$ and it is not the case that $\mathbf{M} \vdash H$. Consider the following proof:

Step 1: $\mathbf{M} \vdash G$	Given
Step 2: $\mathbf{M} \vdash \neg H$	As the 2nd case above
Step 3: $\mathbf{M} \vdash G \rightarrow (\neg H \rightarrow \neg(G \rightarrow H))$	Tautology
Step 4: $\mathbf{M} \vdash \neg H \rightarrow \neg(G \rightarrow H)$	MP on Steps 1 and 3
Step 5: $\mathbf{M} \vdash \neg(G \rightarrow H)$	MP on Steps 2 and 4
Step 6: $\mathbf{M} \vdash \neg F$	Since F is $G \rightarrow H$

  Since $\mathbf{M}$ is consistent, it is not the case that $\mathbf{M} \vdash F$. This contradicts the initial assumption and hence F is true for $\mathfrak{I}$. On the other hand, assume F is true for $\mathfrak{I}$. If it is not the case that $\mathbf{M} \vdash F$, then by the properties of Proposition 3-5, $\mathbf{M} \vdash \neg F$, that is, $\mathbf{M} \vdash G \wedge \neg H$, which means $\mathbf{M} \vdash G$ and $\mathbf{M} \vdash \neg H$. By induction, G is true for $\mathfrak{I}$. Since $\mathbf{M}$ is consistent, it is not the case that $\mathbf{M} \vdash H$. Therefore H is false for $\mathfrak{I}$ and F is false for $\mathfrak{I}$. This again contradicts the initial assumption. Thus F is true for $\mathfrak{I}$.

- Suppose F is $\forall G$ and G is a closed formula. By induction hypothesis, $\vDash G$ if and only if $\mathbf{M} \vdash G$, that is, $\mathbf{M} \vdash G \leftrightarrow \forall x G$, since x is not free in

G. Hence $\mathbf{M} \vdash G$ and $\mathbf{M} \vdash \forall x G$. Furthermore, $\mathbf{M} \vDash G$ if and only if $\mathbf{M} \vDash \forall x G$, since x is not free in G. Hence $\mathbf{M} \vdash \forall x G$ if and only if $\vDash \forall x G$, that is, $\mathbf{M} \vdash F$ if and only if $\vDash F$.

- Suppose F is $\forall x G[x]$ and $G[x]$ is not closed. Since F is closed, x is the only free variable in G. Suppose $\vDash \forall x G[x]$, and assume it is not the case that $\mathbf{M} \vdash \forall G[x]$. Applying Proposition 3-5, $\mathbf{M} \vdash \neg \forall G[x]$, that is, $\mathbf{M} \vdash \exists x \neg G[x]$. From (A) there exists an individual constant a of $\mathcal{P}'$ such that $\mathbf{M} \vdash \neg G[x/a]$. By induction, $\vDash \neg G[x/a]$. Also, $\vDash F$, that is, $\vDash \forall x G[x]$. Therefore, $\vDash \forall x G[x] \to G[x/a]$, that is, $\vDash G[x/a]$, which contradicts the consistency of the interpretation $\mathfrak{I}$. Therefore, $\mathbf{M} \vdash \forall G[x]$. Again, let $\mathbf{M} \vdash \forall x G[x]$ and if possible assume it is not the case that $\vDash \forall x G[x]$. Hence some sequence S does not satisfy $G[x]$. Let t be the i-th component of S that has been substituted for x in showing the unsatisfiability of $G[x]$ wrt S. Then the sequence S does not satisfy $G[x/t]$ either, since the closed term t is mapped to itself under S. Furthermore, $\mathbf{M} \vdash \forall x G[x]$, that is, $\mathbf{M} \vdash \forall G[x/t]$. Hence by our induction hypothesis, $\vDash G[x/t]$. Therefore a contradiction arises and $\vDash \forall x G[x]$, that is, $\vDash F$.

Thus the interpretation $\mathfrak{I}$ is a model for $\mathbf{M}$ and hence a model for $\mathbf{S}_c$. The domain of this interpretation is a one-to-one correspondent with the set of all primitive symbols of $\mathcal{P}$. Hence the proof.

The following theorem is an immediate consequence of Theorem 3-7.

Theorem 3-8: Every consistent theory has a denumerable model.

Theorem 3-9: (Gödel's Completeness Theorem) For each formula F in first-order logic, if $\vDash F$ then $\vdash F$.

Proof: Suppose that $\vDash F$. Then by Proposition 3-6, $\vDash \forall F$ and hence $\neg \forall F$ is not satisfiable. By Theorem 3-7, $\{\neg \forall F\}$ is inconsistent. Therefore, by Proposition 3-4, $\neg \forall F \vdash G \wedge \neg G$, for some formula G. Then, by the weak deduction theorem (Theorem 3-6), $\vdash \neg \forall F \to G \wedge \neg G$, and thus $\vdash \forall F$. Suppose

$x_1, x_2, ..., x_n$ are free variables of F and thus $\vdash \forall x_1 \forall x_2 ... \forall x_n F$. Since a variable is free for itself, axiom A4 gives

$$\vdash \forall x_1 \forall x_2 ... \forall x_n F \rightarrow \forall x_2 ... \forall x_n F$$

By applying MP, $\vdash \forall x_2 ... \forall x_n F$. Continuing this way gives $\vdash F$.

The next two fundamental soundness and completeness theorems follow immediately from the soundness theorem established earlier, and from the above completeness theorem.

Theorem 3-10: (Soundness and Completeness) For each formula F in first-order logic, $\vDash F$ if and only if $\vdash F$.

Theorem 3-11: (Strong Soundness and Completeness) Suppose Γ is any set of formulae and F is a formula in first-order logic. Then $\Gamma \vDash F$ if and only if $\Gamma \vdash F$. In other words, F is a logical consequence of Γ if and only if F is a theorem of Γ.

3.2.3   *Applications*

This section will establish some important theorems by using the soundness and completeness theorem of first-order logic established above. These theorems, especially the compactness theorem, help to establish Herbrand's theorem, which is an important theoretical foundation for the resolution theorem proving.

A theory $\mathcal{K}_1$ is said to be an *extension of a theory* $\mathcal{K}$ if every theorem of $\mathcal{K}$ is a theorem of $\mathcal{K}_1$. A theory $\mathcal{K}$ is *complete* if for any closed formula F of $\mathcal{K}$, either $\vdash F$ or $\vdash \neg F$.

Proposition 3-13: Suppose $\mathcal{K}$ is a consistent theory. Then $\mathcal{K}$ is complete if and only if $\mathcal{K}$ is maximally consistent.

Proposition 3-14: (Lindenbaum's Lemma) If $\mathcal{K}$ is a consistent theory then $\mathcal{K}$ has a consistent, complete extension.

Theorem 3-12: (Compactness Theorem) A set of formulae **S** in first-order logic is satisfiable if and only if every finite subset of **S** is satisfiable.

Proof: The forward part of the theorem follows easily. To prove the converse, suppose every finite subset of **S** is satisfiable. Then every finite subset of **S** is consistent. If not, there is a finite subset $\mathbf{S}_1$ of **S** that is inconsistent. Then, for some formula F, $\mathbf{S}_1 \vdash F$ and $\mathbf{S}_1 \vdash \neg F$. By the soundness theorem, $\mathbf{S}_1 \vDash F$ and $\mathbf{S}_1 \vDash \neg F$. Hence $\mathbf{S}_1$ cannot be satisfiable, which is a contradiction. Since every finite subset of **S** is consistent, **S** itself is consistent. For, if **S** is inconsistent then some finite subset of **S** is inconsistent which is a contradiction. Since **S** is consistent, by Theorem 3-7, **S** is satisfiable.

Theorem 3-13: (Skolem-Löwenheim Theorem) Any theory $\mathcal{K}$ that has a model has a denumerable model.

Proof: If $\mathcal{K}$ has a model, then $\mathcal{K}$ is consistent. For if $\mathcal{K}$ is inconsistent, then $\vdash F \wedge \neg F$ for some formula F. By the soundness theorem, $\vDash F$ and $\vDash \neg F$. If a formula is both true and false then $\mathcal{K}$ cannot have a model, which is a contradiction. By Theorem 3-8, $\mathcal{K}$ has a denumerable model.

3.3   Theorem Proving Procedure

A procedure for determining whether or not a formula is a theorem of a particular theory $\mathcal{K}$ is called a *theorem proving procedure* or *proof procedure* for the theory $\mathcal{K}$. Theorem proving procedures deal with formulae in standard forms, for example, prenex normal form, Skolem conjunctive normal form, and clausal form. This section provides tools for obtaining these forms from given formulae.

3.3.1   *Clausal Form*

A formula is said to be in *prenex normal form* if it is of the form

$$Q_1 x_1 Q_2 x_2 ... Q_n x_n B$$

where each Q_i is either $\forall$ or $\exists$, and the formula B is quantifier free. The formula B is called the *matrix*.   A prenex normal form formula is said to be in *Skolem conjunctive normal form* if it has the form

$$\forall x_1 \forall x_2 ... \forall x_n B$$

where the matrix B is in conjunctive normal form, that is, B is a conjunction of a disjunction of literals (as defined in Section 3.1). Such a Skolem conjunctive normal form formula is said to be a *clause* if it has the form

$$\forall x_1 \forall x_2 ... \forall x_n (L_1 \vee L_2 \vee ... \vee L_m)$$

where each L_i is a literal and $x_1, x_2, ..., x_n$ are the free variables of the disjunction $L_1 \vee L_2 \vee ... \vee L_m$. A formula is said to be in *clausal form* if it is a clause.

For the sake of convenience, a clause is rewritten as the disjunction $L_1 \vee L_2 \vee ... \vee L_m$ of literals without its quantifiers or as the set $\{L_1, L_2, ..., L_m\}$ of literals. Thus when a disjunction $L_1 \vee L_2 \vee ... \vee L_m$ or a set $\{L_1, L_2, ..., L_m\}$ is given as a clause C, where each L_i is a literal, then C is regarded as being of the form $\forall x_1 \forall x_2 ... \forall x_n (L_1 \vee L_2 \vee ... \vee L_m)$, where $x_1, x_2, ..., x_n$ are all the free variables occurring in all the L_i s.

Every formula F can be transformed to some formula G in Skolem conjunctive normal form such that $F \equiv G$ by applying the appropriate transformations of the following steps.

Step 1: (Elimination of $\rightarrow$ and $\leftrightarrow$) Apply the following two conversion rules to any subformula within the given formula:

$F \rightarrow G$ to $\neg F \vee G$

$F \leftrightarrow G$ to $(\neg F \vee G) \wedge (\neg G \vee F)$

Step 2: (Moving $\neg$ inwards) Apply the following conversion rules repeatedly until all negations are immediately to the left of an atomic formula:

$\neg \neg F$ to F

$\neg (F \wedge G)$ to $\neg F \vee \neg G$

$\neg (F \vee G)$ to $\neg F \wedge \neg G$

$\neg \forall x F$ to $\exists x \neg F$

$\neg \exists x F$ to $\forall x \neg F$

Step 3: (Moving quantifiers inwards) Apply the following conversion rules to any subformula within the formula until no rule is applicable:

$\forall x (F \wedge G)$ to $\forall x F \wedge \forall x G$

$\exists x (F \vee G)$ to $\exists x F \vee \exists x G$

$\forall x(F \vee G)$ to $\forall x F \vee G$, provided x is not free in G

$\exists x(F \wedge G)$ to $\exists x F \wedge G$, provided x is not free in G

$\forall x(F \vee G)$ to $F \vee \forall x G$, provided x is not free in F

$\exists x(F \wedge G)$ to $F \wedge \exists x G$, provided x is not free in F

$\forall x F$ to F, provided x is not free in F

$\exists x F$ to F, provided x is not free in F

$\forall x \forall y(F \vee G)$ to $\forall y \forall x(F \vee G)$

$\exists x \exists y(F \wedge G)$ to $\exists y \exists x(F \wedge G)$

The last two transformations in this step should be applied in a restricted manner to avoid infinite computation. These two should be applied to a subformula of a given formula provided y is free in both F and G and x is not free in either F or G.

Step 4: (Variable renaming) Repeat this step until no two quantifiers share the same variable.

When two quantifiers share the same variable, simultaneously rename one of the variables in the quantifier and its associated bound occurrences by a new variable.

Step 5: (Skolemization) Repeat this step until the formula is free from existential quantifiers.

Suppose the formula contains an existential quantifier $\exists x$. The variable x will be within the scope of n universally quantified variables $x_1, x_2, ..., x_n$ (for some $n \geq 0$). At each occurrence of x (other than as the quantifier name), replace x by the term $f(x_1, x_2, ..., x_n)$, where f is an n-ary function symbol that does not occur in the formula. (If $n = 0$ then use a constant symbol instead of $f(x_1, x_2, ..., x_n)$.) This process of removing existential quantifiers from a formula is called *Skolemization* and each newly entered $f(x_1, x_2, ..., x_n)$ is called a *Skolem function instance*.

Step 6: (Rewrite in prenex normal form) Remove all universal quantifiers from the formula and place them at the front of the remaining quantifier free formula (which is the matrix). The resulting formula is now in prenex normal form.

Step 7: (Rewrite in Skolem conjunctive normal form) Apply the following transformations repeatedly to the matrix of the formula until the matrix is transformed to conjunctive normal form:

$$F \vee (G \wedge H) \text{ to } (F \vee G) \wedge (F \vee H)$$

$$(F \wedge G) \vee H \text{ to } (F \vee H) \wedge (G \vee H)$$

Example

This example shows the conversion of a formula to its equivalent Skolem conjunctive normal form. The applicable step from the above transformation procedure is given in the right-hand column.

$\forall x(\neg R(x) \rightarrow P(a) \wedge \neg \exists z \neg Q(z,a)) \wedge \forall x(P(x) \rightarrow \exists y Q(y,x))$	Given
$\forall x(\neg\neg R(x) \vee P(a) \wedge \neg \exists z \neg Q(z,a)) \wedge \forall x(\neg P(x) \vee \exists y Q(y,x))$	Step 1
$\forall x(R(x) \vee P(a) \wedge \forall z \neg\neg Q(z,a)) \wedge \forall x(\neg P(x) \vee \exists y Q(y,x))$	Step 2
$\forall x(R(x) \vee P(a) \wedge \forall z Q(z,a)) \wedge \forall x(\neg P(x) \vee \exists y Q(y,x))$	Step 2
$(\forall x R(x) \vee P(a) \wedge \forall z Q(z,a)) \wedge \forall x(\neg P(x) \vee \exists y Q(y,x))$	Step 3
$(\forall x R(x) \vee P(a) \wedge \forall z Q(z,a)) \wedge \forall x_1(\neg P(x_1) \vee \exists y Q(y,x_1))$	Step 4
$(\forall x R(x) \vee P(a) \wedge \forall z Q(z,a)) \wedge \forall x_1(\neg P(x_1) \vee Q(f(x_1),x_1))$	Step 5
$\forall x \forall z \forall x_1((R(x) \vee P(a) \wedge Q(z,a)) \wedge (\neg P(x_1) \vee Q(f(x_1),x_1)))$	Step 6
$\forall x \forall z \forall x_1((R(x) \vee P(a)) \wedge (R(x) \vee Q(z,a)) \wedge (\neg P(x_1) \vee Q(f(x_1),x_1)))$	Step 7

3.3.2   *Herbrand's theorem*

By the definition of satisfiability, a formula F is unsatisfiable if and only if it is false under all interpretations over all domains. Therefore it is an enormous and almost impossible task to consider all interpretations over all domains to verify the unsatisfiability of F. Hence it is desirable to have one specified domain such that F is unsatisfiable if and only if F is false under all interpretations over this special domain. Fortunately, this kind of special domain exists: it is the Herbrand universe of F, for a given formula F. The power of the Herbrand universe will now be demonstrated by using it to establish some results.

Given a formula F in Skolem conjunctive normal form, the *Herbrand universe* of F, denoted as $\mathcal{HU}(F)$ is inductively defined as follows:

- Any constant symbol occurring in F is a member of $\mathcal{HU}(F)$. If F does not contain any constant symbol, then $\mathcal{HU}(F)$ contains the symbol a.

- If f is an n-ary function symbol occurring in F and $t_1, t_2, \ldots, t_n$ are in $\mathcal{HU}(F)$ then $f(t_1, t_2, \ldots, t_n)$ is a member of $\mathcal{HU}(F)$.

The *Herbrand base* of a formula F, denoted by $\mathcal{HB}(F)$, is the following set:

$$\{\, P(t_1, t_2, \ldots, t_n) \mid P \text{ is an } n\text{-ary predicate symbol occurring in } F$$
$$\text{and } t_1, t_2, \ldots, t_n \in \mathcal{HU}(F) \,\}.$$

A *ground instance* of a formula F in Skolem conjunctive normal form is a formula obtained from the matrix of F by replacing its variables by constants.

Example

Suppose $F = \forall x (Q(x) \wedge (\neg P(x) \vee P(f(x))))$. Then

$$\mathcal{HU}(F) = \{a, f(a), f(f(a)), \ldots\} \text{ and}$$

$$\mathcal{HB}(F) = \{Q(a), Q(f(a)), Q(f(f(a))), \ldots, P(a), P(f(a)), P(f(f(a))), \ldots\}$$

The formula $Q(f(a)) \wedge (\neg P(f(a)) \vee P(f(f(a))))$ is a ground instance of F.

Suppose $F = \forall x (P(a,b) \wedge (\neg P(x,y) \vee Q(x)))$. Then

$$\mathcal{HU}(F) = \{a, b\} \text{ and}$$

$$\mathcal{HB}(F) = \{P(a,a), P(a,b), P(b,a), P(b,b), Q(a), Q(b)\}.$$

Given a formula F in prenex normal form, an interpretation $\mathcal{HI}$ over $\mathcal{HU}(F)$ is a *Herbrand interpretation* if the following conditions are satisfied:

- $\mathcal{HI}$ assigns every constant in $\mathcal{HU}(F)$ to itself.

- Let f be an n-ary function symbol and $t_1, t_2, \ldots, t_n$ be elements of $\mathcal{HU}(F)$. Then $\mathcal{HI}$ assigns f to an n-place function that maps the n-tuple $\langle t_1, t_2, \ldots, t_n \rangle$ to $f(t_1, t_2, \ldots, t_n)$.

Proposition 3-15: Suppose $F = \forall x_1 \forall x_2 ... \forall x_p B[x_1, x_2, ..., x_p]$ is a formula in prenex normal form, where B is the matrix of F. Then F is unsatisfiable if and only if F is false under all Herbrand interpretations.

Proof: If a formula is unsatisfiable then F is false under all interpretations and hence false under all Herbrand interpretations.

Conversely, suppose F is satisfiable. Then there is an interpretation $\mathfrak{I}$ over a domain $\mathbf{D}$ such that $\mathfrak{I}(F) = \top$. Construct a Herbrand interpretation $\mathcal{HI}$ as follows:

> For any n-ary predicate symbol P occurring in F and any $t_1, t_2, ..., t_n \in \mathcal{HU}(F)$, $\langle t_1, t_2, ..., t_n \rangle \in P$ if and only if $\langle \mathfrak{I}(t_1), \mathfrak{I}(t_2), ..., \mathfrak{I}(t_n) \rangle \in \mathfrak{I}(P)$.

For any atomic formula A, $\mathcal{HI}(A) = \mathfrak{I}(A)$. Hence if $p = 0$ in F, that is, if F is a quantifier free formula, $\mathcal{HI}(F) = \mathfrak{I}(F) = \top$. If $p > 0$ in F, then without any loss of generality one can assume that F has the form $\forall x B[x]$, where B is the matrix of F.

Now $\mathfrak{I}(F) = \mathfrak{I}(\forall x B[x]) = \top$.

Therefore $\mathfrak{I}(B[x / \mathfrak{I}(t)]) = \top$, where t is an arbitrary member of $\mathbf{D}$.

Now $\mathcal{HI}_{(x/t)}(B[x]) = \mathfrak{I}_{(x/\mathfrak{I}(t))}(B[x])$.

Since $B[x]$ is quantifier free, $\mathcal{HI}(B[x/t]) = \mathfrak{I}(B[x/t])$.

Therefore $\mathcal{HI}(B[x/t]) = \mathfrak{I}(B[x/\mathfrak{I}(t)]) = \top$.

Thus if a formula F is satisfiable in some interpretation, then F is also satisfiable under the Herbrand interpretation. Hence if F is unsatisfiable under all Herbrand interpretations then F is unsatisfiable.

Suppose $\forall x_1 \forall x_2 ... \forall x_n B[x_1, x_2, ..., x_n]$ is a formula in prenex normal form. An *instance of matrix B* means a formula of the form $B[x_1 / t_1, x_2 / t_2, ..., x_n / t_n]$ where $t_1, t_2, ..., t_n$ are elements of the Herbrand universe of the given formula.

Theorem 3-14: (Herbrand's Theorem) Suppose $F = \forall x_1 \forall x_2 ... \forall x_p B[x_1, x_2, ..., x_p]$ is a formula in prenex normal form. Then F is unsatisfiable if and only if there are finitely many instances of the matrix $B[x_1, x_2, ..., x_p]$ which are unsatisfiable.

Proof: Without any loss of generality, the formula F can be assumed to be of the form $\forall x B[x]$, where B is quantifier free. Suppose $B[x/t_1], B[x/t_2], ..., B[x/t_n]$ are finitely many instances of $B[x]$. Then $\forall x B[x] \vdash B[x/t_1] \wedge B[x/t_2] \wedge ... \wedge B[x/t_n]$. Thus the satisfiability of $\forall x B[x]$ implies the satisfiability of the instances $B[x/t_1], B[x/t_2], ..., B[x/t_n]$. Hence if the instances are unsatisfiable, then the formula $\forall x B[x]$ itself is unsatisfiable.

Conversely, suppose $\forall x B[x]$ is unsatisfiable and, if possible, let any finitely many instances of $B[x]$ be satisfiable. Then, by the compactness theorem, $\{B[x/t] \mid t \in \mathcal{HU}(F)\}$ is satisfiable. Accordingly, $\mathcal{HI}(\forall x B[x]) = \top$, contradicting the assumption that $\forall x B[x]$ is unsatisfiable. Hence there exist finitely many instances of $B[x]$ which are unsatisfiable.

Suppose F is an arbitrary formula. With the aid of the transformation procedure described earlier in this subsection, one can assume that the formula F has the Skolem conjunctive normal form $\forall x_1 \forall x_2 ... \forall x_n (B_1 \wedge B_2 \wedge ... \wedge B_m)$, where each B_i is a disjunction of literals. If $x_1, x_2, ..., x_n$ are all the free variables of the conjunction then, for the sake of convenience, the formula $\forall x_1 \forall x_2 ... \forall x_n (B_1 \wedge B_2 \wedge ... \wedge B_m)$ is rewritten as the set $\{B_1, B_2, ..., B_m\}$.

Thus when a set $\mathbf{S} = \{B_1, B_2, ..., B_m\}$ is given, where each B_i is a disjunction of literals, then $\mathbf{S}$ is regarded as the form $\forall x_1 \forall x_2 ... \forall x_n (B_1 \wedge B_2 \wedge ... \wedge B_m)$ in which $x_1, x_2, ..., x_n$ are all the free variables occurring in all the B_i s. The set $\mathbf{S}$ can also be regarded as the conjunction of clauses $\forall B_1, \forall B_2, ..., \forall B_m$.

<u>Example</u>

Continuing with the previous example, the formula
$$\forall x \forall z \forall x_1 ((R(x) \vee P(a)) \wedge (R(x) \vee Q(z,a)) \wedge (\neg P(x_1) \vee Q(f(x_1), x_1)))$$

can equivalently be written as a set
$$\{\forall x(R(x) \vee P(a)), \forall x \forall z(R(x) \vee Q(z,a)), \forall x_1(\neg P(x_1) \vee Q(f(x_1), x_1))\}$$

of clauses or as a set
$$\{R(x) \vee P(a), R(x) \vee Q(z,a), \neg P(x_1) \vee Q(f(x_1), x_1)\}$$

of clauses when the free variables are assumed to be universally quantified at the front of each clause.

An alternative version of Herbrand's theorem (Theorem 3-14) based on the representation of a formula as a set of clauses is stated below. The proof follows immediately from the original theorem.

Proposition 3-16: A set S of clauses is unsatisfiable if and only if there are finitely many ground instances of clauses of S which are unsatisfiable.

3.3.3　*Implementation of Herbrand's theorem*

Suppose S is a finite, unsatisfiable set of clauses. Then according to Herbrand's theorem, the unsatisfiability of S can be proved in the following manner. First, enumerate a sequence $S_1, S_2, \ldots$ of finite sets of ground instances from S such that $S_1 \subseteq S_2 \subseteq \ldots$ and then test the satisfiability of each S_i. One way of enumerating the sequence $S_1, S_2, \ldots$ is that each member of S_i is of length at most i. (If S_i is the finite set of all ground instances of length $\leq i$, then we eventually have $S \subseteq S_k$ for some k; S_k will be unsatisfiable.) Since each member of S_i is ground, its satisfiability can be checked by a standard method (for example by truth table) of propositional logic.

Gilmore's approach (Gilmore, 1960) was to transform each S_i as it is generated to its equivalent disjunctive normal form D and then to remove from D any conjunction in D containing a complementary pair of literals. If D (disjunction equivalent to S_i, for some i) becomes empty at some stage due to this transformation, then S_i is unsatisfiable and therefore S is unsatisfiable. Gilmore's idea was combinatorially explosive because, for example, if an S_i contains 10 three-literal clauses then there will be 3^{10} conjunctions to be tested.

A more efficient approach to test the unsatisfiability for a set of ground clauses was devised by Davis and Putnam (1960). This method consists of a number of rules which are stated below.

- *Tautology Rule*: Suppose S_1 is obtained from **S** by deleting all clauses from **S** that are tautologies. Then S_1 is unsatisfiable if and only if **S** is unsatisfiable.

- *One-literal Rule*: Suppose there is a unit clause L in **S** and S_1 is obtained from **S** by deleting all clauses from **S** that contains an occurrence of L, and then replacing each clause C from the rest of **S** by a clause obtained from C by removing the occurrence (if any) of the complement of L. Then S_1 is unsatisfiable if and only if **S** is unsatisfiable.

- *Pure-literal Rule*: If a literal L occurs in some clause in **S** but its complement L' does not occur in any clause, then S_1 is obtained from **S** by deleting all clauses containing an occurrence of L. Then **S** is unsatisfiable if and only if S_1 is unsatisfiable. The literal L in this case is called a *pure literal*.

- *Splitting Rule*: Suppose $S = \{C_1 \vee A, ..., C_m \vee A, D_1 \vee \neg A, ..., D_n \vee \neg A, E_1, ..., E_p\}$, where A is an atom and each C_i, D_j, and E_k does not contain any occurrence of A or $\neg A$. Set $S_1 = \{C_1, ..., C_m, E_1, ..., E_p\}$ and $S_2 = \{D_1, ..., D_n, E_1, ..., E_p\}$. Then **S** is unsatisfiable if and only if each S_1 and S_2 is unsatisfiable.

- *Subsumption Rule*: Suppose S_1 is obtained from **S** by deleting every clause D from **S** for which there is another clause C in **S** such that every literal that occurs in C also occurs in D. Then **S** is unsatisfiable if and only if S_1 is unsatisfiable.

The above set of rules is applied repeatedly until no more rules can be applied on the resulting sets. If each of the resulting sets contains the empty clause, then the given set of clauses will be unsatisfiable.

Example

The following steps provide an example:

Step 1: $S = \{\, P(a) \vee \neg P(a) \vee Q(b),\ Q(a),\ Q(a) \vee P(c),\ R(a) \vee P(b),$
$\neg Q(a) \vee S(a) \vee R(b),\ \neg S(b) \vee \neg P(b),\ S(b)\, \}$

Step 2: $\{\, Q(a),\ Q(a) \vee P(c),\ R(a) \vee P(b),\ \neg Q(a) \vee S(a) \vee R(b),$
$\neg S(b) \vee \neg P(b),\ S(b)\, \}$, by the Tautology rule

Step 3: $\{\, R(a) \vee P(b)\,,\; S(a) \vee R(b)\,,\; \neg S(b) \vee \neg P(b)\,,\; S(b)\,\}$, by the One-literal rule

Step 4: $\{\, S(a) \vee R(b)\,,\; \neg S(b) \vee \neg P(b)\,,\; S(b)\,\}$, by the Pure literal rule

Step 5: $\{\, \neg S(b) \vee \neg P(b)\,,\; S(b)\,\}$, by the Pure literal rule

Step 6: $\mathbf{S}_1 = \{\neg S(b), S(b)\}$, $\mathbf{S}_2 = \{\neg P(b)\}$, by the Splitting rule, with $A = S(b)$

Step 7: $\mathbf{S}_1 = \{\neg S(b), S(b)\}$, $\mathbf{S}_2 = \{\}$, by the One-literal rule

The set $\mathbf{S}_1$ is unsatisfiable but $\mathbf{S}_2$ is not. Therefore, $\mathbf{S}$ is satisfiable.

3.4   Resolution Theorem Proving

Theorem proving in a system of first-order logic using the *resolution principle* as the sole inference rule is called *resolution theorem proving*. This style of theorem proving avoids the major combinatorial obstacles to efficiency found in earlier theorem proving methods (Gilmore, 1960; Davis and Putnam, 1960), which used procedures based on Herbrand's fundamental theorem concerning first-order logic. A logic programming system adopts a version of resolution theorem proving as its inference subsystem. Hence it is important to have a clear idea about this kind of theorem proving to understand the internal execution of a logic programming system. This section carries out a detailed discussion of the resolution principle, resolution theorem proving and some important relevant topics including unification and refinements of the resolution principle. Most of the results in this chapter (except the last section) are due to Robinson (1965).

3.4.1   *Resolution principle and unification*

The resolution principle is an inference rule of first-order logic which states that from any two clauses C and D, one can infer a *resolvent* of C and D. The principle idea behind the concept of resolvent, and hence behind the resolution principle, is that of unification. *Unification* is a process of determining whether two expressions can be made identical by some appropriate substitution for their variables. Some definitions and results must be established before formally introducing the concept of unification.

Terms and literals are the only *well-formed expressions* (or simply *expressions*). A *substitution* θ is a finite set of pairs of variables and terms, denoted by $\{x_1 / t_1, x_2 / t_2, ..., x_n / t_n\}$, where x_i s are distinct variables and each t_i is a term different from x_i. The term t_i is called a *binding* for the variable x_i. A substitution θ is called a *ground substitution* if each t_i is a ground term. The substitution given by the empty set is called the *empty substitution* (or *identity substitution*) and is denoted by $\{\}$ or ε. In a substitution $\{x_1 / t_1, x_2 / t_2, ..., x_n / t_n\}$, variables $x_1, x_2, ..., x_n$ are called the *variables of the substitution* and $t_1, t_2, ..., t_n$ are called the *terms of the substitution*.

Let $\theta = \{x_1 / t_1, x_2 / t_2, ..., x_n / t_n\}$ be a substitution and E be an expression. The *application* of θ to E, denoted by $E\theta$, is the expression obtained by simultaneously replacing each occurrence of the variable x_i in E by the term t_i. In this case $E\theta$ is called the *instance* of E by θ. If $E\theta$ is ground then $E\theta$ is called a *ground instance* of E, and E is referred to as a *generalization* of $E\theta$.

Let $\theta = \{x_1 / t_1, x_2 / t_2, ..., x_m / t_m\}$ and $\phi = \{y_1 / s_1, y_2 / s_2, ..., y_n / s_n\}$ be two substitutions. Then the *composition* $\theta\phi$ of θ and ϕ is the substitution obtained from the set

$$\{x_1 / t_1\phi, x_2 / t_2\phi, ..., x_m / t_m\phi, y_1 / s_1, y_2 / s_2, ..., y_n / s_n\}$$

by deleting any binding $x_i / t_i\phi$ for which $x_i = t_i\phi$ and deleting any binding y_i / s_i for which $y_i \in \{x_1, x_2, ..., x_m\}$.

Example

Let $E = p(x, f(x), y, g(a))$ and $\theta = \{x / b, y / h(x)\}$. Then $E\theta = p(b, f(b), h(x), g(a))$. Let $\theta = \{x / b, y / h(z)\}$ and $\phi = \{z / c\}$. Then $\theta\phi = \{x / b, y / h(c), z / c\}$.

The following set of results describes different properties of substitutions and their compositions.

Proposition 3-17: For any substitution θ, $\theta\varepsilon = \varepsilon\theta = \theta$, where ε is the empty substitution.

Proposition 3-18: Let E be any string and α, β be two arbitrary substitutions. Then $(E\alpha)\beta = E(\alpha\beta)$.

Proof: Let $\alpha = \{x_1/e_1, x_2/e_2, ..., x_n/e_n\}$ and $\beta = \{y_1/s_1, y_2/s_2, ..., y_m/s_m\}$. The string E can be assumed to be of the form $E = E_0 x_{i_1} E_1 x_{i_2} ... E_{p-1} x_{i_p} E_p$, where none of the substring E_j of E contains occurrences of variables $x_1, x_2, ..., x_n$, some of E_j are possibly null, and $1 \le i_j \le n$ for $j = 1, 2, ..., p$. Therefore, $E\alpha = E_0 e_{i_1} E_1 e_{i_2} ... E_{p-1} e_{i_p} E_p$ and $(E\alpha)\beta = T_0 t_{i_1} T_1 t_{i_2} ... T_{p-1} t_{i_p} T_p$, where each $t_{i_j} = e_{i_j}\beta$ and $T_j = E_j \gamma$, and γ is a substitution whose variables are not among $x_1, x_2, ..., x_n$. But each element in the composite substitution $\alpha\beta$ is of the form x_i/t_k whenever t_k is different from x_i. Hence $E(\alpha\beta) = T_0 t_{i_1} T_1 t_{i_2} ... T_{p-1} t_{i_p} T_p$.

Proposition 3-19: Let α, β be two arbitrary substitutions. If $E\alpha = E\beta$ for all strings E, then $\alpha = \beta$.

Proof: Let $x_1, x_2, ..., x_n$ be all variables of the two substitutions α and β. Since $E\alpha = E\beta$ for string E and each x_i is also a string, $x_i\alpha = x_i\beta$, $i = 1, 2, ..., n$. Hence the elements of α and β are the same.

Proposition 3-20: The composite operation on substitution is associative, that is, for any substitution α, β, γ, $(\alpha\beta)\gamma = \alpha(\beta\gamma)$. Hence in writing a composition of substitutions, parentheses can be omitted.

Proof: Let E be a string. Then by Proposition 3-18, we have the following:

$$E((\alpha\beta)\gamma) = (E(\alpha\beta))\gamma = ((E\alpha)\beta)\gamma = (E\alpha)(\beta\gamma) = E(\alpha(\beta\gamma)).$$

Hence by Proposition 3-19, $(\alpha\beta)\gamma = \alpha(\beta\gamma)$.

A *unifier* of two expressions E and E' is a substitution θ such that $E\theta$ is syntactically identical to $E'\theta$. If the two atoms do not have a unifier then they are not *unifiable*. A unifier θ is called a *most general unifier* (*mgu*) for the two expressions E and E' if for each unifier α of E and E' there exists a substitution β such that $\alpha = \theta\beta$.

Example

An mgu of two expressions $p(x, f(a, y))$ and $p(b, z)$ is $\{x/b, z/f(a, y)\}$. A unifier of these two expressions is $\{x/b, y/c, z/f(a, c)\}$.

An mgu is unique up to variable renaming. Because of this property the mgu of two expressions is used quite often. The concept of disagreement set for a set of expressions is pertinent here for presenting an algorithm to find an mgu for a set of expressions (if it exists).

Let S be any set $\{E_1, E_2, \ldots E_n\}$ of well-formed expressions. Then the *disagreement set* of S, denoted as $\mathcal{D}(S)$, is obtained by locating the first symbol at which not all E_i have exactly the same symbol, and then extracting from each E_i the subexpression t_i that begins with the symbol. The set $\{t_1, t_2, \ldots t_n\}$ is the disagreement set of S.

Example

Let S be $\{P(a, b, x, y), P(a, b, f(x, y), z), P(a, b, g(h(x)), y)\}$. The string of the first six symbols in each of the expressions in S is "$P(a, b,$". The first symbol position at which not all expressions in S are exactly the same is the seventh position. The extracted subexpressions from each of the expressions in S starting from the seventh position are $x, f(x, y)$, and $g(h(x))$. Hence the disagreement set $\mathcal{D}(S) = \{x, f(x, y), g(h(x))\}$.

Algorithm 3-1: (Unification Algorithm)

Input: S, a set of well-formed expressions.

Step 1: Set $i = 0$, $\theta_0 = \varepsilon$.

Step 2: If $S\theta_i$ is singleton then set $\theta_S = \theta_i$ and return θ_S as an mgu for S.

Step 3: If elements x_i and e_i do not exist in the disagreement set $\mathcal{D}(S\theta_i)$ such that x_i is a variable and x_i does not occur in e_i then stop; S is not unifiable. This check of whether x_i occurs in e_i or not is called the *occur check*.

Step 4: Set $\theta_{i+1} = \theta_i \{x_i / e_i\}$.

Step 5: Set $i = i + 1$ and go to Step 2.

This unification algorithm always terminates for any finite non-empty set of well-formed expressions; otherwise, an infinite sequence $\mathbf{S}\theta_0, \mathbf{S}\theta_1, \ldots$ would be generated. Each such $\mathbf{S}\theta_i$ is a finite non-empty set of well-formed expressions and $\mathbf{S}\theta_{i+1}$ contains one less variable than $\mathbf{S}\theta_i$. So an infinite sequence is not possible, because $\mathbf{S}$ contains only finitely many distinct variables.

Theorem 3-15 below proves that if $\mathbf{S}$ is unifiable, then the algorithm always finds an mgu for $\mathbf{S}$.

Examples

We find an mgu for $\mathbf{S} = \{P(x, g(y), f(g(b))), P(a, z, f(z))\}$ by applying the above unification algorithm.

- $i = 0$, $\theta_0 = \varepsilon$ — By Step 1
- $\mathbf{S}\theta_0 = \mathbf{S}$ and $\mathbf{S}\theta_0$ is not a singleton — By Step 2
- $\mathcal{D}(\mathbf{S}\theta_0) = \{x, a\}$, $x_0 = x$, $e_0 = a$ — By Step 3
- $\theta_1 = \{x/a\}$ — By Step 4
- $i = 1$ — By Step 5
- $\mathbf{S}\theta_1 = \{P(a, g(y), f(g(b))), P(a, z, f(z))\}$ — By Step 2
- $\mathcal{D}(\mathbf{S}\theta_1) = \{g(y), z\}$, $x_1 = z$, $e_1 = g(y)$ — By Step 3
- $\theta_2 = \theta_1 \{z/g(y)\} = \{x/a\} \{z/g(y)\} = \{x/a, z/g(y)\}$ — By Step 4
- $i = 2$ — By Step 5
- $\mathbf{S}\theta_2 = \{P(a, g(y), f(g(b))), P(a, g(y), f(g(y)))\}$ — By Step 2
- $\mathcal{D}(\mathbf{S}\theta_2) = \{y, b\}$, $x_2 = y$, $e_2 = b$ — By Step 3
- $\theta_3 = \theta_2 \{y/b\} = \{x/a, z/g(y)\} \{y/b\} = \{x/a, z/g(b), y/b\}$, By Step 4
- $i = 3$ — By Step 5
- $\mathbf{S}\theta_3 = \{P(a, g(b), f(g(b)))\}$, $\mathbf{S}\theta_3$ is singleton and $\theta_{\mathbf{S}} = \theta_3$, By Step 2

The algorithm terminates here. As another example, the set $\mathbf{S} = \{P(x, x), P(f(a), g(a))\}$ does not have an mgu.

Theorem 3-15: Let $\mathbf{S}$ be any finite non-empty set of well-formed expressions. If $\mathbf{S}$ is unifiable then the above unification algorithm always terminates at Step 2, and $\theta_{\mathbf{S}}$ is an mgu of $\mathbf{S}$.

Proof: Suppose $\mathbf{S}$ is unifiable. To prove that $\theta_{\mathbf{S}}$ is an mgu of $\mathbf{S}$, then for each $i \geq 0$ (until the algorithm terminates) and for any unifier θ of $\mathbf{S}$, $\theta = \theta_i \beta_i$ must hold at Step 2, for some substitution β_i. This is proved by induction on i.

Base Step: $i = 0$, $\theta_0 = \varepsilon$ and hence β_i can be taken as θ.

Induction Step: Assume $\theta = \theta_i \beta_i$ holds for $0 \leq i \leq n$. The unification algorithm stops at Step 2 if $\mathbf{S}\theta_n$ is a singleton then, since $\theta = \theta_n \beta_n$, θ_n is an mgu of $\mathbf{S}$. The unification algorithm will find $\mathcal{D}(\mathbf{S}\theta_n)$ at Step 3 if $\mathbf{S}\theta_n$ is not a singleton then. Since $\theta = \theta_n \beta_n$ and θ is a unifier of $\mathbf{S}$, β_n unifies $\mathbf{S}\theta_n$. Therefore β_n also unifies the disagreement set $\mathcal{D}(\mathbf{S}\theta_n)$. Hence x_n and e_n defined in Step 3 of the unification algorithm satisfy $x_n \beta_n = e_n \beta_n$. Since $\mathcal{D}(\mathbf{S}\theta_n)$ is a disagreement set, at least one well-formed expression in $\mathcal{D}(\mathbf{S}\theta_n)$ begins with a variable. Since a variable is also a well-formed expression, at least one well-formed expression in $\mathcal{D}(\mathbf{S}\theta_n)$ is a variable. Take this variable as x_n and suppose e_n is any other well-formed expression from $\mathcal{D}(\mathbf{S}\theta_n)$. $x_n \beta_n = e_n \beta_n$, since β_n unifies $\mathcal{D}(\mathbf{S}\theta_n)$ and x_n, e_n are members of $\mathcal{D}(\mathbf{S}\theta_n)$. Now if x_n occurs in e_n, $x_n \beta_n$ occurs in $e_n \beta_n$. This is impossible because x_n and e_n are distinct well-formed expressions and $x_n \beta_n = e_n \beta_n$. Therefore x_n does not occur in e_n. Hence the algorithm does not stop at Step 3. Step 4 will set $\theta_{n+1} = \theta_n \{x_n / e_n\}$. Step 5 will set $i = n + 1$ and the control of the algorithm will be back at Step 2.

Set $\beta_{n+1} = \beta_n - \{x_n / x_n \beta_n\}$

$$
\begin{aligned}
\beta_n &= \{x_n / x_n \beta_n\} \cup \beta_{n+1} & \\
&= \{x_n / e_n \beta_n\} \cup \beta_{n+1} & \text{Since } x_n \beta_n = e_n \beta_n \\
&= \{x_n / e_n(\{x_n / x_n \beta_n\} \cup \beta_{n+1})\} \cup \beta_{n+1} & \text{Since } \beta_n = \{x_n / x_n \beta_n\} \cup \beta_{n+1} \\
&= \{x_n / e_n \beta_{n+1}\} \cup \beta_{n+1} & \text{Since } x_n \text{ does not occur in } e_n \\
&= \{x_n / e_n\} \cup \beta_{n+1} & \text{Definition of composite substitution}
\end{aligned}
$$

Thus $\theta = \theta_n \beta_n = \theta_n \{x_n / e_n\} \beta_{n+1} = \theta_{n+1} \beta_{n+1}$.

Hence $\theta = \theta_i \beta_i$, for all $i = 0, 1, \ldots, n+1$.

By the induction principle, for all $i \geq 0$, there is a substitution β_i such that $\theta = \theta_i \beta_i$ until the algorithm terminates in Step 2, for some $i = m$. Furthermore, $\theta_S = \theta_m$ is an mgu for **S**.

Suppose a subset of the set of all literals occurring in a clause C has the same sign (that is, all are negated, or all are not negated) and has an mgu of θ. Then $C\theta$ is called a *factor* of C. If $C\theta$ is a unit clause, then it is called a *unit factor* of C.

Example

Suppose $C = P(x,a) \vee P(f(y),z) \vee Q(y,z)$. Then the literals $P(x,a)$ and $P(f(y),z)$ have the same sign and unify with an mgu $\theta = \{x / f(y), z / a)\}$. Thus $C\theta = p(f(y),a) \vee Q(y,a)$ is a factor of C. If $C = P(x,a) \vee P(f(y),z) \vee P(f(b),a)$, then $P(f(b),a)$ is a unit factor of C.

Suppose C and D are two clauses with no variables in common. Suppose L and M are literals occurring in C and D respectively, where the literals L and M are complementary to each other and their atoms unify with an mgu of θ. Then the clause $(C\theta - \{L\theta\}) \cup (D\theta - \{M\theta\})$, where $X - Y$ means removal of Y from the disjunction X, is called a *binary resolvent* of C and D. The literals L and M are called the *literals resolved upon* and C and D are the *parent clauses* of the resolution operation.

A *resolvent* of clauses C and D is a binary resolvent of C_1 and D_1, where C_1 is either C or a factor of C, and D_1 is either D or a factor of D. Hence a binary resolvent of two clauses C and D is also a resolvent of C and D. The symbol $\square$ denotes an empty resolvent, which is always unsatisfiable.

Example

Suppose $C = P(x,z) \vee Q(f(a)) \vee Q(z)$ and $D = \neg Q(f(y)) \vee \neg R(y)$. Consider L and M as $Q(f(a))$ and $\neg Q(f(y))$ respectively. Then $\theta = \{y / a\}$ and $(C\theta - \{L\theta\}) \cup (D\theta - \{M\theta\}) = P(x,z) \vee Q(z) \vee \neg R(a)$. The resolvent

$P(x, f(a)) \vee \neg R(a)$ of C and D is a binary resolvent between the factor $P(x, f(a)) \vee Q(f(a))$ of C and $\neg Q(f(a)) \vee \neg R(a)$ of D.

This section first establishes a resolution theorem for ground clauses. It is then generalized to establish a resolution theorem for first-order clauses.

The *resolution* of a set **S** of clauses, denoted by $Res(\mathbf{S})$, is the set of all clauses consisting of the members of **S**, together with all the resolvents of all pairs of members of **S**. The *n-th resolution* of a set **S** of clauses, denoted by $Res^n(\mathbf{S})$, is recursively defined as follows:

- $Res^0(\mathbf{S}) = \mathbf{S}$

- $Res^{n+1}(\mathbf{S}) = Res(Res^n(\mathbf{S}))$, $n = 0, 1, \ldots$

It is clear from the above definition that $Res^0(\mathbf{S}) = \mathbf{S} \subseteq Res^1(\mathbf{S}) \subseteq Res^2(\mathbf{S}) \subseteq \ldots$.

Proposition 3-21: If **S** is a finite set of ground clauses then not all inclusions in the above chain are proper.

Proof: Straightforward because resolution does not introduce any new literals in the case of ground clauses.

Theorem 3-16: (Ground Resolution Theorem) If **S** is any finite set of ground clauses then **S** is unsatisfiable if and only if $Res^n(\mathbf{S})$ contains $\square$ (empty clause), for some $n \geq 0$.

Proof: Suppose **S** is unsatisfiable and consider the chain $\mathbf{S} = Res^0(\mathbf{S}) \subseteq Res^1(\mathbf{S}) \subseteq Res^2(\mathbf{S}) \subseteq \ldots$. By Proposition 3-21, this chain terminates. Let $\mathbf{T} = Res^n(\mathbf{S})$ be the terminating set. The set **T** is closed under resolution and $\mathbf{S} \subseteq \mathbf{T}$. Suppose **T** does not contain the empty clause $\square$. Suppose $A_1, \ldots, A_m$ are all the atoms occurring in **T**. Let **M** be a set constructed as follows:

$$\mathbf{M}_0 \quad = \Phi$$

$$\mathbf{M}_j \quad = \mathbf{M}_{j-1} \cup \{A_j\},$$

if there does not exist a clause $M_1 \vee ... \vee M_p$ in $\mathbf{T}$

$\{M_1^c,...,M_p^c\} \subseteq \mathbf{M}_j$, where M_i^c is the complement of M_i

$$= \mathbf{M}_{j-1} \cup \{\neg A_j\}, \qquad \text{otherwise for } j = 1, 2, ..., m.$$

$$\mathbf{M} = \mathbf{M}_m$$

It can now be shown that $\mathbf{M}$ satisfies $\mathbf{T}$. If not, there is a clause C in $\mathbf{T}$ such that the complement of each of the literals occurring in C belongs to $\mathbf{M}_j$, for the least j, $0 \le j \le m$. Hence $\mathbf{M}_j$ is $\mathbf{M}_{j-1} \cup \{\neg A_j\}$. Thus C contains A_j (by the leastness of j). Since $\mathbf{M}_j$ is $\mathbf{M}_{j-1} \cup \{\neg A_j\}$, there exists a clause D in $\mathbf{T}$ such that the complement of each of the literals occurring in D belongs to $\mathbf{M}_{j-1} \cup \{A_j\}$. Thus D contains $\neg A_j$ (by the leastness of j). If the resolvent $(C - \{A_j\}) \cup (D - \{\neg A_j\})$ is non-empty, then the complement of each of the literals occurring in this resolvent is in the set $\mathbf{M}_{j-1}$. By the definition of construction of $\mathbf{T}$, $(C - \{A_j\}) \cup (D - \{\neg A_j\})$ is a member of $\mathbf{T}$ and cannot be empty. This contradicts the leastness of j. Hence $\mathbf{M}$ satisfies $\mathbf{T}$. Since $\mathbf{T}$ is satisfiable and $\mathbf{S} \subseteq \mathbf{T}$, $\mathbf{S}$ is also satisfiable, therefore, it is a contradiction. Thus the original assumption, that $\mathbf{T}$ does not contain empty clause, is false. Hence $\mathbf{T}$ contains $\square$.

To prove the converse of the theorem, suppose $\mathcal{R}es^n(\mathbf{S})$ contains $\square$ and hence $\mathcal{R}es^n(\mathbf{S})$ is unsatisfiable. Therefore $\mathcal{R}es^n(\mathbf{S})$ does not have any model. If C and D are two ground clauses, then any model of $\{C, D\}$ is also a model of $\{C, D, (C - \{L\}) \cup (D - \{L^c\})\}$. Therefore any model of $\mathbf{S}$ is also a model of $\mathcal{R}es^n(\mathbf{S})$. Since $\mathcal{R}es^n(\mathbf{S})$ does not have any model, $\mathbf{S}$ does not have any model. Therefore $\mathbf{S}$ is unsatisfiable.

Proposition 3-22: (Lifting Lemma) If C_1' and C_2' are instances of C_1 and C_2 respectively and C' is a resolvent of C_1' and C_2', then there is a resolvent C of C_1 and C_2 such that C' is an instance of C.

Proof: It can be assumed that there is no common variable between C_1 and C_2. If there is, then the variables are renamed accordingly. Suppose $C' = (C_1'\theta_p - \{L_1'\theta_p\}) \cup (C_2'\theta_p - \{L_2'\theta_p\})$, where θ_p is an mgu of the set of atoms occurring in L_1' and L_2' since C_i' is an instance of C_i, $C_i' = C_i\alpha$, $i = 1, 2$, for some substitution α. Let $L_i^1, ..., L_i^{n_i}$ be the literals of C_i corresponding to L_i', for $i = 1, 2$, that is, $L_i^1\alpha = ... = L_i^{n_i}\alpha = L_i'$.

When $n_i > 1$, suppose β_i is an mgu for $\{L_i^1, ..., L_i^{n_i}\}$ and let $L_i = L_i^1\beta_i = ... = L_i^{n_i}\beta_i$, $i = 1, 2$. Then L_i is a literal in the factor $C_i\beta_i$ of C_i. When $n_i = 1$, let $\beta_i = \varepsilon$ and $L_i = L_i'$. Then L_i' is an instance of L_i. Since the atoms occurring in L_1' and L_2' are unifiable and L_i' is an instance of L_i, the atoms occurring in L_1 and L_2 are unifiable. Let θ be an mgu in this case.

Let $C = ((C_1\beta_1)\theta - \{L_1\theta\}) \cup ((C_2\beta_2)\theta - \{L_2\theta\})$

$$= (C_1(\beta_1\beta_2\theta) - \{L_1^1...L_1^{n_1}\}(\beta_1\beta_2\theta)) - (C_2(\beta_1\beta_2\theta) - \{L_2^1...L_2^{n_2}\}(\beta_1\beta_2\theta))$$

since β_i does not act on L_j^k, $i \neq j$, $i, j = 1, 2$, $1 \leq k \leq n_i$.

Now $C' = (C_1'\theta_p - \{L_1'\theta_p\}) \cup (C_2'\theta_p - \{L_2'\theta_p\})$

$$= (C_1(\alpha\theta_p) - \{L_1^1...L_1^{n_1}\}(\alpha\theta_p)) - (C_2(\alpha\theta_p) - \{L_2^1...L_2^{n_2}\}(\alpha\theta_p))$$

Since $\beta_1\beta_2\theta$ is more general than $\alpha\theta_p$, therefore C' is an instance of C. Hence the lemma.

If **S** is any set of clauses and **P** is any set of terms, then the *saturation* of **S** over **P**, denoted as $Ground_\mathbf{P}(\mathbf{S})$, is the set of all ground clauses obtained from members of S by replacing variables with members of **P**. Occurrences of the same variable in any one clause are replaced by occurrences of the same term. When **P** is $\mathcal{HU}(\mathbf{S})$, then $Ground_\mathbf{P}(\mathbf{S})$ is simply $Ground(\mathbf{S})$.

Proposition 3-23: Let **S** be any set of clauses and **P** be any subset of $\mathcal{HU}(\mathbf{S})$. Then $Res(Ground_\mathbf{P}(\mathbf{S})) \subseteq Ground_\mathbf{P}(Res(\mathbf{S}))$.

Proof: Suppose $C' \in Res(Ground_\mathbf{P}(\mathbf{S}))$. Then there are two clauses C_1 and C_2 in S and their ground instances $C_1\alpha_1$ and $C_2\alpha_2$ such that C' is a resolvent of $C_1\alpha_1$ and $C_2\alpha_2$. Then, by the lifting lemma (Proposition 3-22), C' is an instance of C, where C is a resolvent of C_1 and C_2. Since C is a member of $\mathcal{R}es(\mathbf{S})$ and C' is a ground instance of C, $C' \in \mathcal{G}round_{\mathbf{P}}(\mathcal{R}es(\mathbf{S}))$. Hence $\mathcal{R}es(\mathcal{G}round_{\mathbf{P}}(\mathbf{S})) \subseteq \mathcal{G}round_{\mathbf{P}}(\mathcal{R}es(\mathbf{S}))$.

Example

$$\mathbf{S} = \{P(a) \vee \neg Q(x), Q(b)\}$$
$$\mathbf{P} = \{a\}$$
$$\mathcal{H}\mathcal{U}(\mathbf{S}) = \{a, b\}$$
$$\mathcal{R}es(\mathbf{S}) = \{P(a) \vee \neg Q(x), Q(b), P(a)\}$$
$$\mathcal{G}round_{\mathbf{P}}(\mathcal{R}es(\mathbf{S})) = \{P(a) \vee \neg Q(a), Q(b), P(a)\}$$
$$\mathcal{G}round_{\mathbf{P}}(\mathbf{S}) = \{P(a) \vee \neg Q(a), Q(b)\}$$
$$\mathcal{R}es(\mathcal{G}round_{\mathbf{P}}(\mathbf{S})) = \{P(a) \vee \neg Q(a), Q(b)\}$$
$$\mathcal{R}es(\mathcal{G}round_{\mathbf{P}}(\mathbf{S})) \subseteq \mathcal{G}round_{\mathbf{P}}(\mathcal{R}es(\mathbf{S}))$$
$$\mathcal{R}es(\mathcal{G}round_{\mathbf{P}}(\mathbf{S})) \neq \mathcal{G}round_{\mathbf{P}}(\mathcal{R}es(\mathbf{S}))$$

Hence the converse of Proposition 3-23 is not necessarily true.

Proposition 3-24: If $\mathbf{S}$ is any set of clauses and $\mathbf{P}$ is any subset of $\mathcal{H}\mathcal{U}(\mathbf{S})$, then $\mathcal{R}es^i(\mathcal{G}round_{\mathbf{P}}(\mathbf{S})) \subseteq \mathcal{G}round_{\mathbf{P}}(\mathcal{R}es^i(\mathbf{S}))$, for all $i \geq 0$.

Proof: This proposition is established by induction on i:

Base Step: $i = 0$. Then $\mathcal{R}es^0(\mathcal{G}round_{\mathbf{P}}(\mathbf{S})) = \mathcal{G}round_{\mathbf{P}}(\mathbf{S}) = \mathcal{G}round_{\mathbf{P}}(\mathcal{R}es^0(\mathbf{S}))$.

Induction Step: Suppose the result is true for $i = n$. Then:

$$\mathcal{R}es^{n+1}(\mathcal{G}round_{\mathbf{P}}(\mathbf{S})) = \mathcal{R}es(\mathcal{R}es^n(\mathcal{G}round_{\mathbf{P}}(\mathbf{S})))$$

$$\subseteq \mathcal{R}es(\mathcal{G}round_{\mathbf{P}}(\mathcal{R}es^n(\mathbf{S}))), \qquad \text{for } i = n$$

$$\subseteq \mathcal{G}round_{\mathbf{P}}(\mathcal{R}es(\mathcal{R}es^n(\mathbf{S}))), \qquad \text{Lemma @lemma rtp-res@}$$

$$= \mathcal{G}round_{\mathbf{P}}(\mathcal{R}es^{n+1}(\mathbf{S})), \qquad \text{By definition}$$

Hence the proposition.

The basic version of Herbrand's theorem (Theorem 3-14) can be restated as follows.

Theorem 3-17: If S is any finite set of clauses, then **S** is unsatisfiable if and only if for some finite subset **P** of $\mathcal{HU}(\mathbf{S})$, $\mathit{Ground}_{\mathbf{P}}(\mathbf{S})$ is unsatisfiable.

Theorem 3-18: If **S** is any finite set of clauses, then **S** is unsatisfiable if and only if for some finite subset **P** of $\mathcal{HU}(\mathbf{S})$ and some $n \geq 0$, $\mathcal{R}\!es^{n}(\mathit{Ground}_{\mathbf{P}}(\mathbf{S}))$ contains $\square$.

Proof: Since **S** and **P** are finite, $\mathit{Ground}_{\mathbf{P}}(\mathbf{S})$ is also finite. Hence the theorem follows from Theorem 3-16 and Theorem 3-17.

Theorem 3-19: If **S** is any finite set of clauses, then **S** is unsatisfiable if and only if for some finite subset **P** of $\mathcal{HU}(\mathbf{S})$ and some $n \geq 0$, $\mathit{Ground}_{\mathbf{P}}(\mathcal{R}\!es^{n}(\mathbf{S}))$ contains $\square$.

Proof: Follows from Theorem 3-18 and Proposition 3-24.

Theorem 3-20: (Resolution Theorem) If **S** is any finite set of clauses then **S** is unsatisfiable if and only if $\mathcal{R}\!es^{n}(\mathbf{S})$ contains $\square$, for some $n \geq 0$.

Proof: The set $\mathit{Ground}_{\mathbf{P}}(\mathcal{R}\!es^{n}(\mathbf{S}))$ will contain $\square$ if and only if $\mathcal{R}\!es^{n}(\mathbf{S})$ contains $\square$ (since the replacement of variables by terms cannot produce $\square$ from a non-empty clause in $\mathcal{R}\!es^{n}(\mathbf{S})$). Hence the theorem follows from Theorem 3-19 above.

3.5 Refutation Procedure

This section provides a procedure based on the above resolution theorem (Theorem 3-20) to derive the empty clause from an unsatisfiable set of clauses. Later in this section some techniques are provided to improve the efficiency of the procedure.

Let **S** be a set of clauses (called *input clauses*). A *derivation* (or *deduction*) in **S** is a sequence of clauses $C_1, C_2, \ldots$ such that each C_i is either in **S** or is a resolvent of C_j and C_k, where $1 \leq i \leq n, 1 \leq j, k \leq i$. In the latter case, C_i is a *derived clause*. A derivation is either *finite* or *infinite* according to the length of its sequence.

A *refutation* of $\mathbf{S}$ is a finite derivation $C_1,...,C_n$ in $\mathbf{S}$ such that $C_n = \square$.

The following theorem, which follows immediately from the resolution theorem, is the completeness theorem of a system of logic whose sole inference rule is the resolution principle stated in Section 3.4.

Theorem 3-21: A finite set $\mathbf{S}$ of clauses is unsatisfiable if and only if there is a refutation of $\mathbf{S}$.

One of the ways to find a refutation from a set of unsatisfiable clauses is to compute the sequence $\mathbf{S}, \mathcal{R}es^1(\mathbf{S}), \mathcal{R}es^2(\mathbf{S}),...$ until $\mathcal{R}es^n(\mathbf{S})$ contains $\square$. However, this procedure would be very inefficient. The following example demonstrates this inefficiency.

Example

$\mathbf{S} = \{$

    Clause 1: $\neg P(a)$ Given

    Clause 2: $P(x) \vee \neg Q(x)$ Given

    Clause 3: $P(x) \vee \neg R(f(x))$ Given

    Clause 4: $Q(a) \vee R(f(a))$ Given

    Clause 5: $R(b)$ Given

    Clause 6: $R(c) \vee \neg R(c)$ Given

$\}$

    Clause 7: $\mathcal{R}es(\mathbf{S}) = \mathbf{S} \cup \{$

    Clause 8: $\neg Q(a)$ 1 and 2

    Clause 9: $\neg R(f(a))$ 1 and 3

    Clause 9: $P(a) \vee R(f(a))$ 2 and 4

    Clause 10: $P(a) \vee Q(a)$ 3 and 4

$\}$

$$\mathcal{R}es^2(\mathbf{S}) = \mathcal{R}es(\mathbf{S}) \cup \{$$

Clause 11: $R(f(a))$	1 and 9
Clause 12: $Q(a)$	1 and 10
Clause 13: $P(a)$	2 and 10
Clause 14: $P(a)$	3 and 8
Clause 15: $P(a)$	3 and 9
Clause 16: $R(f(a))$	4 and 7
Clause 17: $Q(a)$	4 and 8

$$\}$$

$$\mathcal{R}es^3(\mathbf{S}) = \mathcal{R}es^2(\mathbf{S}) \cup \{$$

Clause 18: $\square$	1 and 13
Clause 19: ...	

$$\}$$

Many irrelevant, redundant clauses have been generated in the above example. Only the clauses 10 and 13 need to be generated to show that the set of clauses is unsatisfiable. The other clauses are redundant. Clause 6 is a tautology, which is true in any interpretation. Therefore if a tautology is deleted from an unsatisfiable set of clauses, the remaining set of clauses must still be unsatisfiable. Clause $P(a)$ is redundantly generated three times. Satisfiability of the set $\mathbf{S}$ will not depend on clause 5 as there is no clause in $\mathcal{R}es^n(\mathbf{S})$ with an occurrence of a literal $\neg A$ such that A unifies $R(b)$. Hence in this example, the given set $\mathbf{S}$ of clauses is satisfiable if and only if the set of clauses in $\mathbf{S}$ other than clause 5 is satisfiable.

To avoid the inefficiencies that would be caused for the above reasons, the refutation procedure incorporates a number of *search principles*.

- *Purity Principle*: Let $\mathbf{S}$ be any finite set of clauses. A literal L occurring in a clause C in $\mathbf{S}$ is said to be *pure* if there is no clause in D for which a resolvent $(C\theta - L\theta) \cup (D\theta - M\theta)$ exists. The *purity principle* is then stated as follows: From a finite set $\mathbf{S}$ of clauses any clause C containing a pure literal can be deleted from **S**. Then **S** is satisfiable if and only if the resulting set $\mathbf{S} - \{C\}$ is satisfiable.

- *Tautology Principle*: From a finite set **S** of clauses, any clause C which is a tautology can be deleted. Then **S** is satisfiable if and only if the remaining set of clauses $\mathbf{S} - \{C\}$ is satisfiable.

- *Subsumption Principle*: A clause C subsumes a clause D (or D is *subsumed by* C), where $C \neq D$, if there exists a substitution θ such that all the literals that occur in $C\theta$ also occur in D. The *subsumption principle* is then stated as follows: From a finite set **S** of clauses, any clause D which is subsumed by a clause in $\mathbf{S} - \{D\}$ can be deleted. Then **S** is satisfiable if and only if the resulting set $\mathbf{S} - \{D\}$ is satisfiable.

- *Replacement Principle*: The replacement principle can be stated as follows: Suppose C and D are two clauses and R is a resolvent of C and D which subsumes one of C and D. Then in adding R by the resolution principle, one of C and D which R subsumes can be simultaneously deleted.

The above search principles can be used to delete some of the redundant and irrelevant clauses generated during the refutation procedure. In spite of this, there are still many irrelevant clauses which cannot be deleted that are generated during the refutation procedure. *Refinement* of resolution is necessary to achieve an efficient theorem proving procedure. Many refinements of the resolution principles (Chang, 1970) have been proposed.

For example, *input resolution* requires that one of the parent clauses of each resolution operation must be an input clause, that is, not a derived clause. *Unit resolution* requires that at least one of the parent clauses or its factor in each resolution operation be a unit clause. Both unit and input resolutions are incomplete in general. *Linear resolution* is an extension of input resolution in which at least one of the parent clauses to each resolution operation must be either an input clause or an ancestor clause of the parent. Linear resolution is complete. *Linear resolution with selection function* (or *SL resolution*) (Kowalski and Kuehner, 1971) is a restricted form of linear resolution. The main restriction is effected by a selection function which chooses from each clause one single literal to be resolved upon in that clause. In the next chapter, we will make use of linear resolution for inferencing in logic programs.

3.6 Complexity Analysis

Recall that a propositional formula F is satisfiable if there is at least one interpretation I which evaluates the formula F as $\top$ (true). If I satisfies F then I is a model of F. This model checking problem of a formula under an interpretation can be done in polynomial time. On the other hand, the satisfiability problem (that is, to check if there is an interpretation which is a model) is NP-complete. The corresponding Co-NP problem is testing if F is a logical consequence of a set Γ of formulae (i.e. $\Gamma \vDash F$). The best algorithms for solving NP-complete or Co-NP-complete problems require exponential time in the worst case.

Before studying the complexity of satisfiability and logical consequence within the domain of first-order logic, we study first the complexity within an intermediate syntax, called Quantified Boolean Formulae (QBF), which are a generalization of propositional formulae. QBF characterize the complexity class PSPACE, and the complexity results of some systems of propositional and first-order logics can be established by comparing them to QBF.

A QBF is of the form

$$Q_1 X_1 ... Q_n X_n F(X_1,...,X_n)$$

where F is a propositional formula involving the propositional boolean variables $X_1,...,X_n$ and each Q_i is an existential ($\exists$) or universal quantifier ($\forall$). The expression $\exists X_i F$ is read as "there exists a truth assignment to X_i such that F is true," and the expression $\forall X_i F$ is read as "for every truth assignment to X_i, F is true." Following is an example of a true QBF:

$$\forall X_1 \exists X_2 ((X_1 \vee X_2) \wedge (\neg X_1 \vee \neg X_2))$$

On the other hand, $\exists X_1 \forall X_2 ((X_1 \vee X_2) \wedge (\neg X_1 \vee \neg X_2))$ is false. The evaluation problem for a QBF is to decide whether it is true or not. The propositional satisfiability problem for a formula F coincides with the evaluation problem for $\exists X_1 ... \exists X_n F$, where $X_1,...,X_n$ are the propositional variables occurring in F, which is NP-complete. The logical consequence problem for a formula F from a set of formulae Γ coincides with the evaluation problem $\forall X_1 ... \forall X_n (\Gamma \rightarrow F)$, where $X_1,...,X_n$ are the propositional variables occurring in Γ and F, which is Co-NP-complete. In general, the evaluation problem for a QBF is PSPACE-complete. This is because an algorithm can be devised easily that cycles through all possible assignments to the free variables, and substitutions for the quantified variables. In each cycle one needs to store polynomially many values. As the values in each cycle are independent of the previous cycle, we can overwrite them to stay in PSPACE.

Model checking for first-order logic is PSPACE-complete (Stockmeyer 1974, Vardi 1982). Given a finite first-order structure and a first-order formula, just implement an algorithm that cycles through all possible assignments to the free variables, and cycles through possible substitutions for the quantified variables. In each cycle we only need to store polynomially many values. As the values in each cycle are independent of the previous one, we can "overwrite" them, and stay in PSPACE.

The proof is a reduction from the Quantified Boolean Formula (QBF) satisfiability problem. Essentially, evaluating a QBF is shown to be equivalent to model-checking a structure with just two distinguishable elements representing the Boolean values $\top$ (true) or $\bot$ (false). Thus, the complexity of model checking in first-order logic comes purely from large and complicated input formulas.

3.7   Further Readings

There are dozens of good text books on classical logics, of which three excellent ones are (Copi, 1979), (Mendelson, 1987), and (Stoll, 1963). Robinson's paper (1965) is highly recommended as a foundational reading. Chang and Lee (1973) provide a thorough coverage of automated theorem proving. For a more recent account on automated theorem proving, see Gallier (1987 & 2003).

Chapter 4

Logic Programming

This chapter presents the theory and practice of logic programming, a declarative programming paradigm, as opposed to procedural programming such as Java and C++. A logic program is just a collection of logical formulae represented in the form of definite *if-then* rules. A logic program can be regarded as a propositional epistemic model of an agent. Reasoning with such epistemic models, which are represented in a restricted first-order language syntax, not only guarantees decidability and avoids the high complexity involved in theorem proving, but also provides a natural way of expressing an agent's day-to-day decision-making knowledge as arguments.

Program clauses that constitute a logic program are classified according to their syntactic structures. Several categories of logic programs are presented here along with their models and proof-theoretic semantics. Then we describe specialized resolution theorem proving schemes for these categories, namely SLD resolution for implementing backward chaining. Finally, we present the logic programming language Prolog, which can be effectively used to implement the logical and modal epistemic models via "meta-interpreters."

4.1 The Concept

Kowalski (1979b) represented the analysis of an algorithm by the equation

$$Algorithm = Logic + Control$$

that is, given an algorithm for solving a particular problem, the *logic component* specifies the knowledge which can be used in solving the problem and the *control component* determines the way this knowledge can be used to arrive at a solution to the problem. To clarify the concepts of logic and control components, consider the following definition of factorial, which is specified using standard mathematical symbols:

97

$$Fact(n) = 1 \qquad\qquad \text{when } n = 0$$
$$ = n \times Fact(n-1) \qquad\qquad \text{when } n > 0$$

This symbolized definition of factorial constitutes the logic component of the following procedural algorithm for computing a factorial:

Input: n

Step 1: $Fact = 1$

Step 2:   If $n = 0$ then return $Fact$

Step 3: $Fact = Fact \times n$; set $n = n - 1$; go to Step 2

The control component of the above algorithm is embedded within it in the form of branching operators such as "return" and "go to". Unlike procedural programming, in a *logic programming* system, a programmer specifies only the logic component of an algorithm, called the logic program, using the notion of mathematical logic. For example, the definition of factorial as the logic component in a logic programming system is expressed by a well-formed formula in the first-order logic as

$$Fact(0,1) \wedge \forall x_1 \forall x_2 \forall y_1 \forall y_2 (Fact(x_1,y_1) \wedge x_2 = x_1 + 1 \wedge y_2 = x_2 \times y_1$$
$$\rightarrow Fact(x_2,y_2))$$

or, equivalently, as a set of two *definite clauses* as

$$Fact(0,1)$$

$$Fact(x_2,y_2) \leftarrow Fact(x_1,y_1) \wedge x_2 = x_1 + 1 \wedge y_2 = x_2 \times y_1$$

where the interpretation of $Fact(x,y)$ is that number y is the factorial of number x and the other symbols are given their usual meaning ($\leftarrow$ is the same as implication).

The control component in a logic programming system can either be expressed by the programmer through a separate control-specifying language or can be determined entirely by the logic programming system itself. The efficiency of an algorithm can be increased by tuning its control component without changing its logic component. Although a logic program lacks sufficient procedural information for the system to do anything other than make an exhaustive search to find the correct execution path, separating the logic component from the control component has several advantages, including advantages pertaining to the correctness of the algorithm, program improvement, and modification. In a logic programming system, the logic component of an algorithm is specified in mathematical logic. More generally, in a *declarative programming* environment, the programmer specifies the logic component of an algorithm in a standard symbolized manner.

4.2   Program Clauses and Goals

A non-empty clause can be rewritten as

$$M_1 \vee \ldots \vee M_p \vee \neg N_1 \vee \ldots \vee \neg N_q, \qquad p+q \geq 1$$

where M_is and N_js are positive literals, and variables are implicitly, universally quantified over the whole disjunction. Every closed wff can be transformed into this clausal form. The above clause can be written in the form of a *program clause* as

$$M_1 \vee \ldots \vee M_k \leftarrow N_1 \wedge \ldots \wedge N_q \wedge \neg M_{k+1} \ldots \wedge \neg M_p, \quad k \geq 1, p \geq 0, q \geq 0$$

or, in the form of a *goal* as

$$\leftarrow N_1 \wedge \ldots \wedge N_q \wedge \neg M_1 \vee \ldots \vee \neg M_p, \qquad p+q \geq 1$$

Thus, a program clause (or simply *clause*) is a formula of the form

$$A_1 \vee \ldots \vee A_m \leftarrow L_1 \wedge \ldots \wedge L_n, \qquad m \geq 1, n \geq 0$$

where $A_1 \vee \ldots \vee A_m$ is the *head* (or *conclusion* or *consequent*) of the program clause and $L_1 \wedge \ldots \wedge L_n$ is the *body* (or *condition* or *antecedent*) . Each A_i is an atom and each L_j is either an atom (a *positive condition*) or a negated atom (a *negative condition*).   Any variables in $A_1, \ldots, A_m, L_1, \ldots, L_n$ are assumed to be universally quantified over the whole formula. Each A_i (respectively, L_j) is said to have *occurred* in the head (respectively, body) of the clause. The various forms of the clause $A_1 \vee \ldots \vee A_m \leftarrow L_1 \wedge \ldots \wedge L_n$ with values of m and n of interest are the following:

1. $m = 1, n = 0$. That is, the clause has the form

$$A \leftarrow \text{ (or simply } A)$$

   in which the body is empty and the head is a single atom. This clause is called a *unit clause* or *unit assertion*.

2. $m = 1, n \geq 0$, and each L_i is an atom. That is, the clause has the form

$$A \leftarrow B_1 \wedge \ldots \wedge B_n$$

   in which A and B_is are atoms. This clause is called a *definite clause*.

3. $m = 1$, $n \geq 0$. That is, the clause has the form

$$A \leftarrow L_1 \wedge \ldots \wedge L_n$$

in which A is an atom and L_is are literals. This is a *normal clause* (or simply *clause*).

4. $m = 0$, $n \geq 1$, and each L_i is an atom. That is, the clause has the form

$$\leftarrow B_1 \wedge \ldots \wedge B_n$$

in which B_is are atoms. This clause is called a *definite goal*.

5. $m = 0$, $n \geq 1$. That is, the clause has the form

$$\leftarrow L_1 \wedge \ldots \wedge L_n$$

in which L_is are literals. This clause is a *normal goal* (or simply *goal*).

6. $A_1 \vee \ldots \vee A_m \leftarrow L_1 \wedge \ldots \wedge L_n$, $m > 1, n \geq 0$. This clause is called an *indefinite clause*.

Thus a unit clause is a special form of a definite clause; a definite clause is a special form of a normal clause; and a definite goal is a special form of a normal goal. The following example contains various types of clauses involved with the problem of representing numbers.

Example

Clause	Type
C1. $Weather(Eden\,Garden, Wet)$	Unit
C2. $Field(x, y) \leftarrow Weather(x, z) \wedge Causes(z, y)$	Definite
C3. $Game(x) \leftarrow Weather(x, y) \wedge \neg Disruption(y, Transport)$	Normal
C4. $Game(x) \vee Disruption(y, Transport) \leftarrow Weather(x, y)$	Indefinite

Clause $C1$ represents that the weather condition is wet in the Eden Garden area. Clause $C2$ represents that if the weather condition is z in the area of x and z causes the condition y of a field, then the condition of the field x is y. Clause $C3$ represents that if the weather condition in the area of the field x is y and the condition y does not disrupt transport, the game holds at a field x. Clause $C4$ represents that if the weather condition in the area of the field x is y then game holds at x or the condition y disrupts transport. Although clauses $C3$ and $C4$ are logically equivalent, the semantics of a logic program, which includes clause $C3$, may be different if the program replaces $C3$ by $C4$. In such cases, the use of the negation symbol $\neg$ becomes a default rule which may be different from the classical negation studied so far. But it is always clear from the context which particular approach has been taken.

The rest of the chapter does not deal with indefinite clauses with consequent $A_1 \vee ... \vee A_m$ with $m > 1$. The main argument against such considerations is not only to avoid reasoning complexity but also the coarse level of uncertainty that an indefinite clause can represent via its disjunction sign will be subsumed by a more general modeling technique (to be presented in the penultimate chapter) that attaches probabilities to clauses as their degrees of uncertainty.

A *subgoal* of a goal G is a literal occurring in G. A *definite program* is a finite set of definite clauses. A *normal program* (or simply *program*) is a finite set of normal clauses. A *definite program* is a special form of a normal program consists only definite clauses. In writing a set of clauses to represent a program, the same variables may be used in two different clauses. If there are any common variables between two different clauses then they are renamed before they are used.

Example

Program:

C1. $Game(x) \leftarrow Weather(x, y) \wedge \neg Disruption(y, Transport)$

C2. $Field(x, y) \leftarrow Weather(x, z) \wedge Causes(z, y)$

C3. $Commentary(x) \leftarrow Game(x)$

C4. $Weather(Eden\,Garden, Rainy)$

C5. $Disruption(Snow, x)$

C6. $Disruption(Rain, Transport)$

The clause $C1$ makes the above program normal. In this program, the non-ground atom $Disruption(Snow, x)$ represents that the snowy weather disrupts everything.

In the context of the program, you can have a definite goal

$$\leftarrow Commentary(Eden\,Garden)$$

to find whether the radio commentary is being played from Eden Garden or not. Similarly a normal goal of the form

$$\leftarrow Field(x, Wet), \neg Game(x)$$

will find those wet fields where games are not being played.

If $x_1, ..., x_p$ are all the free variables in a normal goal of the form $\leftarrow L_1 \wedge ... \wedge L_n$, then the goal in the context of a program **P** is interpreted as a request for a constructive proof for the formula

$$\exists x_1 ... \exists x_p (L_1 \wedge ... \wedge L_n)$$

This means that one should find a substitution θ for the free variables in the goal such that $(L_1 \wedge ... \wedge L_n)\theta$ is true according to the semantics of the program.

Programs have so far been classified by looking at the structure of the individual clauses. Further subclassifications of programs can be achieved by looking at their constituent set of clauses as a whole. This classification can be facilitated by the use of dependency graphs.

Consider the class of normal programs. The *dependency graph* of a normal program **P** has a node for each predicate symbol occurring in **P** and a directed edge from the node for predicate Q to the node for predicate P whenever predicate Q is in the body of some clause and P is in the head of the same clause. An edge from node Q to node P is *positive* iff there is a clause C in **P** in which P is in the head of C, and Q is the predicate symbol of a positive literal in the body of C. The edge is *negative* if Q is the predicate symbol of a negative literal in the body of C. The *length* of a cycle in a dependency graph is the number of edges occurring in the cycle.

Example

Consider the following program:

$$P(x) \leftarrow Q(x) \wedge \neg R(x)$$
$$R(x) \leftarrow S(x)$$
$$S(x) \leftarrow T(x, y) \wedge P(y)$$
$$Q(a)$$
$$T(a, b)$$

The dependency graph for the above program is shown in Figure 4-1. The set of clauses in this example contains nodes *P*, *Q*, *R*, *S* and *T*. The edges from *Q* to *P*, from *S* to *R*, from *T* to *S* and from *P* to *S* are positive while the edge from *R* to *P* is negative.

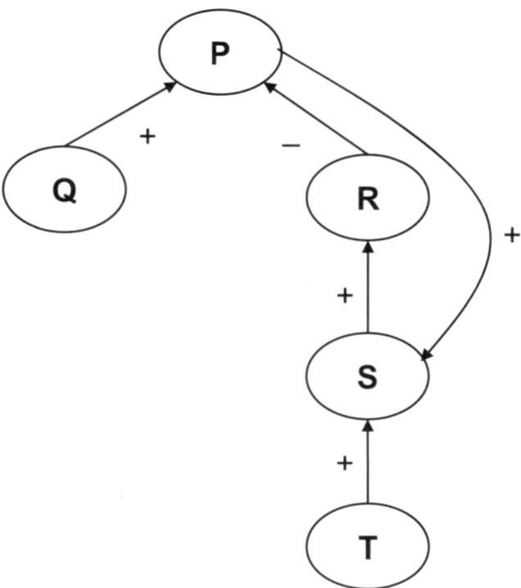

Figure 4-1: Example dependency graph

If the dependency graph for a normal program does not contain any cycles then the program is *hierarchical*.

Having cycles in the dependency graph of a program **P** implies that some predicates of **P** are directly or indirectly defined by themselves. This, of course, complicates the structure of a program and goal evaluations in the context of that program. This class of programs is formally classified through the following definitions.

Let **P** be a normal program and Γ be the dependency graph for **P**. A predicate *P* occurring in **P** is *recursive* if it occurs in a cycle of Γ. Two predicates *P* and *Q* are *mutually recursive* if they both occur in a cycle of Γ. If the dependency graph for a program contains a cycle, then the program is *recursive*.

Example

Consider a program **P** that contains the following clause:

$$Field(x, y) \leftarrow Adjacent(x, z) \wedge Field(z, y)$$

The clause states that the condition of a field x is y provided there exists a field z which is adjacent to x and the condition of z is y. The clause therefore helps to determine the conditions of a field in terms of its adjacent fields. The program **P** is recursive and the predicate *Field* in the program is recursive.

Programs allow negated atoms in their bodies. When this happens, one should be careful about the meaning of the program. Consider, for example, a program **P** that contains the following two clauses:

$$Field\,Open(x) \leftarrow \neg Field\,Closed(x)$$
$$Field\,Closed(x) \leftarrow \neg Field\,Open(x)\overset{\textstyle .}{}$$

The dependency graph of the above program contains a cycle with negative edges, that is, a recursion via negation (so does the program presented earlier whose dependency graph is shown in Figure 4-1). The completion of the program **P**, which is a popular way of defining the semantics of a program, becomes inconsistent. This kind of situation can be prevented by not allowing a recursion via negation.

A normal program is *stratified* (Chandra and Harel, 1985; Apt *et al.*, 1988) if each cycle of the dependency graph is formed by positive edges only (the remainder of the graph may contain negative edges).

Clearly every hierarchical program can be regarded as a specific stratified program and all definite programs are stratified. An alternative definition of stratified programs can be given by introducing the notions of stratification among the definitions of predicate symbols of programs.

Suppose **P** is a program. The *definition of a predicate symbol P* is the subset of **P** consisting of all clauses with P occurring in their heads. Then **P** is regarded as stratified if there is a partition

$$\mathbf{P} = \mathbf{P}_1 \cup ... \cup \mathbf{P}_n$$

such that the following two conditions hold for $i = 1,...,n$:

- If an atom A occurs positively in the body (that is, A occurs in the body) of a clause in $\mathbf{P}_i$ then the definition of its predicate symbol is contained within $\mathbf{P}_j$ with $j \leq i$.

- If an atom A occurs negatively in the body (that is, $\neg A$ occurs in the body) of a clause in $\mathbf{P}_i$ then the definition of its predicate symbol is contained within $\mathbf{P}_j$ with $j < i$.

The program $\mathbf{P}$ is said to be stratified by $\mathbf{P}_1 \cup ... \cup \mathbf{P}_n$ and each $\mathbf{P}_i$ is called a *stratum* of $\mathbf{P}$. The *level* of a predicate symbol is the index of the strata within which it is defined.

A third alternative equivalent definition of stratified programs in terms of its *level mapping*, which is a mapping from its set of predicates to the set of non-negative integers defined as follows:

- The level of a literal is the level of the predicate symbol of the literal.

- The level of a clause is the maximum level of any predicate symbol in the clause.

A program is stratified if it has a level mapping such that, for every clause C, the level of the predicate symbol of every positive condition in C is less than or equal to the level of the predicate symbol in the head of C and the level of the predicate of every negated condition in C is less than the level of the predicate symbol in the head of C.

Example

Consider the following program:

$$C1: \quad P(x) \leftarrow Q(x) \wedge \neg R(x)$$
$$C2: \quad R(x) \leftarrow S(x)$$
$$C3: \quad S(x) \leftarrow T(x,y) \wedge \neg U(y)$$
$$C4: \quad S(x) \leftarrow U(x)$$
$$C5: \quad Q(a)$$
$$C6: \quad R(a)$$
$$C7: \quad T(a,b)$$

The above program is stratified by $\mathbf{P}_1 \cup \mathbf{P}_2 \cup \mathbf{P}_3$, where $\mathbf{P}_3 = \{C_1\}$, $\mathbf{P}_2 = \{C_2, C_3, C_4, C_6\}$ and $\mathbf{P}_1 = \{C_5, C_7\}$. An alternative way the program can be stratified is by $\mathbf{P}_1 \cup \mathbf{P}_2 \cup \mathbf{P}_3 \cup \mathbf{P}_4$, where $\mathbf{P}_4 = \{C_1\}$, $\mathbf{P}_3 = \{C_2, C_6\}$, $\mathbf{P}_3 = \{C_3, C_4, C_7\}$ and $\mathbf{P}_1 = \{C_5\}$.

4.3   Program Semantics

The semantics of a program (van Emden and Kowalski, 1976) deals with its meaning. Three methods for defining program semantics are introduced in this section: declarative, procedural, and fixpoint.

- The *declarative semantics* of a program deals with the interpretation of the program's language, including logical implications, and truth. A program's declarative semantics can be defined by selecting one or more of its models. These models determine which substitution of a given goal is correct in the context of the program.

- The *procedural semantics* (or *operational semantics*) of a program deals with well-formed formulae and their syntax, axioms, rules of inference, and proofs within the program's language. This semantics defines the input/output relations computed by a program in terms of the individual operations evoked by the program inside the machine. Procedural semantics refers to a computational method, called *proof procedure*, for obtaining the meaning of a program.

- The *fixpoint semantics* of a program defines the meaning of a program to be the input/output relation which is the minimal fixpoint of transformation associated with the program **P**. The fixpoint operator builds the intended model of the program in a step-by-step process.

Van Emden and Kowalski first investigated the semantics of Horn logic as a programming language (van Emden and Kowalski, 1976) and compared the declarative semantics of a program with the classical model-theoretic semantics, compared operational semantics with classical proof-theoretic semantics, and compared fixpoint semantics with the fixpoint operator corresponding to a program.

All logical consequences of a program can be considered as positive information. Declarative, procedural, and fixpoint semantics are sometimes concerned only with the positive information about a program that can be derived from that program. The negative information is not explicitly stored in a program and it is assumed by default. This additional implicit negative information cannot be proved from the program by normal inference rules such as modus ponens. *Rules for inferring negative information* are rules that infer this additional implicit negative information. Hence a rule for inferring negative information defines the *semantics for negative informations* in the program and is given in either a declarative or a procedural manner. This rule will be studied alongside the other three kinds of semantics mentioned above.

Example

Consider the following program **P**:

$$Commentary(x) \leftarrow Game(x)$$
$$Game(Eden\,Garden)$$
$$Field(Fenway\,Park, Wet)$$

Several different semantics for the above program can be defined as follows:

- *Declarative semantics*: Facts that are in a minimal model of **P** are taken as true.

- *Procedural semantics*: Facts that can be derived from **P** are taken as true.

- *Semantics for negative information*: Facts that are in $\mathcal{HB}(\mathbf{P})$ and not in at least one minimal model of **P** are taken as false.

Then the declarative and procedural semantics coincide and are represented by the set

$$\{Game(Eden\,Garden), Commentary(Eden\,Garden),$$
$$Field(Fenway\,Park, Wet)\}$$

According to these two semantics, the above set is the set of all true facts in the context of **P**. The semantics for negative information generates the set

$$\mathcal{HB}(\mathbf{P}) - \{Game(Eden\,Garden), Commentary(Eden\,Garden),$$
$$Field(Fenway\,Park, Wet)\}$$

and therefore, $Game(Fenway\,Park)$ is taken as false in the context of **P** according to the semantics for negative information.

Let $\mathcal{R}$ be a rule for inferring negative information for a particular class (such as definite or stratified or normal) of programs and **P** be a program from that class. Then $\mathcal{R}(\mathbf{P})$ will denote the set of negative facts that can be inferred from **P** by the application of $\mathcal{R}$. The following properties (Ross and Topor, 1988) are expected to be satisfied in a typical program **P** of this class:

- *Concise*: Use of $\mathcal{R}$ should enable a useful reduction in the amount of negative information that would otherwise have to be stored as clauses of

P. In other words, the set of facts $\mathcal{R}(\mathbf{P})$ is relatively large compared to **P**. This reflects the property that in a typical program the number of facts that are true is much less than the whole Herbrand base associated with the database. In the context of the above example program **P**, the number of true facts is 3, compared to 15 which is the size of $\mathcal{HB}(\mathbf{P})$.

- *Efficient*: It should be relatively easy to determine whether $\mathcal{R}$ can infer an item of negative information from **P**. In other words, the decision procedure of $\mathcal{R}(\mathbf{P})$ should be relatively efficient.

- *Consistent*: If the program **P** is consistent, then $\mathbf{P} \cup \mathcal{R}(\mathbf{P})$ should be consistent.

Detailed studies of declarative and procedural semantics on individual classes of logic programs are carried out in the next two subsections.

4.4   Definite Programs

This section deals with the semantics of definite programs. Declarative semantics is studied by employing classical model-theoretic semantics and procedural semantics using the SLD-resolution scheme. Unless otherwise stated, in this section the terms "program" and "goal" will always mean "definite program" and "definite goal," respectively. Recall that a definite program is defined as a finite set of definite clauses of the form $A \leftarrow A_1 \wedge ... \wedge A_m$ and a definite goal is defined as $\leftarrow B_1 \wedge ... \wedge B_n$, where $m \geq 0, n \geq 1$, and A, A_i and B_j are atoms.

The only declarative semantics of a definite program studied here is the one given in van Emden and Kowalski (1976), which is formally defined as follows. Let **P** be a definite program. Then the declarative semantics of **P**, denoted as $\mathcal{M}(\mathbf{P})$, is defined as

$$\mathcal{M}(\mathbf{P}) = \{A \mid A \in \mathcal{HB}(\mathbf{P}) \text{ and } \mathbf{P} \vDash A\}$$

which means that $\mathcal{M}(\mathbf{P})$ is the set of all ground atoms of the Herbrand base of **P** that are logical consequences of **P**. The following theorem can be considered as an alternative definition of the semantics in terms of Herbrand models of **P**.

Theorem 4-1: Let **P** be a program and $\mathcal{HM}(\mathbf{P})$ be the set of all Herbrand models of **P**. Then $\mathcal{M}(\mathbf{P}) = \{A \mid A \in \cap \mathcal{HM}(\mathbf{P})\}$.

Proof: $A \in \mathcal{M}(\mathbf{P})$

   Iff $\mathbf{P} \models A$

   iff $\mathbf{P} \cup \{\neg A\}$ has no model

   iff $\mathbf{P} \cup \{\neg A\}$ has no Herbrand model

   iff $\neg A$ is false in all Herbrand models of $\mathbf{P}$

   iff A is true in all Herbrand models of $\mathbf{P}$

   iff $A \in \cap \mathcal{HM}(\mathbf{P})$.

The following theorem establishes the *model intersection property* of a set of Horn clauses.

Theorem 4-2: Suppose $\mathbf{P}$ is a consistent set of Horn clauses and $hm(\mathbf{P})$ is a non-empty set of Herbrand models of $\mathbf{P}$. Then $\cap hm(\mathbf{P})$ is also a model of $\mathbf{P}$.

Proof: If possible, do not let $\cap hm(\mathbf{P})$ be a Herbrand model of $\mathbf{P}$. Then there is a clause C in $\mathbf{P}$ and a ground instance θ of C such that $C\theta$ is false in $\cap hm(\mathbf{P})$. If C has the form $A \leftarrow A_1 \wedge ... \wedge A_n$, then $A\theta \notin \cap hm(\mathbf{P})$ and $A_1\theta,..., A_n\theta \in \cap hm(\mathbf{P})$. Therefore for some $\mathbf{M} \in hm(\mathbf{P})$, $A\theta \notin \mathbf{M}$ and $A_1\theta,..., A_n\theta \in \mathbf{M}$. Hence C is false in $\mathbf{M}$, which contradicts the assumption that $\mathbf{M}$ is a model of $\mathbf{P}$. The other case, when C has the form of a negative Horn clause is similar.

In view of Theorem 4-2, it can be said that $\cap \mathcal{HM}(\mathbf{P})$ is also a Herbrand model and hence the minimal model of $\mathbf{P}$. Thus the declarative semantics of a definite program is equal to its minimal model.

Example

Consider the following program $\mathbf{P}$:

$$P(x,y) \leftarrow Q(x,y)$$
$$P(x,y) \leftarrow Q(x,z) \wedge P(z,y)$$
$$Q(a,b)$$
$$Q(b,c)$$

Then $\mathcal{M}(\mathbf{P}) = \{Q(a,b), Q(b,c), P(a,b), P(b,c), P(a,c)\}$. Any other model of $\mathbf{P}$, for example $\{Q(a,b), Q(b,c), P(a,b), P(b,c), P(a,c), P(b,a), P(a,a)\}$, contains $\mathbf{P}$.

The procedural semantics of a definite program is studied in this section using the *SLD resolution* (SL resolution for Definite clauses) schemes. First, the resolution scheme and its properties are considered.

Let $\mathbf{P}$ be a definite program and G be a goal. An *SLD derivation* of $\mathbf{P} \cup \{G\}$ consists of a sequence $G_0 = G, G_1, \dots$ of goals, a sequence $C_1, C_2, \dots$ of variants of clauses in $\mathbf{P}$ (called the *input clauses* of the derivation) and a sequence $\theta_1, \theta_2, \dots$ of *substitutions*. Each non-empty goal G_i contains one atom, which is the *selected atom* of G_i. The clause G_{i+1} is said to be derived from G_i and C_i with substitutions θ_i and is carried out as follows. Suppose G_i is

$$\leftarrow B_1 \wedge \dots \wedge B_k \wedge \dots \wedge B_n \quad n \geq 1$$

and B_k is the selected atom. Let

$$C_i = A \leftarrow A_1 \wedge \dots \wedge A_m$$

be any clause in $\mathbf{P}$ such that A and B_k are unifiable with the most general unifier θ. Then G_{i+1} is

$$\leftarrow (B_1 \wedge \dots \wedge B_{k-1} \wedge A_1 \wedge \dots \wedge A_m \wedge B_{k+1} \wedge \dots \wedge B_n)\theta$$

and θ_{i+1} is θ. An *SLD refutation* is a derivation ending at an empty clause.

In the definitions of SLD derivation and SLD refutation, the consideration of most general unifiers instead of just one unifier reduces the *search space* for the refutation procedure. A search space for the SLD refutation procedure is an *SLD tree* as defined below.

Let $\mathbf{P}$ be a definite program and G be a goal or an empty clause. An *SLD tree* for $\mathbf{P} \cup \{G\}$ has G as root and each node is either a goal or an empty clause. Suppose

$$\leftarrow B_1 \wedge \dots \wedge B_k \wedge \dots \wedge B_n \quad n \geq 1$$

is a non-empty node with selected atom B_k. Then this node has a descendant for every clause

$$A \leftarrow A_1 \wedge ... \wedge A_m$$

such that A and B_k are unifiable with the most general unifier θ. The descendant is

$$\leftarrow (B_1 \wedge ... \wedge B_{k-1} \wedge A_1 \wedge ... \wedge A_m \wedge B_{k+1} \wedge ... \wedge B_n)\theta$$

Each path in an SLD tree is an SLD derivation and a path ending at an empty clause is an SLD refutation. The search space is the totality of derivations constructed to find an SLD-refutation.

An SLD derivation may be finite or infinite. An SLD refutation is a *successful SLD derivation*. A *failed SLD-derivation* is one that ends in a non-empty goal with the property that the selected atom in this goal does not unify with the head of any program clause. Branches corresponding to successful derivation in an SLD-tree are called *success branches*, branches corresponding to infinite derivations are called *infinite branches* and branches corresponding to failed derivations are called *failed branches*. All branches of a *finitely failed SLD tree* are failed branches.

In general, a given $\mathbf{P} \cup \{G\}$ has different SLD trees depending on which atoms are selected atoms.

Example

Consider the following recursive program presented earlier in this subsection and the goal $\leftarrow P(a, y)$:

$$P(x, y) \leftarrow Q(x, y)$$
$$P(x, y) \leftarrow Q(x, z) \wedge P(z, y)$$
$$Q(a, b)$$
$$Q(b, c)$$

The tree in Figure 4-2 (respectively, Figure 4-3) is based on a literal selection strategy which selects the leftmost (respectively, rightmost) literal from a goal. The tree in Figure 4-3 has an infinite branch whereas all the branches of the tree in Figure 4-2 are finite.

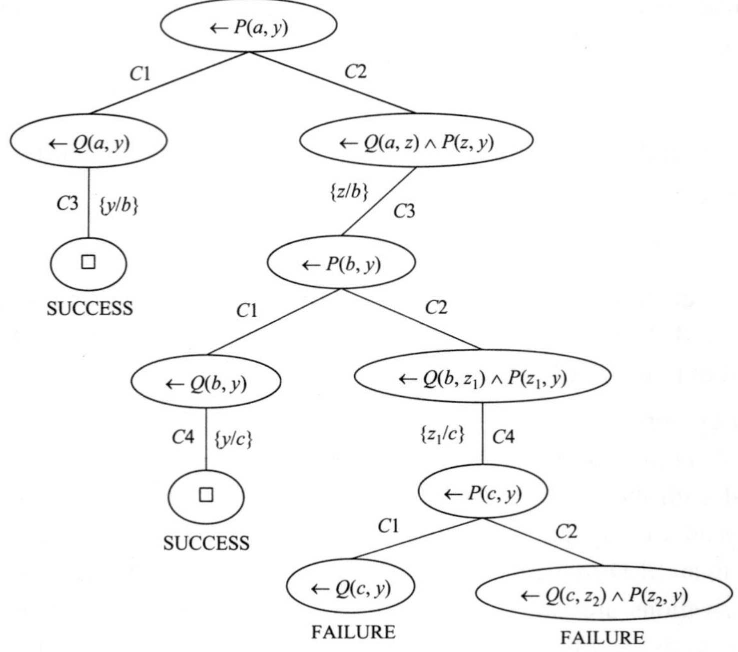

Figure 4-2: An SLD-tree with leftmost literal selection strategy.

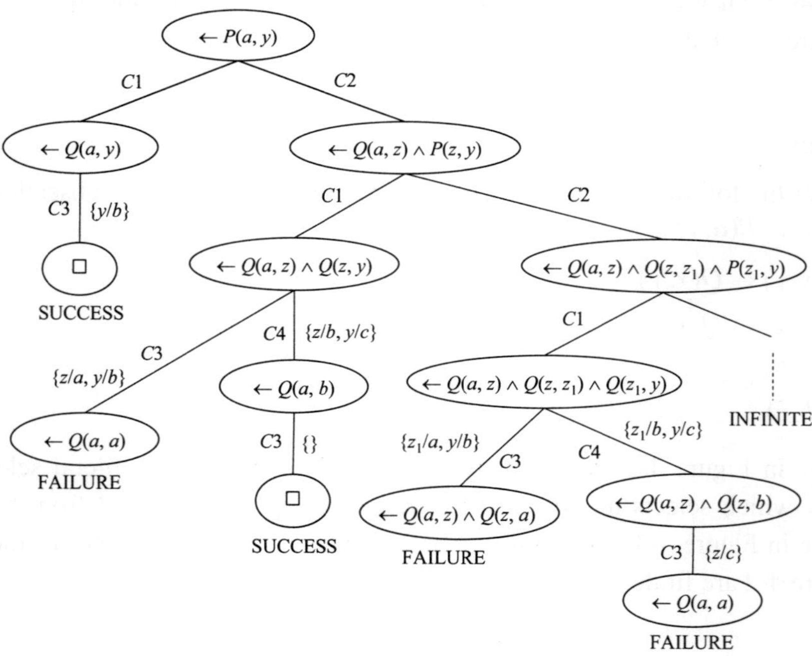

Figure 4-3: An SLD-tree with rightmost literal selection strategy.

The equivalence between declarative and procedural semantics for definite programs is basically the soundness and completeness of SLD-resolution with respect to their minimal models.

Theorem 4-3: (Soundness of SLD Resolution) Let $\mathbf{P}$ be a program and G a goal such that $\mathbf{P} \cup \{G\}$ has an SLD refutation with substitution θ. Then $\mathbf{P} \cup \{G\}$ is inconsistent.

The above soundness theorem also implies that if the goal G is of the form $\leftarrow Q$, then $Q\theta$ is a logical consequence of the program $\mathbf{P}$.

Theorem 4-4: (Completeness of SLD Resolution) Let $\mathbf{P}$ be a program and G be a goal such that $\mathbf{P} \cup \{G\}$ is inconsistent. Then every SLD tree with G as root contains a success branch.

The above completeness theorem also implies that if the goal G is of the form $\leftarrow Q$, then for every substitution θ of $\leftarrow Q$, if $Q\theta$ is a logical consequence of $\mathbf{P}$, then there exists an SLD refutation of $\mathbf{P} \cup \{\leftarrow Q\}$ with answer substitution θ_p such that θ is more general than θ_p.

The completeness theorem establishes the fact that if $\mathbf{P} \cup \{G\}$ is inconsistent, then a refutation can be found if a proper algorithm (for example, breadth-first) is employed to search the tree. The above results also establishes that the set

$$\{A \mid A \in \mathcal{HB}(\mathbf{P}) \text{ and there exists an SLD refutation of } \mathbf{P} \cup \{\leftarrow A\}\}$$

representing the procedural or operational semantics of $\mathbf{P}$ is equal to the declarative semantics $\{A \mid A \in \mathcal{HB}(\mathbf{P}) \text{ and } \mathbf{P} \vDash A\}$ of $\mathbf{P}$.

The above-mentioned declarative and procedural semantics for programs are only concerned with the positive information that can be derived from the program. *Closed World Assumption* (CWA) and *Negation As Failure* (NAF) rules are provided below for studying semantics for negative information in definite programs. Each of these is consistent for the class of definite programs and can be used in conjunction with the other two semantics introduced above to obtain the complete meaning of a definite program.

Given a program $\mathbf{P}$ and $A \in \mathcal{HB}(\mathbf{P})$, one of the ways to say that $\neg A$ may be inferred from $\mathbf{P}$ is if A is not a logical consequence of $\mathbf{P}$. This declarative definition for inferring negative information is what is called the closed world assumption and hence is considered a rule for inferring negative information from a program. Thus

$$CWA(\mathbf{P}) = \{A \mid A \in \mathcal{HB}(\mathbf{P}) \text{ and } A \text{ is not a logical consequence of } \mathbf{P}\}$$

if A is not a logical consequence of $\mathbf{P}$, then each branch of an SLD tree for $\mathbf{P} \cup \{\leftarrow A\}$ is either finitely failed or infinite. This operational aspect of the CWA provides a new rule for inferring negative information from a program and is called the negation as failure rule. The rule can be formally stated as follows. Given a program $\mathbf{P}$ and $A \in \mathcal{HB}(\mathbf{P})$, if $\mathbf{P} \cup \{\leftarrow A\}$ has a finitely failed SLD tree, then A is not a logical consequence of $\mathbf{P}$ and hence $\neg A$ can be inferred from $\mathbf{P}$. Thus

$$NAF(\mathbf{P}) = \{A \mid A \in \mathcal{HB}(\mathbf{P}) \text{ and } \mathbf{P} \cup \{\leftarrow A\} \text{ has a finitely failed SLD tree}\}$$

Naturally CWA is more powerful from the point of view that A may not be a logical consequence of $\mathbf{P}$ but every SLD tree for $\mathbf{P} \cup \{\leftarrow A\}$ has an infinite branch. Consider, for example, the program $\mathbf{P} = \{P(a) \leftarrow P(a)\}$. Clearly $P(a)$ is not a logical consequence of $\mathbf{P}$ and hence CWA infers $\neg P(a)$ from $\mathbf{P}$. But the only branch of the tree for $\mathbf{P} \cup \{\leftarrow P(a)\}$ is infinite. If a system implementing the SLD resolution procedure is not capable of detecting such infinite branches, then a task for finding whether A is a logical consequence or not would be incomplete in that system. As far as the properties of rules for negative information are concerned, both CWA and NAF for definite programs are consistent.

4.5   Normal Programs

The material presented in this section relates to normal programs. The declarative semantics of normal programs is studied through well-known program completion. SLDNF-resolution is also introduced in this section and a relation is established with its declarative counterpart, that is, program completion through the soundness and completeness result. Recall that a normal program has been defined as a finite set of normal clauses of the form

$$A \leftarrow L_1 \wedge \ldots \wedge L_m \qquad\qquad m \geq 0$$

and that a normal goal has been defined as a clause of the form

$$\leftarrow M_1 \wedge \ldots \wedge M_n \quad n \geq 1$$

where A is an atom and L_i and M_j are literals. Unless otherwise mentioned, any reference to programs and goals will always mean normal programs and normal goals.

In the previous section, the declarative semantics of a definite program **P** was defined by the set $\{A \mid A \in \mathcal{HB}(\mathbf{P}) \text{ and } \mathbf{P} \vDash A\}$, and that this semantics coincides with the minimal model of **P**. In contrast, a normal program may not have a unique minimal model. Consider the program $\mathbf{P} = \{Game \leftarrow \neg Rain\}$. The Herbrand base of **P** is $\{Game, Rain\}$. The two minimal models of **P** are $\{Game\}$ and $\{Rain\}$, and their intersection is the empty set, which is not a model of **P**.

While it is said that the program **P** has two minimal models $\{Game\}$ and $\{Rain\}$, the three different program clause representations of $Game \leftarrow \neg Rain$ in **P**, namely $Game \leftarrow \neg Rain$ itself, $Game \vee Rain$ and $Rain \leftarrow \neg Game$, are not distinguished from each other. However, the intent of writing $Game \leftarrow \neg Rain$ is to say that $Game$ is true when $Rain$ is not, and probably the negation of $Rain$ will be inferred by default if $Rain$ cannot be inferred. Neither the reverse $Game \leftarrow \neg Rain$ of $Game \leftarrow \neg Rain$ is intended nor the disjunction $Game \vee Rain$. By distinguishing these representations from each other, the completed program semantics for a normal program is given below.

According to the program completion semantics, the clauses of a program **P** provide the if-parts of the definition of the predicates. The only-if parts in a program are implicitly defined and are obtained by completing each predicate symbol.

<u>Example</u>

Consider a normal program containing only the following two clauses defining the predicate *Weather*:

$$Weather(Fenway\,Park, Wet)$$

$$Weather(x, y) \leftarrow Adjacent(x, z) \wedge Weather(z, y)$$

These two clauses jointly state that the weather in the Fenway Park area is known to be wet, and weather in the area of a field x is y if there exists a field z which is adjacent to x and the weather in the area of z is y. Symbolically,

$$Weather(x, y) \leftarrow$$
$$Weather(Fenway\,Park, Wet) \vee Adjacent(x, z) \wedge Weather(z, y)$$

However, the definition leaves open the possibility that the weather in the area of a field may be known for some other reason. What has been implicitly meant by the program is that the weather in the area of a field x is y only if x is *Fenway Park* and y is *Wet* or there exists a field z which is adjacent to x and the weather in the area of z is y. Symbolically,

$$Weather(x, y) \rightarrow$$
$$Weather(Fenway\,Park, Wet) \vee Adjacent(x, z) \wedge Weather(z, y)$$

The above two clauses jointly give the completed definition of the predicate *Weather*. The formal definition of predicate and program completions are provided below.

Suppose that **P** is a program and

$$P(t_1, ..., t_n) \leftarrow L_1 \wedge ... \wedge L_m$$

is a clause in **P**. If the predicate symbol = is interpreted as the equality (or identity) relation, and $x_1, ..., x_n$ are variables not appearing elsewhere in the clause, then the above clause is equivalent to the clause

$$P(x_1, ..., x_n) \leftarrow x_1 = t_1 \wedge ... \wedge x_n = t_n \wedge L_1 \wedge ... \wedge L_m$$

If $y_1, ..., y_p$ are the variables of the original clause then this can be transformed to the following *general form of the clause*:

$$P(x_1, ..., x_n) \leftarrow \exists y_1 ... \exists y_p (x_1 = t_1 \wedge ... \wedge x_n = t_n \wedge L_1 \wedge ... \wedge L_m)$$

Suppose there are exactly k clauses, $k \geq 0$, in the program **P** defining P (a clause C defines P if C has P in its head). Let

$$P(x_1, ..., x_n) \leftarrow E_1$$
$$...$$
$$P(x_1, ..., x_n) \leftarrow E_k$$

be the k clauses in general form. Then the *completed definition* of P is the formula

$$\forall x_1 ... \forall x_n (P(x_1, ..., x_n) \leftrightarrow E_1 \vee ... \vee E_k)$$

Some predicate symbols in the program may not appear in the head of any program clause. For each such predicate Q, the *completed definition* of Q is the formula

$$\forall x_1 ... \forall x_n \neg Q(x_1,...,x_n)$$

Additional axioms for the equality symbol $=$ are needed as part of the completed definition of a program. The *equality theory* $\mathcal{EQ}$ for a completed program contains these axioms, and they are listed below:

- $c \neq d$, for all pairs c, d of distinct constants (the symbol $\neq$ stands for not equal, that is, $c \neq d$ stands for $\neg(c = d)$).

- $\forall x_1 ... \forall x_n \forall y_1 ... \forall y_m (f(x_1,...,x_n) \neq g(y_1,...,y_m))$, for all pairs f, g of distinct function symbols.

- $\forall x_1 ... \forall x_n (f(x_1,...,x_n) \neq c)$, for each constant c and function symbol f.

- $\forall x(t[x] \neq x)$, for each term $t[x]$ containing x and different from x.

- $\forall x_1 ... \forall x_n \forall y_1 ... \forall y_n ((x_1 \neq y_1) \wedge ... \wedge (x_n \neq y_n) \rightarrow f(x_1,...,x_n) \neq f(y_1,...,y_n))$, for each function symbol f.

- $\forall x_1 ... \forall x_n \forall y_1 ... \forall y_n ((x_1 = y_1) \wedge ... \wedge (x_n = y_n) \rightarrow f(x_1,...,x_n) = f(y_1,...,y_n))$, for each function symbol f.

- $\forall x(x = x)$

- $\forall x_1 ... \forall x_n \forall y_1 ... \forall y_n ((x_1 = y_1) \wedge ... \wedge (x_n = y_n) \rightarrow (P(x_1,...,x_n) \rightarrow P(y_1,...,y_n)))$, for each predicate symbol P (including $=$).

The fact that $=$ is an equivalence relation is implied by the above axioms. The *completion* of **P** (Clark, 1978; Lloyd, 1987), denoted by *Comp*(**P**), is the collection of completed definitions of predicate symbols in **P** together with the equality theory.

Example

Let the predicate symbol P be defined by the following clauses:

$$P(a,x) \leftarrow Q(x,y) \wedge \neg R(y)$$
$$P(b,c)$$

Then the completed definition of P is.

$$\forall x_1 \forall x_2 (P(x_1, x_2) \leftrightarrow$$

$$(\exists x \exists y ((x_1 = a) \wedge (x_2 = x) \wedge Q(x, y) \wedge \neg R(y)) \vee ((x_1 = b) \wedge (x_2 = c))))$$

and the completed definitions of Q and R are $\forall x_1 \forall x_2 \neg Q(x_1, x_2)$ and $\forall x_1 \neg R(x_1)$ respectively.

Consider the program **P** that has only one clause $P \leftarrow \neg P$. Apart from the equality axioms, *Comp*(**P**) is $\{P \leftrightarrow \neg P\}$, which is inconsistent. Therefore, the completion of a program may not be consistent. But, *Comp*(**P**) is consistent when **P** is stratified. The program $\{P \leftrightarrow \neg P\}$ is of course not stratified.

SLDNF resolution is thought of as a procedural counterpart of the declarative semantics given in terms of completed programs. This resolution scheme is, to a large extent, the basis of present-day logic programming. The resolution scheme is essentially SLD resolution augmented by the negation as failure (NAF) inference rule.

Let **P** be a program and G be a goal. An *SLDNF derivation* of $\mathbf{P} \cup \{G\}$ consists of a sequence $G_0 = G, G_1, ...$ of goals, a sequence $C_1, C_2, ...$ of variants of clauses of **P** or negative literals, and a sequence $\theta_1, \theta_2, ...$ of substitutions satisfying the following:

- Consider the case when G_i is $\leftarrow L_1 \wedge ... \wedge L_m$, $m \geq 1$, and the selected literal L_k is positive. Suppose $A \leftarrow M_1 \wedge ... \wedge M_n$ is the input clause C_{i+1} such that L_k and A are unifiable with the mgu θ_{i+1}. Then the derived goal G_{i+1} is

$$\leftarrow (L_1 \wedge ... \wedge L_{k-1} \wedge M_1 \wedge ... \wedge M_n \wedge L_{k+1} \wedge ... \wedge L_m)\theta_{i+1}$$

- Consider the case when G_i is $\leftarrow L_1 \wedge ... \wedge L_m$, $m \geq 1$, and the selected literal L_k is a ground negative literal of the form $\neg A$. If there is a finitely failed SLDNF tree for $\mathbf{P} \cup \{\leftarrow A\}$, then θ_{i+1} is the identity substitution, C_{i+1} is $\neg A$, and G_{i+1} is

$$\leftarrow L_1 \wedge ... \wedge L_{k-1} \wedge L_{k+1} \wedge ... \wedge L_n$$

- If the sequence $G_0, G_1, ...$ is finite, then either

  - the last goal is empty, or

- the last goal is $\leftarrow L_1 \wedge ... \wedge L_m$, $m \geq 1$, the selected literal L_k is positive, and there is no program clause whose head unifies with L_k, or

- the last goal is $\leftarrow L_1 \wedge ... \wedge L_m$, $m \geq 1$, the selected literal L_k is a ground negative literal $\neg A$, and there is an SLDNF refutation of $\mathbf{P} \cup \{\leftarrow A\}$.

An SLDNF derivation is *finite* if it consists of a finite sequence of goals; otherwise it is *infinite*. An SLDNF derivation is *successful* if it is finite and the last goal is the empty goal. A successful SLDNF derivation is an *SLDNF refutation*. An SLDNF-derivation is failed if it is finite and the last goal is not the empty goal.

Let $\mathbf{P}$ be a normal program and G be a goal. Then an *SLDNF tree* for $\mathbf{P} \cup \{G\}$ is defined as follows:

- Each node of the tree is a goal.

- The root node is G.

- Let $\leftarrow L_1 \wedge ... \wedge L_m$, $m \geq 1$, be a node in the tree and suppose that the selected literal L_k is positive. Then this node has a descendant for each input clause $A \leftarrow M_1 \wedge ... \wedge M_n$ such that L_k and A are unifiable. The descendant is

$$\leftarrow (L_1 \wedge ... \wedge L_{k-1} \wedge M_1 \wedge ... \wedge M_n \wedge L_{k+1} \wedge ... \wedge L_m)\theta$$

where θ is the most general unifier of L_k and A.

- Let $\leftarrow L_1 \wedge ... \wedge L_m$, $m \geq 1$, be a node in the tree and suppose that the selected literal L_k is a ground negative literal of the form $\neg A$. If for every branch of an SLDNF tree for $\mathbf{P} \cup \{\leftarrow A\}$ the terminating node is a non-empty goal, then the single descendant of the node is

$$\leftarrow L_1 \wedge ... \wedge L_{k-1} \wedge L_{k+1} \wedge ... \wedge L_n$$

If there exists a branch of an SLDNF tree for $\mathbf{P} \cup \{\leftarrow A\}$ for which the terminating node is an empty goal, then the node has no descendant.

- A node which is the empty goal has no descendants.

In an SLDNF tree, a branch which terminates at an empty goal is a *success branch*, a branch which does not terminate is an *infinite branch* and a branch which terminates at a non-empty goal is a *failed branch*. An SLDNF tree for which every branch is a failed branch is indeed a *finitely failed SLDNF tree*. Each branch of an SLDNF-tree corresponds to an SLDNF-derivation.

According to the definition of SLDNF resolution, a computation cannot proceed (or *flounders*) further from a goal with only non-ground negative literals left. The allowedness restriction on programs and goals are introduced in order to prevent floundering of any computation. Formally, a clause $A \leftarrow M_1 \wedge ... \wedge M_n$ in a program **P** is *allowed* or *range-restricted* (Clark, 1978; Lloyd, 1987; Gallaire *et al.*, 1984) if every variable that occurs in the clause occurs in a positive literal of the body $M_1 \wedge ... \wedge M_n$. The whole program **P** is allowed if each of its clauses is allowed. A goal $\leftarrow L_1 \wedge ... \wedge L_m$ is allowed if every variable that occurs in the goal occurs in a positive literal of the goal. If a program is range-restricted, then by definition, each of its assertions (that is, a clause with empty body) is ground.

Theorem 4-5: (Soundness of SLDNF Resolution) Let **P** be a program and $G = \leftarrow L_1 \wedge ... \wedge L_n$ be a goal. If $\mathbf{P} \cup \{G\}$ has an SLDNF refutation with substitutions $\theta_1, ..., \theta_m$, then $\forall((L_1 \wedge ... \wedge L_n)\theta_1 ... \theta_m)$ is a logical consequence of $Comp(\mathbf{P})$.

Although SLDNF resolution is sound with respect to the completion, the resolution is not necessarily complete even when the completion is consistent. For example, if **P** is $\{P \leftarrow Q, P \leftarrow \neg Q, Q \leftarrow Q\}$, then **P** is stratified and hence $Comp(\mathbf{P})$ is consistent. The atom P is a logical consequence of $Comp(\mathbf{P})$, but $\mathbf{P} \cup \{\leftarrow P\}$ does not have an SLDNF refutation. This example also proves that the SLDNF-resolution is not even complete for stratified programs. Hence one should look for a class of programs which is less general than the stratified class of programs.

In his paper, Clark (1978) proved that the SLDNF-resolution is complete for an *allowed hierarchical class* of programs and goals. A further completeness result was also established for *structured programs* introduced by Barbuti and Martelli (1986). The completeness result has been pushed up to the class of allowed *strict stratified programs* (Cavedon and Lloyd, 1989; Baratelli and File, 1988).

4.6   Prolog

This chapter presents the logic programming language Prolog. The main topics discussed in this chapter include the syntax of Prolog programs and goals, the style of their execution, semantics, and usefulness of extralogical features and meta-programming.

Prolog is a logic programming language and a *Prolog program* is a normal logic program, which is a finite set of normal clauses, and a *Prolog goal* is a normal goal. Some additional extralogical features from outside the framework of first-order logic have been incorporated within normal Prolog programs and goals. These extra features allow (among other things) the achievement of the following:

- Performance enhancement by reducing the search space, that is, by pruning the redundant branches of search trees (for example, using the *cut* symbol)

- Increasing readability (for example, using *if-then*/*or* constructs)

- Input/output operations (for example, using *read*/*write* predicates)

- Clause management (for example, using *assert*/*retract* predicates)

- Set-type evaluation in addition to the resolution-based, tuple-at-a-time approach (for example, using *setof*/*bagof* predicates)

- Meta-programming (for example, using *clause*, *call* predicates)

To incorporate these features, the syntax of Prolog programs (and Prolog goals) is extended using some special symbols, special predicates, and special constructs.

4.6.1   Prolog Syntax

A standard Prolog type of syntax is now introduced for dealing with the normal Prolog programs and goals in the rest of the book. Extralogical symbols, predicates, and constructs are introduced in the relevant sections.

- In general, constants, functions and predicate symbols of a Prolog program begin with a lower-case letter (or the dollar symbol, $) or may comprise any string of characters enclosed in single quotes. The rest of a line starts with % is treated as comments.

- Variables begin with at least one upper-case letter (or the underline symbol _). The symbol _ on its own is a special type of variable called the "don't know" variable. Discrete occurrences of this particular variable in a program clause or in a goal are considered different from each other.

- The symbols for $\wedge$ (and sign), $\vee$ (or sign), $\leftarrow$ (implication) and $\neg$ (negation) are `,`, `;`, `:-`, and `not` respectively.

- Arithmetic function symbols (for example, `+`, `-`, `*`, `/`) and some comparison predicates (such as `<`, `>`) may be written in their infix forms. For example, the addition of two numbers represented by variables `X` and `Y` is usually expressed as `X+Y` rather than `+(X,Y)` (although this is also perfectly correct syntax).

- Arithmetic assignment is achieved by the predicate `is` and the arithmetic equality symbol is represented by the symbol `=:=`.

- The unification of two terms is performed using the symbol `=` whereas to check whether or not two terms are syntactically equal (that is, the equality symbol) the symbol `==` is used.

In describing a goal, subgoal, or a piece of code, the occurrence of an expression of the form ⟨type⟩ or ⟨type-n⟩ can be replaced by an arbitrary expression which is of type `type`. So, for example, a subgoal `square(`⟨integer-1⟩`,`⟨integer-2⟩`)` means the arguments of the predicate `square` are arbitrary integers. With these conventions, a Prolog clause has the following form:

⟨atom⟩`:-` ⟨literal-1⟩`,` ⟨literal-2⟩`,` `...,` ⟨literal-m⟩`.`

where $m \geq 0$. Every clause in a Prolog program or a goal is terminated with the dot symbol (`.`).

Example

The normal clause

$$Rain(x) \leftarrow Condition(x, Wet) \wedge \neg Sprinkler(x)$$

encodes the statement that if a field's condition is wet and the sprinkler system of the field is not on then it is raining at the field. In the syntax of first-order logic, this clause is expressed in Prolog syntax as

```
rain(X):- condition(X, wet), not sprinkler(X).
```

A Prolog program's goal is expressed by the symbol ?- followed by a conjunction of literals and then terminated with (.). Hence a normal Prolog goal has the form

```
?- ⟨literal-1⟩, ⟨literal-2⟩, ..., ⟨literal-n⟩.
```

Example

A Prolog goal to find all fields where it is raining and not closed: is

```
?- rain(X), not closed(X).
```

Commands to a Prolog interpreter are in the form of goals.

4.6.2 Theoretical Background

The inference mechanism adopted in a standard Prolog system (considered as a Prolog interpreter throughout the text) is the same as SLDNF-resolution with the following restrictions:

- The computation rule in standard Prolog systems always selects the leftmost literal in a goal (it follows a leftmost literal selection strategy).

- Standard Prolog uses the order of clauses in a program as the fixed order in which clauses are to be tried, that is, the search tree is searched depth-first.

- Prolog omits the occur check condition for unifying two expressions.

Therefore, the search trees in the context of the execution of normal Prolog programs and goals are similar in structure to SLDNF-trees. Given a Prolog program and a goal, the search tree is unique because of the fixed computation rule and the fixed clause order in which they are fetched for the purpose of resolution. For your convenience, the definition of search tree is provided below in the terminology of Prolog.

Suppose $\mathbf{P}$ is a normal Prolog program and G is a Prolog goal. Then the *search tree* T for $\mathbf{P} \cup \{G\}$ (or the search tree for G with respect to $\mathbf{P}$) is defined as follows:

- The root of T is G.

- A leaf node of T is either a goal (a failure node) or an empty clause (a success node).

- Suppose the node N is

```
?- ⟨literal-1⟩, ⟨literal-2⟩, ..., ⟨literal-n⟩
```

and the leftmost literal ⟨literal-1⟩ is positive. Suppose

```
⟨atom-1⟩:-⟨literal-conjunction-1⟩
...
⟨atom-p⟩:-⟨literal-conjunction-p⟩
```

are the only clauses of **P** such that ⟨literal-1⟩ and each of ⟨atom-i⟩ unifies with an mgu ⟨substitution-i⟩. The order shown for these clauses is the order in which they will be tried for resolution with the goal of node N. Then N has p descendants and they are (from left to right)

```
?- (⟨literal-conjunction-1⟩, ⟨literal-2⟩, ...,
        ⟨literal-n⟩)⟨substitution-1⟩
...
?- (⟨literal-conjunction_p⟩, ⟨literal-2⟩, ...,
        ⟨literal-n⟩)⟨substitution-p⟩.
```

If for some i, ⟨literal-conjunction-i⟩ is true, (that is, the clause ⟨atom-i⟩:-⟨literal-conjunction-i⟩ is a fact), then the corresponding descendant is

```
?- (⟨literal-2⟩, ..., ⟨literal-n⟩)⟨substitution-i⟩.
```

This becomes the empty clause □ when $n = 0$.

- Suppose the node N is

```
?- ⟨literal-1⟩, ⟨literal-2⟩, ..., ⟨literal-n⟩
```

and the leftmost literal ⟨literal-1⟩ is a negative literal of the form not ⟨atom⟩ (The search tree construction does not stop even if the leftmost literal is non-ground and the answer in this case may be incorrect.) Then a recursive process is established to apply the negation as failure (NAF) rule, that is, to find the search tree for the goal ?- ⟨atom⟩ with respect to **P**. If all the branches of this tree result in failure, then the only descendant of N is

```
?- ⟨literal-2⟩, ..., ⟨literal-n⟩
```

without any substitution of its variables; otherwise, if the system is able to find a success node of the search tree without going into an infinite loop, then N is considered a failure node.

In the above definition, a success branch corresponds to a successful Prolog derivation which causes some instantiations to the variables of *G* and is an answer to the goal *G*. The following example constructs a Prolog search tree.

Example

Consider the following program **P** and goal *G*:

Program:

```
R1. p(X,Y):- q(X,Y),not r(Y).
R2. p(X,Y):- s(X),r(Y).
F1. q(a,b).
F2. q(a,c).
F3. r(c).
F4. s(a).
F5. t(a).
```

Goal:

```
?- p(X,Y),t(X).
```

The Prolog search tree for $\mathbf{P} \cup \{G\}$ is given in Figure 4-4. The leftmost branch is a successful derivation and is given below (selected literals are underlined).

Goals	Input Clauses	Substitutions
?- <u>p(X,Y)</u>,t(X)	R1	{X/X, Y/Y}
?- <u>q(X,Y)</u>,not r(Y),t(X)	F1	{X/a, Y/b}
?- <u>not r(b)</u>,t(a)	not r(b)	{}
?- <u>t(a)</u>	t(a)	{}
□		

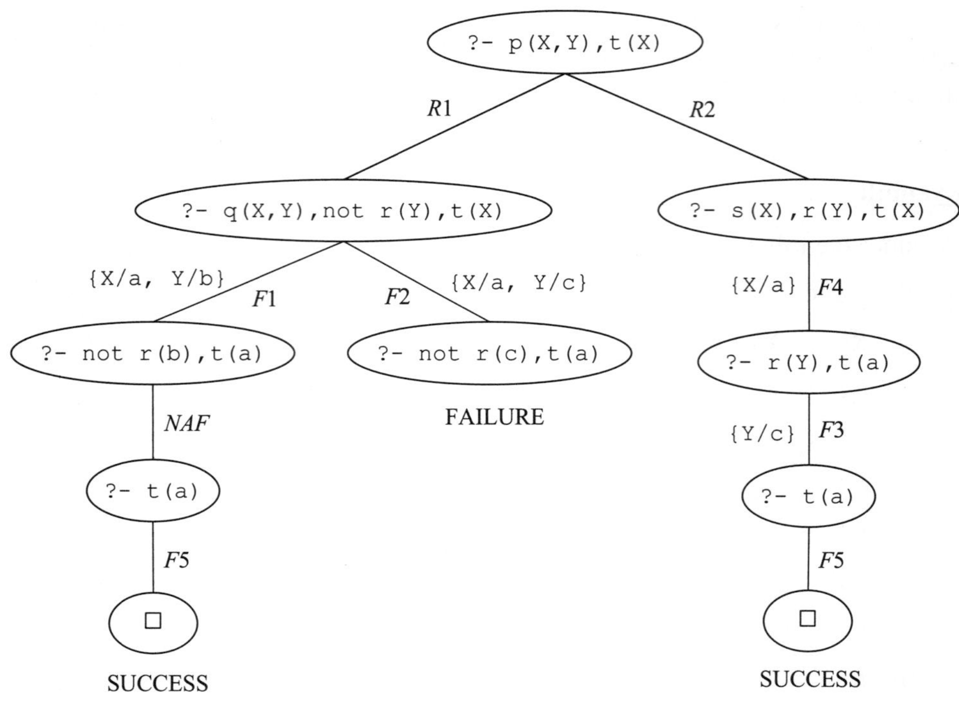

Figure 4-4: A Prolog search tree.

4.6.3 *Backtracking*

When a complete search tree for a goal with respect to a program is generated without going into a loop, all success branches of the tree generate all answers to the goal *G*. In the context of the example presented in the previous sub-section, there are two answers to the goal ?-p(X,Y),t(X) and they correspond to the two success branches of Figure 4-4. These two answers are {X/a, Y/b} and {X/a, Y/c}. Since the search tree is generated depth-first, the leftmost branch is generated first and therefore the answer {X/a, Y/b} results.

To try to find the next alternative answer, the user can initiate *backtracking*, which is a request for an alternative solution. Backtracking after detecting a failed branch is automatic. Hence, after the user initiates backtracking, all subsequent backtrackings are automatic until the interpreter finds an alternative answer to the goal. In the case of the above example, when backtracking is initiated after the first answer has been generated, the system must try to find an alternative solution for the last subgoal t (a) to find an alternative answer for the original goal. Since there is no more t (a) in the program, the last but one subgoal is tried. This is the negative subgoal not r (b) which cannot generate any new answers.

Backtracking further, the subgoal q (X, Y) is resolved against the second fact q (a, c) under the predicate q and another descendant ?- not r (c), t (a) is created from the node ?-q (X, Y), not r (Y), t (X). The selected subgoal not r (c) from this node does not succeed and hence this node is considered a failure node. The interpreter backtracks automatically to the root node and generates the rightmost success branch eventually.

4.6.4 *The Cut*

Prolog provides an extralogical system predicate called the *cut* (!) to reduce the search space. This reduction is achieved by dynamically pruning redundant branches of the search tree. A normal clause with one cut looks like

```
⟨atom⟩:- ⟨literal-1⟩, ..., ⟨literal-i⟩, !,
                ⟨literal-i+1⟩, ..., ⟨literal-n⟩.
```

On finding a solution, the cut prunes all alternative solutions to the conjunction ⟨literal-1⟩, ..., ⟨literal-i⟩, but does not affect the conjunction ⟨literal-i+1⟩, ..., ⟨literal-n⟩. To illustrate this, consider the following example.

Example

Refer again to the program **P** which generated the search tree in Figure 4-4 for the goal ?- p (X, Y), t (X). The search tree for **P** ∪ {*G*}, where *G* is the goal ?- p (X, Y), !, t (X), is same as the one in Figure 4-4. Due to the presence of the cut symbol in the goal the alternative solution p (X, Y) (with the help of *R*2) is pruned. Again refer to the same program **P**. If the rule *R*1 is replaced by p (X, Y) :- q (X, Y), !, not r (b), then the tree contains only the leftmost branch of Figure 4-4.

Cuts are divided into two different categories. *Green cuts* prune only those branches of the search tree that do not lead to any new solutions. Hence removal of those cuts from a program does not change the declarative meaning of the program.

Example

Consider the following program which computes the absolute value of an integer X and illustrates the use of a green cut in a program:

```
absolute(X, X):- X >= 0, !.
absolute(X, Y):- X < 0, Y is -X.
```

Without using the cut in the first clause a redundant check would be performed when calculating the absolute value of a non-negative integer using a goal such as ?- absolute(3, Z). By using a cut this redundant check is eliminated.

All other cuts are *red cuts* and they do affect the declarative meaning of a program. Red cuts should be used in a program with special care.

Example

Consider the following example illustrating the use of a red cut:

```
absolute(X, X):- X >= 0,  !.
absolute(X, Y):- Y is -X.
distance(X, Y, Z):- Z1 is X-Y, absolute(Z1, Z).
```

The cut in the above program is absolutely essential. In its absence a goal of the form ?-distance(2, -3, Z) would instantiate Z to 5 first and then to –5 upon backtracking which is obviously incorrect.

4.6.5   *Special Constructs and Connectives*

To increase the readability and, to some extent, the efficiency of a Prolog program, a few extra connectives and constructs have been introduced. When the or-connective ; is used in program clauses (and also in goals) as

```
⟨atom⟩:- ⟨lit-conj-1⟩, (lit_conj-2;
                    lit-conj_3), lit-conj_4.
```

this single clause is interpreted as a combination of two clauses

```
⟨atom⟩:- ⟨lit-conj-1⟩, lit-conj-2, lit-conj-4
⟨atom⟩:- ⟨lit-conj-1⟩, lit-conj-3, lit-conj-4
```

where `lit-conj` is an abbreviation of `literal-conjunction`. In the body of a Prolog program clause an occurrence of an expression of the form

$$\langle \texttt{lit-conj-1}\rangle \ \texttt{->}\ \langle \texttt{lit-conj-2}\rangle;\ \langle \texttt{lit-conj-3}\rangle$$

using the special conditional if-then construct (`->`) can be thought of as an alternative form of an atom (not a first-order atom)

$$\langle \texttt{predicate-symbol}\rangle(\langle \texttt{lit-conj-1}\rangle,$$
$$\langle \texttt{lit-conj-2}\rangle,\ \langle \texttt{lit-conj-3}\rangle)$$

where the definition of $\langle \texttt{predicate-symbol}\rangle$ is

```
⟨predicate-symbol⟩(⟨lit-conj1⟩, ⟨lit-conj-2⟩, _):-
    ⟨lit-conj-1⟩, !, ⟨lit-conj-2⟩.
⟨predicate-symbol⟩(_, _, ⟨lit-conj-3⟩):-
    ⟨lit-conj-3⟩.
```

The above use of the cut is red.

Example

An example of the use of the construct `->` is to define the absolute value of a number `X` as in the program

```
absolute(X, Y):-
    X >= 0 -> Y = X; Y is -X.
```

where `Y` gives the absolute value of `X`.

4.6.6 Negation

Following SLDNF-resolution, Prolog provides a form of negation based on the negation as failure principle. The negation of $\langle \texttt{atom}\rangle$ is written in Prolog as `not` $\langle \texttt{atom}\rangle$ and can be implemented using the cut as follows:

```
not X :- X, !, fail.
not X.
```

The first of the above clauses prunes the other branches of the SLDNF-tree for the goal "`?- X`" when a success branch is found and marks the original goal as failed. This way of inferring negation may fail to generate all answers as is evident from the following example.

 Decision-Making Agents

Example

Program:

```
p(a).
q(b).
```

Goals:

```
?- p(X), not q(X).
?- not q(X), p(X).
```

The above pair of goals is the same except for the order of the literals. With leftmost literal selection strategy, the first goal succeeds and returns X = a as an answer. However, the second goal has a finitely failed search tree with respect to the program and therefore does not generate any answers.

As was pointed out in the description of SLDNF-resolution, negative literals should be selected only when they are ground. If at some stage of a computation the goal contains only non-ground negative literals, then the computation *flounders*. When programs and goals are considered to be range-restricted, a literal selection strategy can always be found such that computations never flounder. Under Prolog's leftmost literal selection strategy, even if programs and goals are range-restricted, a computation may flounder. This can be demonstrated by considering the above program and its second goal. However, by rearranging the literals in the goals as well as in the bodies of the program clauses, it is always possible to make the negative literals ground before their selection. One of the ways to achieve this is by placing all negative literals in a goal or in the body of a clause after all the positive literals. By applying this strategy to the above program and the second goal, it is reduced to the first one.

It is again worth mentioning here that if a non-ground negative literal "not A" is chosen from a goal and if the goal "?- A" has a finitely failed search tree with respect to the program, then the computation can be continued without fear of losing any answers.

4.6.7 Equality

The symbol = in Prolog is treated as an unifying relation. That is, a subgoal of the form

$$\langle \text{expression-1} \rangle = \langle \text{expression-2} \rangle$$

succeeds when ⟨expression-1⟩ and ⟨expression-2⟩ unify. Such a success will cause instantiation to the variables of ⟨expression-1⟩ and ⟨expression-2⟩ giving an mgu of ⟨expression-1⟩ and ⟨expression-2⟩. For example, the goal

```
?- f(X,b) = f(a,Y).
```

will succeed with variables X and Y instantiated to a and b respectively, that is, the goal returns {X/a, Y/b} as an mgu.

When the two arguments of the predicate = are equal, the mgu in this case can be taken as an identity substitution. The predicate \= is just the negation of the predicate = and is defined using the cut and fail combination as follows:

```
⟨expression-1⟩ \= ⟨expression-2⟩:-
        ⟨expression-1⟩ = ⟨expression-2⟩, !, fail.
_ \= _.
```

Since Prolog omits the occur check condition when unifying two expressions, goals of the form

```
?- X = f(X).
```

do not fail although the normal unification procedure would, of course, fail to unify the two expressions X and f(X). Prolog, however, would instantiate X to the infinite string f(f(f(f(f(....

The predicate == is used to check whether or not two terms are syntactically equal. The subgoal

```
?- f(X,b) == f(a,Y).
```

fails as its constituent terms f(X,b) and f(a,Y) are not syntactically equal (though the two terms are unifiable). The predicate \== is the negation of == and can be defined using the cut and fail combination in the same way as \= was defined by =.

4.6.8   List

A list structure is a sequence of any finite number of term structures. In Prolog it is represented as

```
[⟨expression-1⟩, ..., ⟨expression-n⟩]
```

where each ⟨expression-1⟩, ..., ⟨expression-n⟩ is a member of the list. The expression ⟨expression-1⟩ is called the *head* of the list and the list [⟨expression-2⟩, ..., ⟨expression-n⟩] is called the *tail* of the list.

The *empty list*, denoted as `[]`, does not have any members. A list in Prolog is abbreviated as [⟨expression⟩|⟨list⟩] in which ⟨expression⟩ is the head and ⟨list⟩ is the tail. Thus the goal

$$?- \ [H|T] = [a, \ b, \ c, \ d].$$

would succeed with instantiation {H/a, T/[b, c, d]}.

A list structure can be thought of as a functional value *list*(*H*,*T*) which is when the one-to-one function symbol *list* is applied on its two arguments *H* and *T*. The functional value *list*(*H*,*T*) is also a list in which *H* is the head and *T* is the tail.

Example

As an application of list structure, consider the following Prolog program which checks whether a given structure is a member of a list or not:

```
member(X, [Y|_]):-X == Y.
member(X, [_|T]):-member(X, T).
```

The subgoal `member(X, L)` succeeds when `X` is a member of `L`. Another useful operation on lists is to join two lists together to form another. The following Prolog program serves this purpose:

```
append([], L, L).
append([X|L1], L2, [X|L3]):-append(L1, L2, L3).
```

The subgoal `append(L1, L2, L3)` succeeds when the list `L3` is obtained by appending `L2` to `L1`.

4.6.9   Arithmetic

The arithmetic function symbols (`*`, `/`, `+`, `-`, `mod`, `div`) are interpreted function symbols in Prolog. The assignment of the value of an arithmetic expression to a variable is achieved by the binary predicate symbol `is`. A subgoal of the form

⟨variable⟩ is ⟨arithmetic-expression⟩

causes an evaluation of ⟨arithmetic-expression⟩, giving a value. Then ⟨variable⟩ is instantiated to this value. For example, the subgoal

$$?- \ X \ is \ 2+3*5.$$

succeeds and instantiates X to 17. Equality testing between the values of two arithmetic expressions is carried out in Prolog by means of the predicate = : =. Hence a subgoal of the form

$$\langle \texttt{arithmetic-expression-1} \rangle \ \texttt{=:=}$$
$$\langle \texttt{arithmetic-expression-2} \rangle$$

succeeds when the value of $\langle \texttt{arithmetic-expression-1} \rangle$ is equal to that of $\langle \texttt{arithmetic-expression-2} \rangle$. Thus the goal

```
?- 2+3*5 =:= 57/3-2.
```

succeeds. Other arithmetic binary predicates are < (strictly less than), > (strictly greater than), <= (less than or equal to), >= (greater than or equal to). The arguments of these predicates and also the predicate = : = should be ground.

4.6.10 Input/Output

A Prolog program can read data from several input files (called *input streams*) and can output data to several output files (called *output streams*). However only one input stream and one output stream can be active at any time during the execution of the program. Reading input from a user's terminal and outputing data to a user's terminal are considered as input and output to a special file named *user*. At the beginning of the execution of a program these two streams are open. The subgoal

```
see (⟨file-name⟩)
```

succeeds if the file ⟨file-name⟩ exists. The subgoal will cause the current input stream to switch to ⟨file-name⟩. The existence of a file can be tested by the predicate *exists*. The subgoal

```
exists (⟨file-name⟩)
```

succeeds if the file ⟨file-name⟩ exists in the current directory. In dealing with output streams, the subgoal

```
tell (⟨file-name⟩)
```

always succeeds and causes the current output stream to switch to ⟨file-name⟩.

Reading a term from the current input stream is accomplished by the unary predicate *read* and writing a term to the current output stream is done by the predicate *write*. Hence the subgoal

```
read (⟨variable⟩)
```

succeeds with an instantiation of the variable ⟨variable⟩ to a term from a current input file. When the end of the current input file is reached, ⟨variable⟩ is instantiated to the constant `end_of_file`. Each term of the input file must be followed by . and a carriage return. In a similar way, the subgoal

$$\text{write}(\langle\text{term}\rangle)$$

succeeds and causes the term ⟨term⟩ to be written on the current output stream. The subgoal

$$\text{tab}(\langle\text{integer}\rangle)$$

succeeds and writes ⟨integer⟩ number of spaces on the current output stream. Also the nullary predicate `nl` causes the start of a new line on the current output stream. Current input and output streams can be closed by the nullary predicates `seen` and `told` respectively.

Example

Consider the following program to display the contents of a file on the terminal (assuming the current output stream is `user`) using the predicates `read` and `write`:

```
display_file(FileName):-
    exists(FileName),
    see(FileName),
    repeat,
        read(Term),
        (Term = end_of_file -> !, seen ;
        write( Term ), write('.'), nl, fail).
```

A goal of the form

```
?- display_file(⟨file-name⟩).
```

displays the contents of the file ⟨file-name⟩ on the terminal but the variable names occurring in any terms therein are replaced by suitable internal variable names.

As opposed to reading and writing terms, single characters from the input or output streams can also be read or written with the help of the unary predicates `get0` and `put`. For example, a single character can be read from an input stream by using a subgoal of the form

```
get0(⟨variable⟩)
```

and ⟨variable⟩ will be instantiated to the ASCII character code for the character read. A subgoal of the form

```
put(⟨variable⟩)
```

displays the character whose ASCII character code is ⟨variable⟩.

Example

The above program to display the contents of a file can be rewritten using the two predicates `get0` and `put` as follows:

```
display_file(FileName):-
   exists(FileName),
   see(FileName),
   repeat,
      get0(Char),
      (Char = 26 -> % ASCII code for control-Z
      !, seen ;
      put(Char), fail).
```

In this case the variable names will be displayed as they were in the input file.

4.6.11 Clause Management

To read clauses from a file the predicate `consult` is used. Hence a command

```
?- consult(⟨file-name⟩).
```

reads the clauses from the file ⟨file-name⟩ into the Prolog interpreter. An alternative syntax for this is

```
?- [⟨file-name⟩].
```

A number of different files can be consulted as

```
?- [⟨file-name-1⟩, ..., ⟨file-name-n⟩].
```

This is useful when different parts of a program are held in a number of different files. A command of the form

```
?- reconsult(⟨file-name⟩).
```

results in the clauses in the file being added to the existing clause set. At the same time, any clauses with the same predicate symbol and arity in their heads as those in ⟨file-name⟩ are deleted from the existing clause set. An alternative syntax for reconsulting the file is

$$?-\ [-\langle\text{file-name}\rangle].$$

A sequence of files can be consulted and reconsulted. As an example, consider the command

$$?-\ [\langle\text{file-name-1}\rangle,\ -\langle\text{file-name-2}\rangle,\ \langle\text{file-name-3}\rangle].$$

which causes ⟨file-name-1⟩ to be consulted, then ⟨file-name-2⟩ to be reconsulted on the resultant set of clauses and then ⟨file-name-3⟩ to be consulted on the resultant set of clauses. Consulting or reconsulting the special file *user* causes the system to read in clauses from the terminal until the predicate symbol end_of_file is entered.

An individual clause can be added to an existing set of clauses by the command

$$?-\ \text{assert}(\langle\text{clause}\rangle).$$

There are two different versions of the assert predicate, namely asserta and assertz. The former causes the clause to be placed towards the beginning of the existing set of clauses before any other clauses with same head predicate and the latter will place it at the end after any clauses with the same head predicate. Deleting a clause from an existing set of clauses is achieved by

$$?-\ \text{retract}(\langle\text{clause}\rangle).$$

Considering clause as a term, Prolog attempts to unify it with an existing clause. The first clause for which this unification succeed is deleted from the existing clause set. The command

$$?-\ \text{retractall}(\langle\text{clause}\rangle).$$

deletes all the clauses whose heads unify with clause. The command

$$?-\ \text{abolish}(\langle\text{predicate-symbol}\rangle,\ \langle\text{integer}\rangle).$$

causes the deletion of all clauses whose head predicate symbol is ⟨predicate-symbol⟩ with arity ⟨integer⟩.

4.6.12 Set Evaluation

The evaluation strategy for producing answers in pure Prolog is basically tuple-at-a-time, that is, branches of the search tree are explored one after another using a depth-first strategy with backtracking. Every success branch produces an answer. When Prolog tries to find another success branch through backtracking, all information about the previous branch of the computation is lost. Therefore, in Prolog, there is no connection between the computation of answers coming from two different branches of the search tree.

The predicate `setof` enables you to accumulate different answers to a query and thus provides a set-at-a-time evaluation strategy. The syntax of this predicate is outside the framework of first-order logic and can be given as

```
setof(⟨term⟩, ⟨conjunction⟩, ⟨list⟩)
```

The above subgoal is true when ⟨list⟩ is the sorted set of instances of ⟨term⟩ for which the goal ⟨conjunction⟩ succeeds. A slightly different version of the `setof` predicate is `bagof` whose syntax is similar to the above, that is

```
bagof(⟨term⟩, ⟨conjunction⟩, ⟨list⟩)
```

The semantics of the above subgoal is given as true when ⟨list⟩ is the multiset (that is, a well-defined collection of elements, not necessarily distinct) of all instances of ⟨term⟩ for which ?- ⟨conjunction⟩ succeeds.

Example

Consider a program containing the following clauses:

```
event_type(fenway_park, baseball).
event_type(fenway_park, concert).
event_type(fenway_park, baseball).
event_type(eden_garden, cricket).
event_type(eden_garden, soccer).
event_type(eden_garden, cricket).
event_type(mcc_lords, cricket).
```

A goal to find all the types of event held at Fenway Park can be given as

```
?- setof(X, event_type(fenway_park, X), L).
```

which causes an instantiation of L as

```
L = [baseball,concert]
```

On the other hand, the goal

```
?- bagof(X, event_type(fenway_park, X), L).
```

causes an instantiation of L to

```
    L = [baseball,concert,baseball]
```

The predicate `setof` can have multiple solutions. The goal

```
    ?- setof(X, event_type(Y, X), L).
```

causes an instantiation of Y and L as

```
    L = [baseball,concert]
    Y = fenway_park
```

respectively. Upon backtracking, the next instantiation of Y and L would be

```
    L = [cricket,soccer]
    Y = eden_garden
```

and so on. The same goal can be interpreted to find all Xs such that `father(Y, X)` is true for some Y and is expressed as

```
    ?- setof(X, Y^event_type(Y, X), L).
```

where Y^ can be interpreted as an existential quantifier. The instantiation to L would be

```
    L = [baseball,concert,cricket,soccer]
```

The predicate `bagof` has similar interpretations in the above two cases but does not discard any duplicate elements from a list.

The system predicate `setof` can be implemented by using some of the predicates already introduced.

<u>Example</u>

Consider one such alternative implementation to find all solutions without backtracking:

```
    set_of(Term, Goal, Instance):-
       assert(term_goal(Term, Goal)),
       set_of([], UInstance),
       sort(UInstance, Instance),
       retract(term_goal(_,_)).
    set_of(L, UInstance):-
       term_goal(Term, Goal),
       Goal,
       not member(Term, L), !,
```

```
   set_of([Term|L], UInstance).
   set_of(UInstance, UInstance).
member(H, [H|_]):-!.
member(H, [_|T]):-member(H, T).
```

A major drawback in the above implementation is that each time a new solution is generated, the search tree corresponding to the goal `Goal` is traversed from the beginning and not from the point where the last solution is found. A further alternative implementation can be achieved by using the `assert` and `retract` predicates. This implementation would be particularly difficult in the presence of rules.

4.6.13 *Meta Programming*

A *meta-program* analyzes, transforms, and simulates other programs by treating them as data. Meta-programming is inherent in the nature of logic programming and Prolog is no exception. Some meta-programming examples are as follows: a translator (or a compiler) for a particular language is a meta-program which takes programs written in that language as input, the output for the translator (or compiler) is another program in another language, and an editor can also be viewed as a meta-program which treats other programs as data.

An interpreter for a language, called a *meta-interpreter*, is also a meta-program. If the language of a meta-interpreter $\mathcal{M}$, that is, the language in which $\mathcal{M}$ is written, coincides with the language for which $\mathcal{M}$ is written, then the meta-interpreter $\mathcal{M}$ is called a *meta-circular interpreter*. Since program and data are uniformly expressed as clauses in logic programming environments, writing a meta-circular interpreter is an inherent feature.

Example

Consider an example of a meta-circular interpreter written in Prolog which simulates Prolog-like execution by selecting the rightmost atom from a goal:

```
solve(true).
solve((B, A)):- solve(B), solve(A).
solve(not A):- not solve(A).
solve(A):- clause(A, B), solve(B).
```

Another example is the meta-program which transforms a set of definite rules (with no occurrence of constant symbols) to another program by *saturating* each non-recursive predicate occurring in the body of a rule. For example, the following Prolog version of some definite program:

```prolog
a(X, Y):- p(X, Y).
a(X, Y):- p(X, Z), a(Z, Y).
p(X, Y):- f(X, Y).
p(X, Y):- m(X, Y).
m(X, Y):- f(Z, Y), h(Z, X).
```

can be transformed to the following Prolog program:

```prolog
a(X, Y):- f(X, Y).
a(X, Y):- f(Z, Y), h(Z, X).
a(X, Y):- f(X, Z), a(Z, Y).
a(X, Y):- f(Z1, Z), h(Z1, X), a(Z, Y).
p(X, Y):- f(X, Y).
p(X, Y):- f(Z, Y), h(Z, X).
m(X, Y):- f(Z, Y), h(Z, X).
```

To write this meta-program, suppose the program clauses are stored under a unary predicate `rule` in the form of `rule(⟨rule⟩)`. Then, in the following meta-program, the subgoal `saturate(⟨rule-1⟩, ⟨rule-2⟩)` succeeds only when ⟨rule-2⟩ is obtained from ⟨rule-1⟩ by the saturation process described above:

```prolog
base(P):- not rule((P:-_)).
recursive(P):- rule((P:-B)), recursive(P, B), !.
recursive(H, H).
recursive(H, (B,Bs)):-
    recursive(H, B); recursive(H, Bs).
recursive(H, H1):-
    rule((H1:-B)), recursive( H, B ).
expand((B,Bs), (ExpB,ExpBs)):-!,
expand(B, ExpB), expand(Bs, ExpBs).
expand(H, H):- (recursive( H ); base(H)), !.
expand(H, ExpB):- rule((H:-B)), expand(B, ExpB).
```

4.7   Prolog Systems

SICStus Prolog (http://www.sics.se/isl/sicstus.html) is an ISO standard compliant (http://www.logic-programming.org/prolog_std.html) Prolog development system with constraint solver. Logic Programming Associates (http://www.lpa.co.uk/) offers Prolog compilers on Windos, DOS, and Macintosh platforms. GNU Prolog (http://pauillac.inria.fr/~diaz/gnu-prolog/) is a free and ISO standard compliant Prolog compiler with constraint solving over finite domains. Amzi! (http://www.amzi.com/) offers Prolog + Logic servers on various platforms, including Windows, Linux, and Solaris.

The Prolog system E5 (Environment for 5th Generation Application), being developed by the author, has three modes of operations: Prolog, Lisp, and Expert System. It is an interpreter developed fully in Visual C++ 6.0 and therefore runs only under the Microsoft Windows environment. The system provides an interactive, multi-document, and graphical user interface. The system is a full implementation of ANSI Prolog and supports various features, including libraries for graphical user interface development, routines for integrating ODBC compliant data sources as facts, and a debugging facility. The interpreter is also COM compliant so that it can be called by other applications. E5 is an ideal platform for rapid prototyping of complex reasoning processes. Please contact the author for more details or to obtain an evaluation version.

4.8   Complexity Analysis

Complexities of various classes of logic programming are given in terms of polynomial hierarchy. The classes Σ_k^p, Π_k^p, and Δ_k^p of the polynomial hierarchy are recursively defined as follows:

$$\Sigma_0^p = \Pi_0^p = \Delta_0^p = P$$

$$\Sigma_{k+1}^p = NP^{\Sigma_k^p}, \quad \Pi_{k+1}^p = co - \Sigma_{k+1}^p, \quad \Delta_{k+1}^p = P^{\Sigma_k^p}, \text{ for all } k \geq 0$$

Therefore, $\Sigma_1^p = \text{NP}, \Pi_1^p = \text{co-NP}, \Delta_1^p = \text{P}$. The class Σ_2^p contains all problems solvable in non-deterministic polynomial time, provided it is possible to freely use a subroutine for a problem in NP. Testing the validity of a QBF of the form

$$\exists X_1...\exists X_n \forall Y_1...\forall Y_m F$$

is Σ_2^p-complete. On the other hand, testing the validity of a QBF of the form

$$\forall X_1...\forall X_n \exists Y_1...\exists Y_m F$$

is Π_2^p-complete.

We study here breifly the complexity of the following query problem in logic programming: given a program **P** and a goal or query of the form $\leftarrow L_1 \wedge ... \wedge L_n$, determine from the completion of the program **P** whether there is a proof for the formula

$$\exists x_1 ... \exists x_p (L_1 \wedge ... \wedge L_n)$$

where $x_1, ..., x_p$ are all the free variables in $\leftarrow L_1 \wedge ... \wedge L_n$. Note that there is no difference between data complexity and program complexity, as defined in (Vardi, 1982), since the program can be considered as part of the query; by complexity here we mean the expression or the combined complexity. If **P** is non-recursive without function symbol then the complexity is PSPACE-complete. For stratified non-recursive programs without function symbol, the complexity is still PSPACE-complete, and NEXP-complete (if P = NP then NEXP = EXP) in the case of recursive programs. In the case of arbitrary function symbols, the complexity for programs with n levels of stratification is Σ_n^p-complete. See (Dantsin et al., 2001) for a detailed discussion on the complexity of various classes of logic programs.

4.9   Further Readings

The author's own book (Das, 1992) presents a detailed overview of logic programming and its related field deductive databases. Kowalski's book (1979a) is must for foundational reading. A theoretical account of logic programming can be found in (Lloyd, 1987). (Russell and Norvig, 2002) is an excellent book that discusses, among other areas, logic, logic programming, and Prolog from the artificial intelligence perspective. There are a number of good books on Prolog, including (Bratko, 2000), (Clocksin and Mellish, 2003), (O'Keefe, 1990), and (Sterling and Shapiro, 1994).

Chapter 5

Logical Rules for Making Decisions

This chapter provides a foundation for building propositional epistemic models for decision-making agents by describing the use of a special subset of classical logics that take the form of conditional *if-then* type rules. The if-then type syntax provides domain experts a natural way of expressing their knowledge, be it as arguments for or against carrying out actions or simple conditional statements for deriving implicit knowledge from observations. Unlike the definite rules handled in the last chapter, the syntax of rules considered in this chapter allows negative consequents for representing arguments against certain options. Moreover, to break the coarse grain granularity of Boolean type truth-value, we embed numbers from the dense set [0,1] to represent degrees of uncertainty with two semantics in mind: 1) classical probability; and 2) mass values, as in Dempster-Shafer theory. To determine ranking among a set of decision options by aggregating incoming evidence, we apply the Bayesian formula to compute posterior probabilities in the case of probability semantics, and apply the Dempster combination rule to compute overall degrees of belief in the case of belief semantics.

The embellishment of rules with degrees of uncertainty really combines the pure propositional epistemic model with the probabilistic epistemic model. The adoption of the Dempster-Shafer theory is a slight departure from the probability thrust of this book, but it fits naturally within the realm of possible world semantics to be adopted later, and also generalizes the theory of probability.

When multiple, disparate, and uncertain knowledge sources, be they subject matter experts or fielded sensors, are involved in making decisions, some kind of quantitative criterion is useful for measuring the quality of the consensus generated by pooling evidence from these knowledge sources. Formally, consensus in such a disparate, group-decision-making context refers to agreement on some decision by all members (experts and sensors) of the group, rather than a majority, and the consensus process is what a group goes through to reach this agreement. (The consensus process provides a way to focus on areas where experts disagree to help initiate conflict-resolving discussions.)

In this chapter, we present a criterion for measuring consensus among disparate sources when the underlying mechanism for handling uncertainty is the Dempster-Shafer theory (applicable to the Bayesian theory as a special case). The two approaches for combining knowledge sources described in this chapter implicitly assume that all sources are equally credible. This is simply not the case when there are experts of various degrees of expertise and sensors with various degrees of reliability. This chapter also presents a method for taking into account such confidence factors during the handling of uncertainty.

We begin the presentation of this chapter with an introduction of how the if-then type rule syntax evolved from the syntax of classical logics.

5.1   Evolution of Rules

In Chapter 3, on the classical logics, we presented a procedure for converting an arbitrary first-order sentence to its equivalent conjunctive normal form, where each conjunct is a clause or in clausal form. Such a clause is then interpreted as various types of program clauses or if-then type of rules. Figure 5-1 shows the overall evolution along with examples.

Expert system rules and Prolog rules are just the definite version of general if-then type rules with attached uncertainty values. Rules normally used in rule-based expert systems to represent expert knowledge are restricted to propositional syntax to avoid reasoning complexity and to enhance efficiency. Each such rule has an associated degree of uncertainty represented by elements from dictionaries (for example, certainty factor values in MYCIN expert system are drawn from the dictionary $[-1,+1]$ (Heckerman and Shortliffe, 1991)). Arguments are more general than definite rules since the syntax of an argument against an option requires negative consequent.

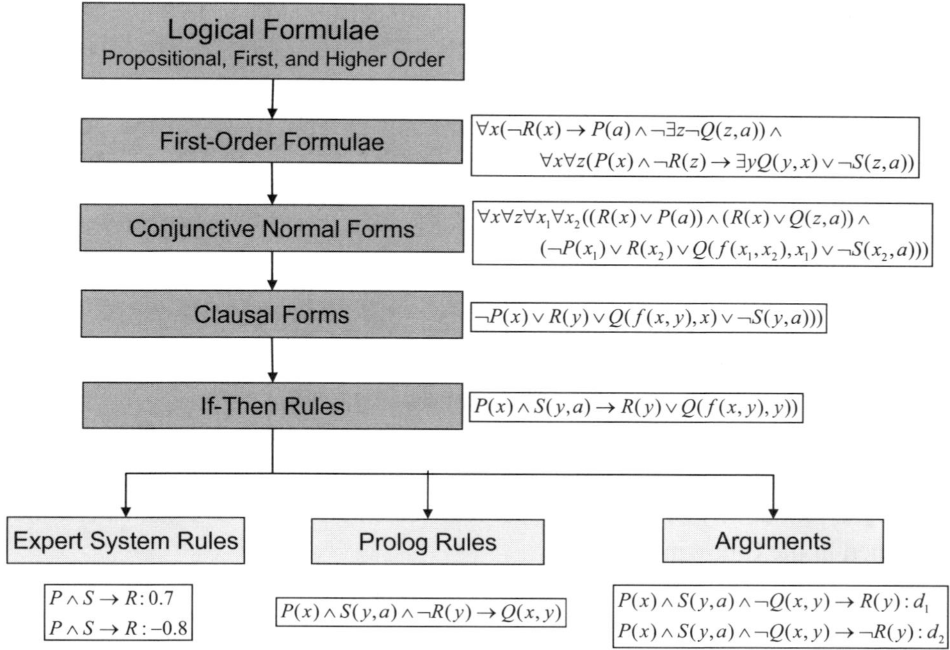

Figure 5-1: Evolution of rules with examples

The if-then type rules in the decision-making context we will be dealing with in this chapter have the following general form:

IF	*Events*
THEN	*Hypothesis* (*D*)

The above rule is interpreted as "If *Events* occur, then *Hypothesis* follows with a certain degree of uncertainty *D*." In the absence of *D*, the rule becomes a definite rule, and reasoning with such rules can be performed using any of the resolution principles without the need to handle uncertainty. If *D* is a probability value, then the rule is interpreted as "If *Events* occur, then the probability that *Hypothesis* will follow is *D*." *Events* is a conjunction of propositional symbols, each representing an event, and *Hypothesis* can be a property symbol as in

IF *Rainy Weather* AND *Players Injured* THEN *Game On* (−0.8)

or an action symbol as in

IF *Rainy Weather* AND *Players Injured* THEN *Cancel Game* (0.8)

But the logical reasoning presented here does not distinguish between a property symbol and an action symbol. Degrees of uncertainty can be drawn from a quantitative dictionary (for example, $[0,1]$ or $[-1,+1]$) or a qualitative dictionary (for example, {*High, Medium, Low*}). Our objective here is to interpret rules upon receiving evidence (for example, evidence on the state of weather) and then compute aggregated evidence on individual decision options (for example, *Game On* and *Cancel Game*) by handling uncertainty appropriately. In the next two sections, we present two approaches to handling uncertainty: Bayesian probability theory and the Dempster-Shafer theory of belief function.

5.2   Bayesian Probability Theory for Handling Uncertainty

In this approach, degrees of uncertainty are represented as a probability value from [0,1]. This approach describes the decision options from an application as a set of possible outcomes, termed "hypotheses." Bayesian inference requires an initial (prior) probability for each hypothesis in the problem space. The inference scheme then updates probabilities using evidence. Each piece of evidence may update the probability of a set of hypotheses calculated via Bayesian rule, which is defined in the following:

$$p(A\,|\,B) = \frac{p(B\,|\,A)p(A)}{p(B)}$$

where A and B are events that are not necessarily mutually exclusive, $p(A\,|\,B)$ is the conditional probability of event A occurring given that event B has occurred, $p(B\,|\,A)$ is the conditional probability of event B occurring given that event A has occurred, $p(A)$ is the probability of event A occurring, and $p(B)$ is the probability of event B occurring. In general, if there are m mutually exclusive and exhaustive hypotheses $H_1,...,H_m$ (that is, $\sum_{i=1}^{m} p(H_i) = 1$ and n possible events $E_1,...,E_n$ that can occur, then the probability of a hypothesis given some evidence is computed as follows:

$$p(H_i\,|\,E_j) = \frac{p(E_j\,|\,H_i)p(H_i)}{\sum_{k=1}^{m} p(E_j\,|\,H_k)p(H_k)}$$

The Bayesian inference mechanism is illustrated in the following example.

<u>Example</u>

Consider a knowledge base consisting of the following three rules describing the chances of canceling a game based on a variety of evidence:.

IF *Heavy Rain* THEN *Cancelled* (0.7)

IF *Players Injured* THEN *Cancelled* (0.8)

IF *Terrorist Threat* THEN *Cancelled* (0.9)

The two mutually exclusive and exhaustive hypotheses are:

$H_1 = Cancelled$

$H_2 = \neg Cancelled$

Three independent events on which evidence can be gathered are:

$E_1 = Heavy\ Rain$

$E_2 = Players\ Injured$

$E_3 = Terrorist\ Threat$

Below, we illustrate how posterior probabilities of the hypotheses are updated as evidence on rain and player injury are gathered.

Initial State

Conditional probabilities $p(H_i | E_j)$ for the hypotheses are inferred as follows from the three rules of the knowledge base and the relation $p(H_1 | E_j) + p(H_2 | E_j) = 1$:

$p(H_1 | E_1) = p(Cancelled | Heavy\ Rain) = 0.7$

$p(H_2 | E_1) = p(\neg Cancelled | Heavy\ Rain) = 0.3$

$p(H_1 | E_2) = p(Cancelled | Players\ Injured) = 0.8$

$p(H_2 | E_2) = p(\neg Cancelled | Players\ Injured) = 0.2$

$p(H_1 | E_3) = p(Cancelled | Terroist\ Threat) = 0.9$

$p(H_2 | E_3) = p(\neg Cancelled | Terroist\ Threat) = 0.1$

Given prior probabilities $p(E_j)$ and $p(H_i)$ are listed below:

$p(E_1) = p(Heavy\ Rain) = 0.20$ $p(H_1) = p(Cancelled) = 0.2$

$p(E_2) = p(Players\ Injured) = 0.10$ $p(H_2) = p(\neg Cancelled) = 0.8$

$p(E_3) = p(Terroist\ Threat) = 0.01$

The following conditional probabilities $p(E_j | H_i)$ for evidence are computed via the application of Bayesian rule:

$$p(E_1 \mid H_1) = p(Heavy\ Rain \mid Cancelled) = 0.7$$
$$p(E_1 \mid H_2) = p(Heavy\ Rain \mid \neg Cancelled) = 0.075$$

$$p(E_2 \mid H_1) = p(Players\ Injured \mid Cancelled) = 0.4$$
$$p(E_2 \mid H_2) = p(Players\ Injured \mid \neg Cancelled) = 0.025$$

$$p(E_3 \mid H_1) = p(Terroist\ Threat \mid Cancelled) = 0.045$$
$$p(E_3 \mid H_2) = p(Terroist\ Threat \mid \neg Cancelled) = 0.001$$

Evidence of Heavy Rain

Assume that we first receive confirming evidence e on E_1 (that is, heavy rain). Then the probability $p(Cancelled \mid Heavy\ Rain)$ directly provides the posterior probability 0.7 for the game being cancelled, which is a significant increase from the prior probability $p(Cancelled) = 0.2$. If you are not completely sure about heavy rain, then soft evidence e can be encoded as the following likelihood:

$$\begin{bmatrix} p(e \mid Heavy\ Rain) \\ p(e \mid \neg Heavy\ Rain) \end{bmatrix} = \begin{bmatrix} 0.95 \\ 0.05 \end{bmatrix}$$

The posterior probability of E_1 upon receiving e is computed as shown below (α is the normalizing constant):

$$p(Heavy\ Rain \mid e) = \alpha \times p(e \mid Heavy\ Rain) \times p(HeavyRain)$$
$$= \alpha \times \begin{bmatrix} 0.95 \\ 0.05 \end{bmatrix} \times \begin{bmatrix} 0.2 \\ 0.8 \end{bmatrix} = \begin{bmatrix} 0.83 \\ 0.17 \end{bmatrix}$$

You can then compute the posterior probability of the hypotheses as follows:

$$p(Cancelled \mid e) = p(Cancelled \mid Heavy\ Rain) \times p(HeavyRain \mid e) +$$
$$p(Cancelled \mid \neg Heavy\ Rain) \times p(\neg HeavyRain \mid e)$$

But we have

$$p(Cancelled \mid \neg Heavy\ Rain)$$

$$= \frac{p(\neg Heavy\ Rain \mid Cancelled) \times p(Cancelled)}{p(\neg Heavy\ Rain)}$$

$$= \frac{(1 - p(Heavy\ Rain \mid Cancelled)) \times p(Cancelled)}{(1 - p(Heavy\ Rain))}$$

$$= \frac{(1 - 0.7) \times 0.2}{1 - 0.2} = 0.07$$

Therefore,

$$p(Cancelled \mid e) = 0.7 \times 0.83 + 0.07 \times 0.17 = 0.59$$

$$p(\neg Cancelled \mid e) = 0.41$$

Note that the probability of the game being cancelled has increased significantly from the earlier prior value 0.2, but not as much as to 0.7 when evidence on E_1 was certain.

Evidence of Player Injury

In addition to the confirming evidence on E_1, suppose now we observe confirming evidence on E_2 (that is, player injury). The posterior probabilities are computed using the following formulae:

$$p(H_i \mid E_1 E_2) = \frac{p(E_1 E_2 \mid H_i) \times p(H_i)}{\sum\limits_{j=1}^{2} p(E_1 E_2 \mid H_j) \times p(H_j)}, \qquad i = 1, 2$$

Since E_1 and E_2 are independent, $p(E_1 E_2 \mid H_i) = p(E_1 \mid H_i) \times p(E_2 \mid H_i)$. Therefore,

$$p(Cancelled \mid Heavy\ Rain, Players\ Injured)$$

$$= \frac{0.7 \times 0.4 \times 0.2}{0.7 \times 0.4 \times 0.2 + 0.075 \times 0.025 \times 0.8} = 0.97$$

$$p(\neg Cancelled \mid Heavy\ Rain, Players\ Injured)$$

$$= \frac{0.075 \times 0.025 \times 0.8}{0.7 \times 0.4 \times 0.2 + 0.075 \times 0.025 \times 0.8} = 0.03$$

Note that the posterior probability of the game being cancelled has increased further due to evidence of both heavy rain and player injury. This process of probability revision continues as evidence arrives.

When the requisite initial assumptions (for example, prior probabilities, event independence) are fairly accurate, the Bayesian approach typically provides optimum results and is difficult to beat. However, there is always some question as to how accurate our *a priori* assumptions are for any given situation we are modeling. Under such circumstances, where a priori assumptions are inaccurate, Bayesian methods may perform poorly. The Dempster-Shafer theory was specifically developed to mitigate these weaknesses.

5.3 Dempster-Shafer Theory for Handling Uncertainty

The theory of belief functions (Shafer, 1976), also known as Dempster-Shafer theory, is a generalization of the Bayesian theory of subjective probability (mainly by virtue of its explicit definition of the concept of ignorance) to combine accumulative evidence or to change prior opinions in the light of new evidence. Whereas the Bayesian theory requires probabilities for each question of interest, belief functions allow us to base degrees of belief for one question (for example, whether the game is on) on probabilities for a related question. Arthur P. Dempster set out the basic ideas of the theory (Dempster, 1966) and then Glenn Shafer developed the theory further (Shafer, 1976). Briefly, the theory may be summarized as follows.

Suppose expert X (for example, a weatherman, traffic cop, or grounds man) says that the game is not on due to heavy rain. The decision maker's subjective probabilities for expert X being reliable or unreliable are 0.7 and 0.3. Now, expert X's statement must be true if reliable, but not necessarily false if unreliable. The expert's testimony therefore justifies 0.7 "degrees of belief" that the game is not on, but only a zero (not 0.3) degree of belief that the game is on. The numbers 0.7 and 0 together constitute a belief function.

Suppose subjective probabilities were based on the decision maker's knowledge of the frequency with which experts like X are reliable witnesses. 70% of statements made would be true by reliable witnesses, n% would be true by unreliable ones, and $(30 - n)$% would be false statements by unreliable witnesses. 0.7 and 0 are the lower bounds of true probabilities $(70 + n)/100$ and $(30 - n)/100$ respectively. Thus, a single belief function is always a consistent system of probability bounds, but may represent contradictory opinions from various experts. For example, consider the belief function 0.7 and 0 from expert X's opinion of the game not being on, and 0.8 and 0 from expert Y's opinion of the game being on. The lower bound of the true probability for the game not being on in the first case is 0.7, but the upper bound is 0.2 in the second case, yielding contradiction.

Let Ω be a finite set of mutually exclusive and exhaustive propositions, called the *frame-of-discernment*, about some problem domain ($\Omega = \{Cancelled, \neg Cancelled\}$ in our example decision making problem) and $\Pi(\Omega)$ is be the power set of Ω. A *basic probability assignment* (BPA) or *mass function* is the mapping

$$m : \Pi(\Omega) \rightarrow [0,1]$$

which is used to quantify the belief committed to a particular subset A of the frame of discernment given certain evidence. The probability number $m(A)$, the *mass* of A, says how much belief there is that some member of A is in fact the case, where

$$m(\Phi) = 0 \text{ and } \sum_{A \subseteq \Omega} m(A) = 1$$

The value 0 indicates no belief and the value 1 indicates total belief, and any values between these two limits indicate partial beliefs. If the probability number p for only a partial set A of hypotheses is known then the residual complementary probability number $1 - p$ is assigned to the frame-of-discernment, thus allowing the representation of ignorance. A basic probability assignment m is Bayesian if $m(A) = 0$ for every non-singleton set A. For any set $A \subseteq \Omega$ for which $m(A) \neq 0$, A is called a *focal element*.

The measure of total belief committed to $A \subseteq \Omega$ can be obtained by computing the *belief function Bel* for $A \subseteq \Omega$ which simply adds the mass of all the subsets of A:

$$Bel(A) = \sum_{B \subseteq A} m(B)$$

A *single* belief function represents the lower limit of the true probability and the following plausibility function provides the upper limit of the probability:

$$Pl(A) = \sum_{B \cap A \neq \Phi} m(B) = 1 - Bel(A^c)$$

Mass can be recovered from belief function as follows:

$$m(B) = \sum_{A \subseteq B} (-1)^{|A-B|} Bel(A)$$

So there is a one-to-one correspondence between the two functions m and Bel. Two independent evidences expressed as two basic probability assignments m_1 and m_2 can be combined into a single joined basic assignment $m_{1,2}$ by Dempster's rule of combination:

$$m_{1,2}(A) = \begin{cases} \dfrac{\displaystyle\sum_{B \cap C = A} m_1(B)m_2(C)}{1 - \displaystyle\sum_{B \cap C = \Phi} m_1(B)m_2(C)} & A \neq \Phi \\ \\ 0 & A = \Phi \end{cases}$$

Example

In order to illustrate the Dempster-Shafer theory in the context of our example, we consider only the following three different expert rules for canceling the game along with the degrees of reliability on experts from whom the rules have been acquired:

IF *Heavy Rain* THEN *Game Cancelled*

Expert: *Grounds Man*; Reliability: 0.7

IF *Players Injured* THEN *Game Cancelled*

Expert: Club Physiotherapist; Reliability: 0.8

IF *Attack Threat* THEN *Game Cancelled*

Expert: Homeland Security; Reliability: 0.9

Note that Dempster-Shafer theory requires that evidences to be combined are independent. In the above set of rules the potential usable evidences (rainy condition, player injury, and terrorist attack threat) are essentially so.

In case of rain evidence, the values 0.7 and 0 together constitute a belief function. The focal element is $\{Cancelled\}$ and the mass distribution is $m_1(\{Cancelled\}) = 0.7$. We know nothing about the remaining probability so it is allocated to the whole frame of discernment as $m_1(\Omega) = 0.3$, where $\Omega = \{Cancelled, \neg Cancelled\}$. Evidence of player injury provides the focal element $\{Cancelled\}$ other than Ω, with $m_2(\{Cancelled\}) = 0.8$. The remaining probability, as before, is allocated to the whole frame of discernment as $m_2(\Omega) = 0.2$. Dempster's rule can then be used to combine the masses as follows:

$Can-Cancelled$	$m_2(\{Can\}) = 0.8$	$m_2(\Omega) = 0.2$
$m_1(\{Can\}) = 0.7$	$m_{1,2}(\{Can\}) = 0.56$	$m_{1,2}(\{Can\}) = 0.14$
$m_1(\Omega) = 0.3$	$m_{1,2}(\{Can\}) = 0.24$	$m_{1,2}(\Omega) = 0.06$

Now, $Bel(\{Cancelled\}) = 0.56 + 0.24 + 0.14 = 0.94$. Therefore, the combined belief and plausibility are computed in the following table:

Focal Element (A)	$Bel(A)$	$Pl(A)$
$\{Cancelled\}$	0.94	1.0
Ω	1.0	1.0

Example

Here is a more interesting example with a third decision option for the game to be delayed. Therefore, the *frame of discernment* is $\Omega = \{On, Cancelled, Delayed\}$. We consider the following set of expert rules along with the degrees of reliability:

IF *Heavy Rain* THEN NOT *On*

Expert: *Grounds Man*; Reliability: 0.7

IF *Players Injured* THEN *Cancelled*

Expert: *Club Physiotherapist*; Reliability: 0.8

IF *Financial Crisis* THEN NOT *Cancelled*

Expert: *Club Administrator*; Reliability: 0.6

Here also the potential usable evidences (rainy condition, player injury, and club financial situation) are essentially independent.

In case of rain evidence, the game not being on (that is, cancelled or delayed) and the values 0.7 and 0 together constitute a belief function. The focal element other than Ω is $\{Cancelled, Delayed\}$ and the mass is distributed to it as $m_1(\{Cancelled, Delayed\}) = 0.7$. We know nothing about the remaining probability so it is allocated to the whole frame of discernment as $m_1(\Omega) = 0.3$.

There is also evidence that the current financial situation of the club is bad, resulting in 0.6 subjective probability that the game will not be cancelled in this situation. The new evidence suggests the focal element $\{On, Delayed\}$ other than Ω with $m_2(\{On, Delayed\}) = 0.6$. The remaining probability, as before, is allocated to the whole frame of discernment as $m_2(\Omega) = 0.4$. Considering that the rainy condition and the current financial situation are independent of each other, Dempster's rule can then be used to combine the masses as follows:

$Can - Cancelled$ $Del - Delayed$	$m_2(\{On, Del\}) = 0.6$	$m_2(\Omega) = 0.4$
$m_1(\{Can, Del\}) = 0.7$	$m_{1,2}(\{Del\}) = 0.42$	$m_{1,2}(\{Can, Del\}) = 0.28$
$m_1(\Omega) = 0.3$	$m_{1,2}(\{On, Del\}) = 0.18$	$m_{1,2}(\Omega) = 0.12$

Therefore the combined belief and plausibility are computed in the following table:

Focal Element (A)	$Bel(A)$	$Pl(A)$
$\{Delayed\}$	0.42	1.0
$\{On, Delayed\}$	0.60	1.0
$\{Cancelled, Delayed\}$	0.70	1.0
Ω	1.0	1.0

The basic probability assignments m_1 and m_2 are different but consistent, and therefore the degrees of belief in both $\{Cancelled, Delayed\}$ and $\{On, Delayed\}$ being true (that is, the game is delayed) is the product of $m_1(\{Cancelled, Delayed\})$ and $m_2(\{On, Delayed\})$, or 0.42.

Finally, the player injury situation suggests the focal element *{Cancelled}* and Ω with $m_3(\{Cancelled\}) = 0.8$ and $m_3(\Omega) = 0.2$. The Dempster rule of combination applies as before, but with one modification. When the evidence is inconsistent, their products of masses are assigned to a single measure of inconsistency, say k.

$Can - Cancelled$ $Del - Delayed$	$m_3(\{Can\}) = 0.8$	$m_3(\Omega) = 0.2$
$m_{1,2}(\{Del\}) = 0.42$	$k = 0.336$	$m(\{Del\}) = 0.084$
$m_{1,2}(\{On, Del\}) = 0.18$	$k = 0.144$	$m(\{On, Del\}) = 0.036$
$m_{1,2}(\{Can, Del\}) = 0.28$	$m(\{Can\}) = 0.224$	$m(\{Can, Del\}) = 0.056$
$m_{1,2}(\Omega) = 0.12$	$m(\{Can\}) = 0.096$	$m(\Omega) = 0.024$

The total mass of evidence assigned to inconsistency k is $0.336 + 0.144 = 0.48$. The normalizing factor is $1 - k = 0.52$. The resulting masses of evidence are as follows:

$$m(\{Cancelled\}) = (0.224 + 0.096)/0.52 = 0.62$$
$$m(\{Delayed\}) = 0.084/0.52 = 0.16$$
$$m(\{On, Delayed\}) = 0.036/0.52 = 0.07$$
$$m(\{Cancelled, Delayed\}) = 0.056/0.52 = 0.11$$
$$m(\Omega) = 0.024/0.52 = 0.04$$

Therefore, the combined belief and plausibility are computed in the following table:

Focal Element (A)	$Bel(A)$	$Pl(A)$
$\{Cancelled\}$	0.62	0.77
$\{Delayed\}$	0.16	0.38
$\{On, Delayed\}$	0.23	0.38
$\{Cancelled, Delayed\}$	0.89	1.0
Ω	1.0	1.0

Hence, the most likely hypothesis is the cancellatation of the game.

Let us consider two examples to illustrate two special cases for evidence aggregation.

Example

Hypothetically, consider the case when the set of focal elements of the basic probability distribution m_2 is exactly the same as m_1. The evidence combination table is shown below:

$Can - Cancelled$ $Del - Delayed$	$m_2(\{Can, Del\}) = 0.6$	$m_2(\Omega) = 0.4$
$m_1(\{Can, Del\}) = 0.7$	$m_{1,2}(\{Can, Del\}) = 0.42$	$m_{1,2}(\{Can, Del\}) = 0.28$
$m_1(\Omega) = 0.3$	$m_{1,2}(\{Can, Del\}) = 0.18$	$m_{1,2}(\Omega) = 0.12$

Now,

$$Bel(\{Cancelled, Delayed\}) = 0.42 + 0.18 + 0.28$$
$$= 0.88$$
$$= 0.6 + 0.7 - 0.6 \times 0.7$$

In general, when two mass distributions m_1 and m_2 agree on focal elements, then the combined degree of belief on a common focal element is $p_1 + p_2 - p_1 \times p_2$, where p_1 and p_2 are mass assignments on the focal element by the two distributions.

As opposed to agreeing on focal elements, if m_2 is contradictory to m_1 then an example evidence combination is shown below:

$Can - Cancelled$ $Del - Delayed$	$m_2(\{On\}) = 0.6$	$m_2(\Omega) = 0.4$
$m_1(\{Can, Del\}) = 0.7$	$k = 0.42$	$m_{1,2}(\{Can, Del\}) = 0.28$

$m_1(\Omega) = 0.3$	$m_{1,2}(On) = 0.18$	$m_{1,2}(\Omega) = 0.12$

In this case,

$$Bel(\{Cancelled, Delayed\})$$
$$= 0.28/(1-0.42) = 0.7(1-0.6)/(1-0.42)$$

In general, when two mass distributions m_1 and m_2 are contradictory, then the combined degree of belief on the focal element for m_1 is $p_1(1-p_2)/(1-p_1 \times p_2)$ and the combined degree of belief on the focal element for m_2 is $p_2(1-p_1)/(1-p_1 \times p_2)$, where p_1 and p_2 are mass assignments on the focal element by the two distributions.

5.4 Measuring Consensus

The need to measure consensus in a decision-making context with the Dempster-Shafer theory as the underlying theoretical foundation is first explained in this section with the help of a concrete example.

<u>Example</u>

Suppose some source of evidence heavily suggests against the game being cancelled (that is, the game is either on or delayed) and the values 0.9 and 0 together constitute a belief function. The focal element is $\{On, Delayed\}$ and the mass distribution is $m_1(\{On, Delayed\}) = 0.9$. We know nothing about the remaining probability 0.1 so it is allocated to the whole frame of discernment $\Omega = \{On, Cancelled, Delayed\}$. Another source of evidence heavily suggests against the game being on. The mass distribution in this case is $m_2(\{Cancelled, Delayed\}) = 0.9$. The remaining probability, as before, is allocated to the whole frame of discernment. Dempster's rule can then be used to combine the masses as follows:

Can – Cancelled *Del – Delayed*	$m_2(\{Can, Del\}) = 0.9$	$m_2(\Omega) = 0.1$
$m_1(\{On, Del\}) = 0.9$	$m_{1,2}(\{Del\}) = 0.81$	$m_{1,2}(\{On, Del\}) = 0.09$

$m_1(\Omega) = 0.1$	$m_{1,2}(\{Can, Del\}) = 0.09$	$m_{1,2}(\Omega) = 0.01$

Therefore, the combined belief and plausibility are computed in the following table:

Focal Element (A)	$Bel(A)$	$Pl(A)$
$\{Delayed\}$	0.81	1.0
$\{On, Delayed\}$	0.09	0.91
$\{Cancelled, Delayed\}$	0.09	0.91
Ω	1.0	1.0

The above result is counterintuitive, given that none of the two sources of evidence explicitly supported the game being delayed. Moreover, the table above does not reflect the underlying high degree of disagreement. The consensus metric provided below will be able to highlight this case. The "entropy"-based algorithm presented here generates a very low (2.3 out of 3) degree of consensus (zero or minimum entropy is the highest consensus). The consensus-measuring criterion is based on the generalization of the concept of entropy from point function to set function (Stephanou and Lu, 1988). This generalization is composed of three measurements: belief entropy, core entropy, and partial ignorance.

Suppose Ω is a frame of discernment, m is a basic probability assignment, and $\mathbf{F} = \{A_1,..., A_n\}$ is the set of focal elements.

The *belief entropy* of the pair $\langle m, \mathbf{F} \rangle$ is defined as follows (log is base 2):

$$E_b(m, \mathbf{F}) = -\sum_{i=1}^{n} m(A_i) \log m(A_i)$$

The belief entropy is a measure of the degree of confusion in the decision-maker's knowledge about the exact fraction of belief that should be committed to each focal element in $\mathbf{F}$. Thus, the belief entropy is naturally equal to zero if the entire belief is committed to a single focal element, that is, $m(A_k) = 1$, for some k. The maximum belief entropy occurs when belief is divided in equal fractions among the focal elements.

The *core entropy* of the pair $\langle m, \mathbf{F} \rangle$ is defined as follows:

$$E_c(m,\mathbf{F}) = -\sum_{i=1}^{n} r_i \log r_i$$

where

$$r_i = \frac{\|A_i\| - 1}{\sum_{j=1}^{n}(\|A_i\| - 1)}$$

and $\|A_i\|$ is the cardinality of A_i ($0\log 0$ is considered 0). The core entropy is a measure of the degree of confusion in the decision-maker's knowledge of which focal elements the true value might be in. If belief is committed to a single focal element A_k, then r_k is one, and therefore the core entropy is equal to zero. The core entropy is maximum when belief is divided among a number of focal elements with the same cardinality.

The *partial ignorance* of the pair $\langle m, \mathbf{F} \rangle$ is defined as follows:

$$I(m,\mathbf{F}) = \sum_{i=1}^{n} m(A_i) s_i(A_i)$$

where $s_i(A_i) = \dfrac{\|A_i\| - 1}{\|\Omega\| - 1}$ and $\|A_i\|$ is the cardinality of A_i, and the assumption is that the frame of discernment has more than one element. Partial ignorance is a measure of the inability to confine the true value within a small-sized focal element. Consequently, partial ignorance is large when large belief is committed to large focal elements. It is zero when the entire belief is committed to a singleton.

The *generalized entropy* of the pair $\langle m, \mathbf{F} \rangle$ is then defined as follows:

$$E(m,\mathbf{F}) = E_b(m,\mathbf{F}) + E_c(m,\mathbf{F}) + I(m,\mathbf{F})$$

Now suppose that we have two basic probability assignments m_1 and m_2, corresponding to two different expert knowledge sources, defining focal elements $\{A, \Omega\}$ and $\{B, \Omega\}$ respectively, where $A \subset \Omega$ and $B \subset \Omega$. Let us consider four different cases as shown in Figure 5-2 (Stephanou and Lu, 1988).

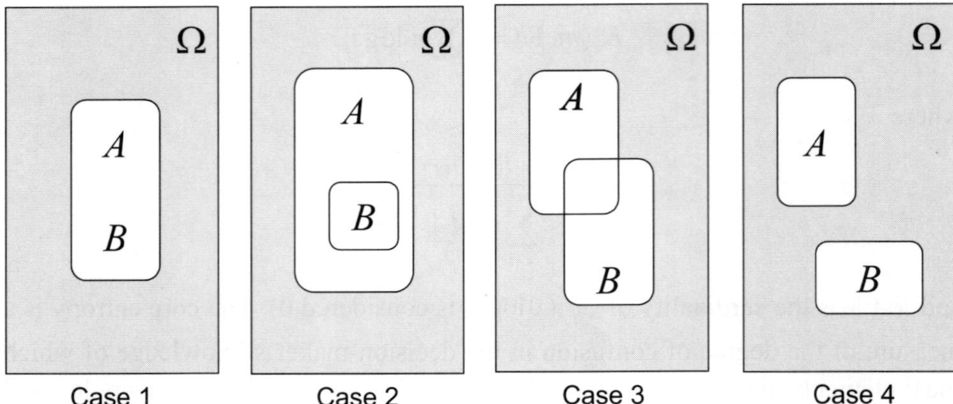

Figure 5-2: Combining expert knowledge sources

It can be shown that the generalized entropy of the knowledge sources via Dempster's rule of combination for each of the first three cases is smaller than the sum of the individual entropies of the knowledge sources. In other words, the following relation holds for the first three cases:

$$E(m_{12},\mathbf{F}) \leq E(m_1,\mathbf{F}) + E(m_2,\mathbf{F})$$

The above implies that the pooling of two concordant bodies of evidence reduces entropy. The disjointedness between A and B in the last case indicates that there are two bodies of evidence supporting two conflicting propositions, and hence the combined generalized entropy does not necessarily get reduced.

Example

To illustrate reduction of entropy, consider the frame of discernment $\Omega = \{On, Cancelled, Delayed\}$ the following two focal elements, and their corresponding basic probability assignments:

$A = \{Cancelled, Delayed\}$ $B = \{On, Delayed\}$

$\mathbf{F}_1 = \{A, \Omega\}$ $\mathbf{F}_2 = \{B, \Omega\}$

$m_1(A) = 0.7, m_1(\Omega) = 0.3$ $m_2(B) = 0.6, m_1(\Omega) = 0.4$

The entropies and ignorance are computed as follows:

Belief Entropy:

$$E_b(m_1, \mathbf{F}_1) = -m_1(A)\log m_1(A) - m_1(\Omega)\log m_1(\Omega)$$
$$= -0.7\log 0.7 - 0.3\log 0.3 = 0.88$$
$$E_b(m_2, \mathbf{F}_2) = -m_2(A)\log m_2(A) - m_2(\Omega)\log m_2(\Omega)$$
$$= -0.6\log 0.6 - 0.4\log 0.4 = 0.97$$

Core Entropy:

$$E_c(m_1, \mathbf{F}_1) = -r_1(A)\log r_1(A) - r_1(\Omega)\log r_1(\Omega)$$
$$= -\frac{1}{3}\log\frac{1}{3} - \frac{2}{3}\log\frac{2}{3} = 0.92$$
$$E_c(m_2, \mathbf{F}_2) = -r_2(A)\log r_2(A) - r_2(\Omega)\log r_2(\Omega)$$
$$= -\frac{1}{3}\log\frac{1}{3} - \frac{2}{3}\log\frac{2}{3} = 0.92$$

Partial Ignorance:

$$I(m_1, \mathbf{F}_1) = m_1(A)s_1(A) + m_1(\Omega)s_2(\Omega)$$
$$= 0.7\cdot\frac{1}{2} + 0.3\cdot\frac{2}{2} = 0.65$$
$$I(m_2, \mathbf{F}_2) = m_2(B)s_1(B) + m_2(\Omega)s_2(\Omega)$$
$$= 0.6\cdot\frac{1}{2} + 0.4\cdot\frac{2}{2} = 0.7$$

Generalized Entropy:

$$E(m_1, \mathbf{F}_1) = 0.88 + 0.92 + 0.65 = 2.45$$
$$E(m_2, \mathbf{F}_2) = 0.97 + 0.92 + 0.70 = 2.59$$

Now, consider Case 3 in Figure 5-1. The Dempster combination rule yields a new set of focal elements and basic probability assignments as follows:

$$A \cap B = \{Delayed\}$$
$$\mathbf{F} = \{A, B, A \cap B, \Omega\}$$
$$m_{1,2}(A) = 0.18, m_{1,2}(B) = 0.28, m_{1,2}(A \cap B) = 0.42, m_{1,2}(\Omega) = 0.12$$

The entropies and ignorance for the combined set of focal elements are computed as follows:

Belief Entropy:

$$E_b(m_{1,2},\mathbf{F}) = -0.18\log 0.18 - 0.28\log 0.28$$
$$-0.42\log 0.42 - 0.12\log 0.12 = 1.85$$

Core Entropy:

$$E_c(m_{1,2},\mathbf{F}) = -\frac{1}{4}\log\frac{1}{4} - \frac{1}{4}\log\frac{1}{4} - \frac{2}{4}\log\frac{2}{4} = 1.5$$

Partial Ignorance:

$$I(m_{1,2},\mathbf{F}) = 0.18\cdot\frac{1}{2} + 0.28\cdot\frac{1}{2} + 0.42\cdot\frac{0}{2} + 0.12\cdot\frac{2}{2} = 0.35$$

Generalized Entropy:

$$E(m_{1,2},\mathbf{F}) = 1.85 + 1.50 + 0.35 = 3.70$$

Thus we observe that the relation $E(m_{12},\mathbf{F}) \leq E(m_1,\mathbf{F}) + E(m_2,\mathbf{F})$ holds.

5.5   Combining Sources of Varying Confidence

Here we present a credibility transformation function (Yager, 2004) for combining sources with various confidences. This approach discounts the evidence with a credibility factor α and distributes remaining evidence $1-\alpha$ equally among the n elements of the frame of discernment. The transformed credibility function takes the following form:

$$m_\alpha(A) = m(A)\cdot\alpha + \frac{1-\alpha}{n}$$

In the Bayesian formalism where each focal element A is a singleton set, we distribute the remaining evidence $1-\alpha$ as per their prior probabilities. In other words, we modify the credibility function as the following:

$$p_\alpha(A) = p(A)\cdot\alpha + p_0(A)\cdot\frac{1-\alpha}{n}$$

When prior probabilities are uniformly distributed among elements of the frame of discernment, this case becomes a special case of the credibility function for the Dempster-Shafer case. Moreover, if the prior probability of A is zero, then it remains zero after the transformation via the credibility function.

Example

$$\Omega = \{On, Cancelled, Delayed\}$$
$$A = \{On, Cancelled\}$$
$$m_1(A) = 0.7$$

If the reliability of the source of m_1 is 0.8, then

$$m_{0.8}(A) = 0.7 \cdot 0.8 + \frac{1-0.8}{2} = 0.66$$

$$m_{0.8}(\Omega) = 0.3 \cdot 0.8 + \frac{1-0.8}{2} = 0.34$$

5.6   Advantages and Disadvantages of Rule-Based Systems

Subject matter experts usually express their predictive and actionable domain knowledge in terms of expressions such as "this action should be taken under that condition," "this situation means that," and so on. Such knowledge, with clearly separated antecedent and consequent parts, can be easily transformed to rules. Moreover, each rule is a self-documented, independent piece of knowledge that can be maintained and acquired in a local context without analyzing its global impact.

Rule-based systems provide both data-driven reasoning (that is, forward chaining) and goal-driven reasoning (that is, backward chaining as in SLD resolution presented in the last chapter). A derivation via each such reasoning method is transparent in the sense that derivation steps can be laid out in sequence to provide a basis for explaining an answer to a query. Rules can be attached with supports drawn from various types of qualitative and quantitative dictionaries, other than just probabilities, provided appropriate algebras exist for manipulating dictionary terms (for example, mass values in Dempster-Shafer, certainty factors). Rules can also be learned, given sufficient training instances, using standard machine learning techniques, such as decision trees for learning propositional rules and inductive logic programming for learning first-order rules (Mitchell, 1997).

Since each rule is self-documented, a special reasoning procedure is required to implement abductive or diagnostic reasoning (this kind of reasoning is an integral part of evidence propagation algorithms for belief networks). Moreover, a knowledge base consisting of many isolated rules does not provide a global view of the domain that it is modeling in terms of links and relationships among variables.

5.7   Background and Further Readings

Maturity and high-level of interests in the area of rule-based expert system have culminated dozens of text books. (Jackson, 2002) provides a good background of the subject. Shafer's own book (1976) and the edited volume by Yager et al. (1994) are good sources on Dempster-Shafer theory of belief function. The paper by Stephanou and Lu (1988) is the source of the entropy-based consensus measurement presented here. See (Smets, 1991) for various other formalisms for handling uncertainty, including fuzzy logic (Zadeh, 1965) and possibility theory (Zadeh, 1978; Dubois and Prade, 1988), and their relations to the theory of belief function.

Chapter 6

Bayesian Belief Networks

This chapter presents the Bayesian belief network (or simply *belief network*; also known as *causal network* or *probabilistic network*) technology for developing the probabilistic foundation of epistemic models for decision-making agents. A straightforward Bayesian approach to probabilistic epistemic modeling via a set of random variables is too inefficient, because problems with a large number of variables require maintaining large joint distributions to compute posterior probabilities based on evidence. The belief network technology avoids this via certain independence assumptions based on which application network models are constructed.

In this chapter, we start by defining belief networks and illustrating the concept of conditional independence within belief networks. Then we presents the evidence, belief, and likelihood concepts. Next, we present two propagation algorithms to compute variables' posterior probabilities based on evidence: one for the class of networks without cycles, and another for a class of networks with restrictive cycles, and analyze the complexities of these two algorithms.

The task of constructing a belief network can be divided into two subtasks: 1) specification of the causal structure among variables of the network; and 2) specification of prior and conditional probabilities. Usually, automated learning of the structure from the data is much harder than learning probabilities, although techniques exist for automated learning in each of these areas. However, the structure is much easier to elicit from experts than probabilities, especially when nodes have several parents (in which case their conditional probabilities become large and complex). We provide some guidance for acquiring network probabilities, including the noisy-or technique. Finally, we conclude the chapter by summarizing the overall advantages and disadvantages of belief network technology and tools.

165

6.1   Bayesian Belief Networks

A Bayesian belief network is a network (that is, a graph consisting of nodes and links) with the following interpretation:

- Each node in the network represents a random variable that can take on multiple discrete values; these values are mutually exclusive events constituting the sample space over which the random variable is defined. The terms *node* and *variable* are synonymous in the context of a belief network and are often used interchangeably.

- Each link in the network represents a relation or *conditional dependence* between the two variables it connects, and an associated *conditional probability table* (CPT) quantifies the relation associated with the link.

- There is a notion of causality between two linked variables, so links with explicit direction (represented by arrows) are drawn from "cause" nodes to "effect" nodes.

- The network is based on some *marginal* and *conditional independence* assumptions among nodes. These assumptions are described in the next section.

- The state of a node is called a *belief*, and reflects the posterior probability distribution of the values associated with that node, given all the a priori evidence.

The probabilities in a CPT are typically acquired from subject matter experts in the domain, but can also be learned automatically given a large enough number of training instances. The causality restriction can sometimes be relaxed by allowing links between a pair of nodes that are simply correlated and the direction of the arrow between the two nodes is decided based on the ease of acquisition of the probabilities in the CPT. A typical causal relationship between two variables inevitably brings the temporal dimension into the modeling problem. We avoid building any formal dynamic or temporal belief networks within which the state of a variable is allowed to change over time by modeling a simple snapshot of the problem domain at a particular time.

Example

An example belief network is shown in Figure 6-1. It illustrates causal influences on whether or not a game is going to be played, given the current weather and other related sources of evidence. Note that there are many ways to approach modeling a decision-making problem with belief networks. The selection of random variables and their granularities and interdependences are largely subjective, but should be driven by the problem solving requirements. (Please note that we are not emphasizing any specific modeling methodology; the specific belief network in Figure 6-1 is constructed to illustrate the technology via a single unified example.)

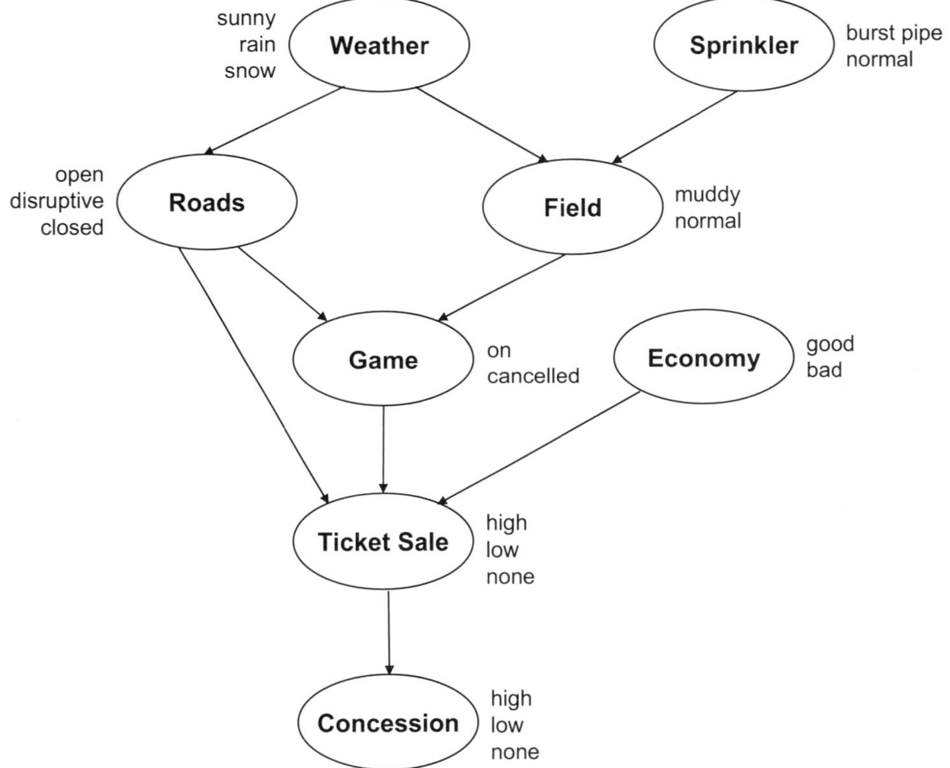

Figure 6-1: An example belief network

Each node in the belief network in Figure 6-1 and its associated mutually exclusive states (possible values of the random variable representing the node) are described below (from top to bottom):

- *Weather*: Overall weather condition during the game – {*sunny, rain, snow*}

- *Sprinkler*: Current state of the water sprinkler – {*normal, burst pipe*}

- *Roads*: Condition of city roads before the game – {*open, disruptive, closed*}

- *Field*: Overall condition of the field for the game – {*muddy, normal*}
- *Game*: Whether the game is on or cancelled – {*on, cancelled*}
- *Economy*: Current economic condition of the country – {*good, bad*}
- *Ticket Sale*: Level of ticket sale at the gate – {*high, low, none*}
- *Concession*: Level of concession (for example, snacks and drinks) inside the stadium – {*high, low, none*}

For example, all states in the random variable *Weather* are mutually exclusive, that is, the overall weather condition cannot be both sunny and rain at the same time. But a probability density function f of the variable *Weather* with the domain {*sunny, rain, snow*} can be defined and interpreted as " $f(x)$ is the probability that *Weather* will assume the value x." One such function can be defined as follows:

$$f(sunny) = 0.55, f(rain) = 0.15, f(snow) = 0.3$$

The causal influences by numbers (as shown in Figure 6-2) along the directions of the arrows are defined as follows:

- Weather condition determines the condition of the roads. For example, snowy weather is likely to cause roads to be closed.

- Weather condition and the sprinkler state together determine the suitability of the field condition. For example, a rainy day is likely to cause the field to be muddy and therefore unsuitable for a game. A burst pipe is likely to have the same effect.

- Road and field conditions together determine the status of a game, that is, whether it is going to be played or not. For example, the disruptive condition of the roads is likely to cause the game to be cancelled as is an unsuitable field condition.

- Condition of the roads, game status, and the country's economy together determine the ticket sale volume. For example, when the game is on, the ticket sale is likely to be low due to disruptive road conditions, since these conditions might also cause low attendance (not modeled explicitly). The volume of ticket sale is likely to be even lower if the economy is bad.

- Ticket sale determines the concession volume, since the presence of fewer customers (low ticket sales) is likely to cause a lower volume of concession sales.

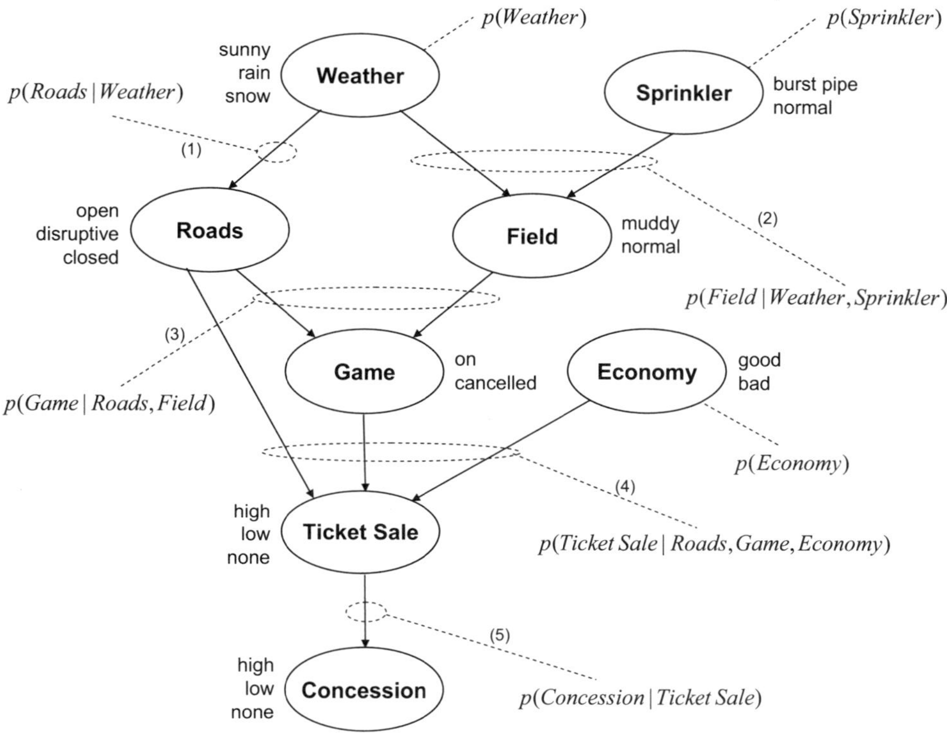

Figure 6-2: Belief network with prior and conditional probabilities

The three prior probabilities (corresponding to the three nodes without parents) and five conditional probabilities of the belief network in Figure 6-2 are provided in the following tables.

	sunny	0.55
Weather	rain	0.15
	snow	0.3

Table 6-1: *p(Weather)*

	burst pipe	0.01
Sprinkler	normal	0.99

Table 6-2: *p(Sprinkler)*

Economy	good	0.6
	bad	0.4

Table 6-3: $p(Economy)$

Weather		sunny	rain	snow
	open	0.9	0.7	0.8
Roads	disruptive	0.09	0.2	0.15
	closed	0.01	0.1	0.05

Table 6-4: $p(Roads \mid Weather)$

Weather		sunny		rain		snow	
Sprinkler		burst pipe	normal	burst pipe	normal	burst pipe	normal
Field	muddy	0.6	0.95	0.1	0.3	0.2	0.4
	normal	0.4	0.05	0.9	0.7	0.8	0.6

Table 6-5: $p(Field \mid Weather, Sprinkler)$

Roads		open		disruptive		closed	
Field		muddy	normal	muddy	normal	muddy	normal
Game	on	0.99	0.1	0.8	0.1	0.01	0
	cancelled	0.01	0.9	0.2	0.9	0.09	1

Table 6-6: $p(Game \mid Roads, Field)$

Roads		open				disruptive				closed			
Game		on		cancelled		on		cancelled		on		cancelled	
Economy		good	bad	good	bad	good	bad	good	bad	good	bad	good	bad
Ticket Sale	high	.9	.8	0	0	.2	.1	0	0	.09	0	0	0
	low	.1	.2	0	0	.8	.9	0	0	.9	.9	0	0
	none	0	0	1	1	0	0	1	1	.01	.1	1	1

Table 6-7: $p(Ticket\ Sale \mid Roads, Game, Economy)$

Ticket Sale		high	low	none
Concession	high	1	0	0
	low	0	1	0
	none	0	0	1

Table 6-8: $p(Concession \mid Ticket\ Sale)$

This belief network and its associated CPTs shown above are used throughout the rest of this chapter to illustrate algorithms and other related concepts. Note that the CPT in Table 6-8 is in the form of an identity matrix, and guarantees the perfect causal relationship between the states of *Concession* and *Ticket Sale* variables.

6.2 Conditional Independence in Belief Networks

Two random variables Y and Z are said to be (marginally) *independent*, denoted as $Y \perp Z$, if

$$p(Y,Z) = p(Y)p(Z)$$

for any combination of values for the variables Y and Z. The variable Y is *conditionally independent* of Z given another variable X, denoted as $Y \perp Z \mid X$, if

$$p(Y,Z \mid X) = p(Y \mid X)p(Z \mid X).$$

Therefore,

$$p(Y \mid Z,X) = \frac{p(Y,Z \mid X)}{p(Z \mid X)} = \frac{p(Y \mid X)p(Z \mid X)}{p(Z \mid X)} = p(Y \mid X)$$

Similarly, $p(Z \mid Y,X) = p(Z \mid X)$. Note that marginal independence (no conditioning) does not imply conditional independence; nor does conditional independence imply marginal independence.

Figure 6-3 represents conditional independence in a chain fragment of a belief network where a node X is between two other nodes Y and Z. We factorize the joint probability distribution of the variables X, Y, and Z as follows:

$$p(X,Y,Z) = p(Z \mid X,Y)p(X,Y) = p(Z \mid X)p(X \mid Y)p(Y)$$

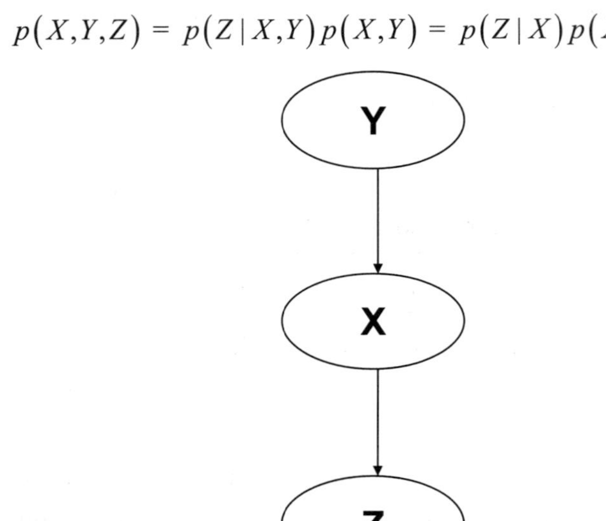

Figure 6-3: Conditional independence in chain fragment: Z is conditionally independent of Y given X

Example

Figure 6-4 shows an example instantiation of conditional independence in a chain network fragment as shown in Figure 6-3. The variables X, Y, and Z represent respectively the sprinkler state, field condition, and the status of the game, respectively. If we observe, with complete certainty, that the field condition is unsuitable, then the probability of the game being cancelled is determined.

Therefore, confirmation that the sprinkler is normal or that it has a burst pipe will not change the probability of the game status, and vice versa.

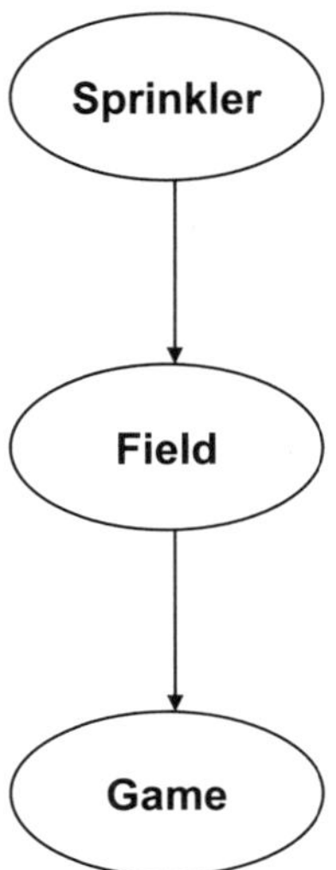

Figure 6-4: Example conditional independence in a chain network fragment: *Sprinkler* and *Game* are conditionally independent given *Field*

Figure 6-5 represents conditional independence in a tree network fragment of a belief network where the node X is the parent of two other nodes Y and Z. In this case, we factorize the joint probability distribution of the variables X, Y, and Z as follows:

$$p(X,Y,Z) = p(Z\,|\,X,Y)p(X,Y) = p(Z\,|\,X)p(Y\,|\,X)p(X)$$

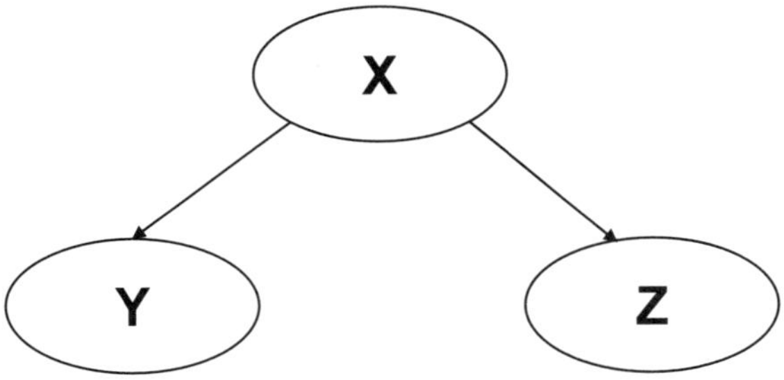

Figure 6-5: Conditional independence in tree network fragment: Z is conditionally independent of Y given X

Example

Figure 6-6 shows an example instantiation of conditional independence in a tree network fragment. The variables X, Y and Z represent the weather, roads, and field conditions, respectively. If we observe rainy weather then the probabilities of the road condition being disruptive and the field condition unsuitable are determined, and the confirmation of the road condition being disruptive will not change the probability the field being unsuitable, and vice versa.

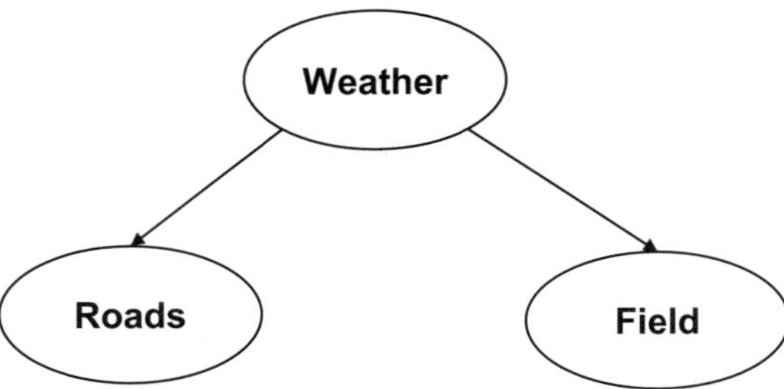

Figure 6-6: Example conditional independence in a tree network fragment: *Roads* and *Field* are conditionally independent given *Weather*

Figure 6-7 shows conditional dependence in a polytree network fragment between the nodes Y and Z, given that we know about X; the two variables are marginally independent if we know nothing about X. For a polytree fragment as shown in Figure 6-7, the probability distribution of the variables Y, Z, and X can be factorized as follows:

$$p(X,Y,Z) = p(X|Y,Z)p(Y,Z) = p(X|Y,Z)p(Y)p(Z)$$

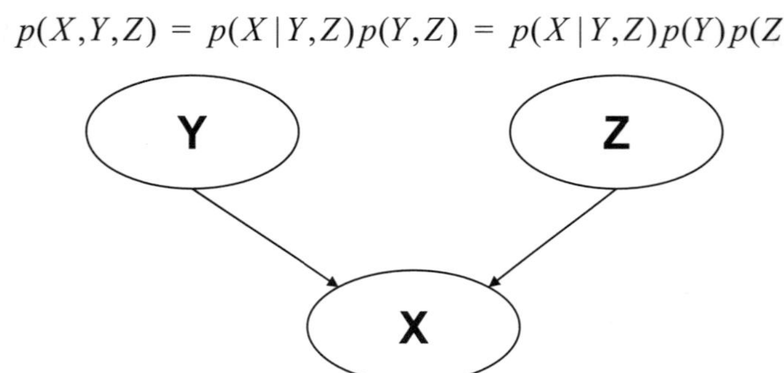

Figure 6-7: Conditional dependence in a polytree fragment: *Y* is conditionally dependent on *Z* given *X*

Example

Figure 6-8 shows an example instantiation of the conditional dependence in a polytree network fragment. Both the weather condition and the sprinkler status can affect the field condition. Before any observation is made on the field condition, the probability of the weather condition to be in a particular state is independent of the probability of the sprinkler being normal or having a burst pipe. However, once a particular field condition, say the unsuitable condition, is observed, the weather condition may influence the sprinkler. For example, observation of rainy weather (thus explaining why the field is unsuitable) may decrease the probability of a burst pipe in the sprinkler system. This phenomenon is termed as *explaining away*. In other words, observation of the status of one parent "explains away" the other, given a value of the child node.

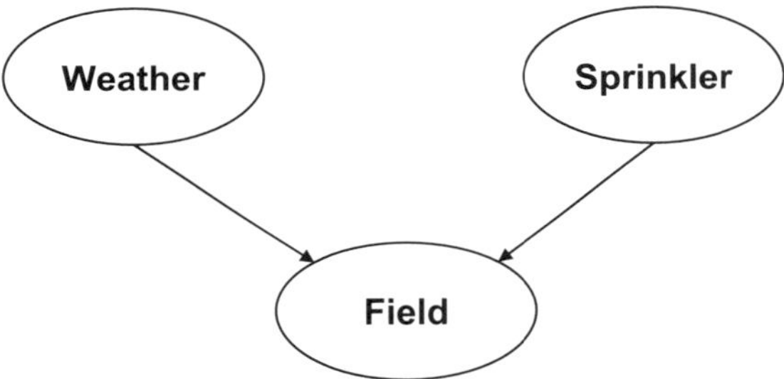

Figure 6-8: Example conditional independence in a polytree network fragment: *Weather* is conditionally dependent on *Sprinkler* given *Field*

In view of the joint distribution formulae for these three types of network fragments, the joint probability distribution in a directed acyclic graph (DAG) ("directed" means the links have an explicit direction represented by arrows, and "acyclic" means that the arrows may not form a directional cycle in the network) type of network can be factored into conditional probabilities, where each factor involves only a node and its parents. This is stated through the following result:

Proposition 6-1: Consider a network consisting of variables $X_1, X_2, ..., X_n$. The joint probability distribution $p(X_1, X_2, ..., X_n)$ is the product of all conditional probabilities specified in the network:

$$p(X_1, X_2, \ldots, X_n) = \prod_{i=1}^{n} p(X_i \mid pa(X_i))$$

where $pa(X_i)$ denotes the parent variables of X_i.

Example

Consider the network shown in Figure 6-1. For any combination of values w, s, r, f, g, e, t, c of the variables *Weather*, *Sprinkler*, *Roads*, *Field*, *Game*, *Economy*, *Ticket Sale*, and *Concession*, respectively, the joint probability is

$$p(w,s,r,f,g,e,t,c) =$$

$$p(w)p(s)p(r \mid w)p(f \mid w,s)p(g \mid r,f)p(t \mid r,g,e)p(c \mid t)$$

Influence in a belief network is only allowed to flow along the links given in the network. Therefore, independence between two nodes is represented by the absence or blocking of links between the two nodes. Whether a link between a pair of nodes exists or not is determined by a property called *d-separation*. Before we formally define *d*-separation, we need to introduce three kinds of connections between a node X and two of its neighbors Y and Z. The three possibilities are shown in the three figures Figure 6-3, Figure 6-5, and Figure 6-7. Their example instantiations are shown in Figure 6-4, Figure 6-6, and Figure 6-8 respectively.

In Figure 6-3, there are links from Y to X and from X to Z. In other words, Y has an influence on X, which in turn influences Z. The connection or path between Y and Z is called *linear*. In this case, causal or deductive evidence from Y will change the belief of X, which then changes the belief of Z. Similarly, diagnostic or abductive evidence from Z will change the belief of X, which then changes the belief of Y. But if the state of X is known, then the connection between Y and Z are blocked and cannot influence each other any more. Thus, Y and Z become independent given X. We say that Y and Z are *d-separated* given X.

In Figure 6-5, there are links from X to Y as well as from X to Z. In other words, X has influence on both Y and Z, and Y and Z will influence each other via X. The connection or path between Y and Z is called *diverging* and the node X is said to have diverging arrows. As in the linear case, if the state of X is known then Y and Z cannot influence each other any more, and we say that Y and Z are *d-separated* given X.

The third case, shown in Figure 6-7, is opposite to the previous case. In this case, there are links from Y to X as well as from Z to X. In other words, both Y and Z have influence on X. The connection or path in this case is called *converging* and the node X is said to have converging arrows. In this case, if nothing is known about X then Y and Z are independent, and therefore cannot influence each other. But if the state of X is known then Y and Z can influence each other. In other words, Y and Z are already *d-separated*, but not when X is given.

In general, two nodes Y and Z in a DAG type of network are *d-separated* if for all paths between Y and Z, there is an intermediate node X such that either:

- The path between Y and Z is serial or diverging at node X and the state of X is known, or

- The path between Y and Z is converging at node X and neither X nor any of its descendants have received evidence

Two nodes Y and Z in a DAG are *d-connected* if they are not *d*-separated. The following proposition establishes a connection between conditional independence and *d*-separateness.

Proposition 6-2: If any two nodes Y and Z in a DAG are *d*-separated with evidence e entered, then Y and Z are conditionally independent given e (i.e. $Y \perp Z \mid e$ or $p(Y \mid Z, e) = p(Y \mid e)$).

Example

Consider the network shown in Figure 6-1. Let $Y = Roads$ and $Z = Sprinkler$. The two paths between the two nodes Y and Z are:

$$Roads \leftarrow Weather \rightarrow Field \leftarrow Sprinkler$$
$$Roads \rightarrow Game \leftarrow Field \leftarrow Sprinkler$$

The first path contains a diverging node (*Weather*) and the second path contains a converging node (*Game*). If the state of the variable *Weather* is known and the variable *Game* and its descendents *Ticket Sale*, and *Concession* have not received evidence, then the nodes *Roads* and *Sprinkler* are *d*-separated. Alternatively, if the variable *Field* and its descendents *Game, Ticket Sale* and *Concession* have not received evidence, then the nodes *Roads* and *Sprinkler* are *d*-separated.

The above definition of *d*-separation between two nodes takes into account the evidence entered into the network. Here we present a more generalized definition of *d*-separation that identifies a set of nodes, instead of a single node, that could potentially separate two nodes in a network. Moreover, the definition provided here is between two sets of nodes rather than between two nodes.

For any three disjoint subsets S_X, S_Y, and S_Z of nodes, S_X is said to *d-separate* S_Y and S_Z if every path between a node in S_Y to a node in S_Z there is a node X satisfying one of the following two conditions:

- X has converging arrows and none of X or its descendants are in S_X, or

- X does not have a converging arrows and X is in S_X

Example

Consider the network shown in Figure 6-1. Let $S_Y = \{Roads\}$ and $S_Z = \{Sprinkler\}$. The set of all paths from a node in S_Y to a node in S_Z is:

$$Roads \leftarrow Weather \rightarrow Field \leftarrow Sprinkler$$
$$Roads \rightarrow Game \leftarrow Field \leftarrow Sprinkler$$

Suppose $S_X = \Phi$. The first path contains node *Field* with converging arrows and none of its three descendants *Game, Ticket Sale*, and *Concession* is in S_X. The second path contains the node *Game* with converging arrows and none of its two descendants *Ticket Sale* and *Concession* is in S_X. Therefore, S_Y and S_Z are *d*-separated by the empty set. But, if we consider $S_X = \{Ticket\ Sale\}$, the first path contains the node *Field* with converging arrows and its descendant *Ticket Sale* is in S_X. Also, the second path contains the node *Game* with converging arrows and its descendants *Ticket Sale* is in S_X. Although, the first path contains the node *Weather* without converging arrows, the node does not belong to S_X. Therefore, $\{Ticket\ Sale\}$ does not *d*-separate S_Y and S_Z. Note that Φ does not *d*-

separate $\{Roads\}$ and $\{Field\}$, but $\{Weather\}$ does, and so does $\{Weather, Ticket\, Sale\}$.

The generalized set-theoretic definition of *d*-separation above yields the following proposition.

Proposition 6-3: For any three disjoint subsets S_X, S_Y, and S_Z of variables in a DAG, S_X *d*-separates S_Y and S_Z if and only if S_Y and S_Z are conditionally independent given S_X (that is, $S_Y \perp S_Z \mid S_X$).

Once we have built belief networks, either with the help of domain experts or via automated learning from past observations, we need to reason with them, that is, examine how the variables in a network change their beliefs when observations are propagated into the network as evidence.

6.3 Evidence, Belief, and Likelihood

Evidence on a variable is a statement of certainties of its states based on certain observation. Since the states of a belief network variable are mutually exclusive, such a statement of certainty of a state variable is usually made with a percentage that represents the chance of being in that state. If the statement constituting evidence for a variable gives the exact state of the variable (that is, 100%), then it is *hard* evidence (which is also called *instantiation*); otherwise, the evidence is called *soft*. As an example, consider the variable *Weather* whose states are *sunny*, *rain*, and *snow*. If the evidence e is based on someone's direct observation of the weather and states that the weather is sunny, then it is hard evidence and is denoted by *Weather = sunny*. In general, if $E = \{X_1, ..., X_n\}$ is the set of all variables whose values are known as $X_1 = a_1, ..., X_n = a_n$ then

$$e = \{X_1 = a_1, ..., X_n = a_n\}$$

where each a_i is hard evidence of the state of X_i. For example, if $E = \{Weather, Roads, Game\}$ and the evidence states that the weather is sunny, the roads are open, and the game is on, then

$$e = \{Weather = sunny, Roads = open, Game = on\}$$

On the other hand, consider the situation when the source of evidence on the variable *Weather* is based on the observation of output from a sensor (the sensor is not explicitly modeled as a belief network variable). The statement constituting evidence states that there is a 80% chance that the weather is sunny, 15% chance that the weather is rainy, and 5% chance that the weather is snowy. The evidence in this case is inexact and therefore *soft*. Evidence on a variable X yields a *likelihood vector*, denoted as $\lambda(X)$, expressed in terms of probability measures.

For example, the above soft evidence on the variable *Weather* yields the likelihood vector

$$\lambda(Weather) = \begin{bmatrix} 0.80 \\ 0.15 \\ 0.05 \end{bmatrix}$$

The hard evidence $e = \{Weather = sunny\}$ yields the likelihood vector $\begin{bmatrix} 1 \\ 0 \\ 0 \end{bmatrix}$.

Usually, total evidence accumulated on the states of a variable is expected to be more or less than 100%. An example of such type of evidence obtained by observing output from a sensor (for example, the sensor displays green for sunny weather, blue for rainy, and red for snowy; ambiguities occur because a faulty sensor or a noisy environment can cause display of different lights for a given weather state) states that there is a 70% chance that the weather is sunny, a 50% chance that the weather is rainy, and a 15% chance that the weather is snowy. The likelihood vector for this evidence is the following:

$$\lambda(Weather) = \begin{bmatrix} 0.70 \\ 0.50 \\ 0.15 \end{bmatrix}$$

The above evidence states that if all 100 weather circumstances similar to the current one are sunny, the sensor prediction is likely to be correct (that is, displays green) 70 times; if all are rainy, it is likely to be correct (blue) 50 times; and if all are snowy, it is likely to be correct (red) 15 times.

How do we then relate evidence to probability? Observe that 70:50:15 is the ratio of the number of times the sensor is likely to produce $e_{Weather}$ if all 100 weather circumstances are sunny, to the number of times it is likely to produce

$e_{Weather}$ if all 100 are rainy, to the number of times it is likely to produce $e_{Weather}$ if all 100 are snowy. This relation yields the following likelihood ratio:

$$p(e_{Weather} \mid Weather = sunny) : p(e_{Weather} \mid Weather = rainy) :$$
$$p(e_{Weather} \mid Weather = snowy) = 70 : 50 : 15$$

This ratio gives the following likelihood:

$$\lambda(Weather = sunny) = p(e_{Weather} \mid Weather = sunny) = 0.70$$

$$\lambda(Weather = rainy) = p(e_{Weather} \mid Weather = rain) = 0.50$$

$$\lambda(Weather = snowy) = p(e_{Weather} \mid Weather = snow) = 0.15$$

The likelihood vector $\lambda(Weather)$ is $p(e_{Weather} \mid Weather)$, and we therefore have the following:

$$\lambda(Weather) = p(e_{Weather} \mid Weather) = \begin{bmatrix} 0.70 \\ 0.50 \\ 0.15 \end{bmatrix}$$

If the sensor always displays green when the weather is sunny, then the above likelihood vector changes to

$$\lambda(Weather) = \begin{bmatrix} 1.0 \\ 0.5 \\ 0.15 \end{bmatrix}$$

But, conversely, the green display does not necessarily mean the weather is sunny though it certainly indicates a high probability of sunny weather.

The CPTs of a belief network remain unchanged upon the arrival of evidence. When evidence is *posted* to the designated node to compute posterior probabilities of the nodes in the networks, the node state certainties, or probability distribution, change. After receiving evidence e, the posterior probability of node X is $p(X \mid e)$. The *belief* of the node X of a belief network, denoted as $Bel(X)$, is the overall belief of the node X contributed by all evidence so far received. Therefore, if e is the evidence received so far then $Bel(X) = p(X \mid e)$.

Consider the network fragment as shown in Figure 6-9. Suppose e_X^+ and e_X^- are the total evidence connected to X through its parents and children, respectively. In other words, e_X^+ and e_X^- are the evidence contained in the upper and lower sub-networks with respect to the node X. We then define the following two π and λ vectors:

$$\pi(X) = p(X \mid e_X^+)$$

$$\lambda(X) = p(e_X^- \mid X)$$

The vectors $\pi(X)$ and $\lambda(X)$ represent the distributions of the total supports among the states of X through its parents and children respectively. If the network that contains the fragment is a tree, then the vectors $\pi(X)$ and $\lambda(X)$ represent the distributions of the total causal and diagnostic supports among the states of X by all its ancestors and descendants, respectively.

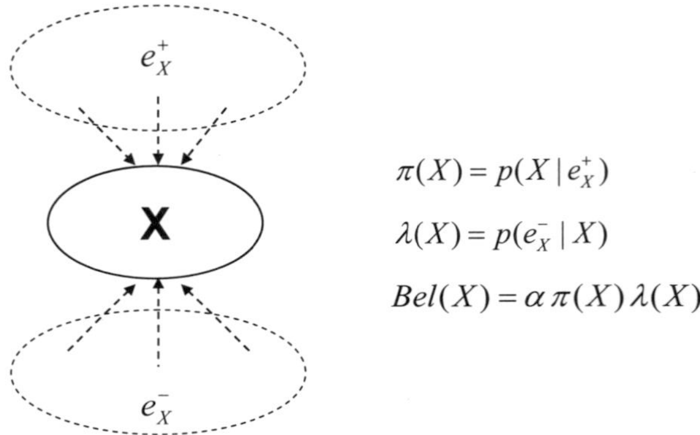

Figure 6-9: Network fragment containing node X

6.4   Prior Probabilities in Networks without Evidence

Evidence propagation and belief updating starts with "fresh" networks without any observed evidence. Then the π and λ vectors of the variables, and hence the belief vectors, are updated incrementally as evidence is accumulated. In this section, we detail how these initial vectors of the variables in a fresh network are computed.

If no evidence has yet been propagated in a network, then, for every variable X in the network, $\pi(X)$ is $p(X)$, since e_X^+ is empty. Therefore, $\pi(X)$ is simply the prior probability of the variable X. Since e_X^- is the empty set, $\lambda(X)$ is $p(\Phi \mid X)$. Since Φ is a constant, each $p(\Phi \mid x)$ is equal to $\dfrac{1}{n}$, where n is the number of states of X. For the purpose of simplicity, we will write an initial λ vector simply as a non-normalized n-vector $(1,1,...,1)$.

Recall that the relationships among the variables in a network are quantified via CPTs of the form $p(X \mid U_1, U_2, ..., U_n)$ for a variable X with parents $U_1, U_2, ..., U_n$. Therefore, if X has no parent (that is, is a *root node*) then its CPT is just $p(X)$, which is its prior probability. With this in mind, we present a simple recursive routine to compute the beliefs and the π and λ vectors in a fresh network.

First, mark all the root nodes. Then recursively compute $p(X)$ of a node X, each of whose parents is already marked, and mark the node X itself. If X has parents $U_1, U_2, ..., U_n$ then

$$p(X) = \sum_{U_1, ..., U_n} p(X \mid U_1, ..., U_n) p(U_1, ..., U_n)$$

$$= \sum_{U_1, ..., U_n} p(X \mid U_1, ..., U_n) p(U_1 \mid U_2, ..., U_n) ... p(U_{n-1} \mid U_n) p(U_n)$$

Since $U_1, U_2, ..., U_n$ are marginally independent

$$p(X) = \sum_{U_1, ..., U_n} p(X \mid U_1, ..., U_n) \prod_{i=1}^{n} p(U_i)$$

Thus, $p(X)$ can be computed using its CPT and the prior probabilities of its parents.

Example

Consider the network in Figure 6-10, along with the prior probabilities $p(Roads)$ and $p(Field)$ of the root nodes *Roads* and *Field*, respectively, and the two CPTs $p(Game \mid Roads, Field)$ and $p(Ticket\,Sale \mid Game)$ of the two other nodes of the network. The network also shows the initial π and λ vectors, and hence belief vectors, of each of the two root nodes.

The prior probability of the node *Game* and then of the node *Ticket Sale* are computed as follows:

$$p(Game) = \sum_{Roads, Field} p(Game \mid Roads, Field) p(Roads) p(Field) = \begin{bmatrix} 0.67 \\ 0.33 \end{bmatrix}$$

$$p(Ticket\,Sale) = \sum_{Game} p(Ticket\,Sale \mid Game) p(Game) = \begin{bmatrix} 0.50 \\ 0.17 \\ 0.33 \end{bmatrix}$$

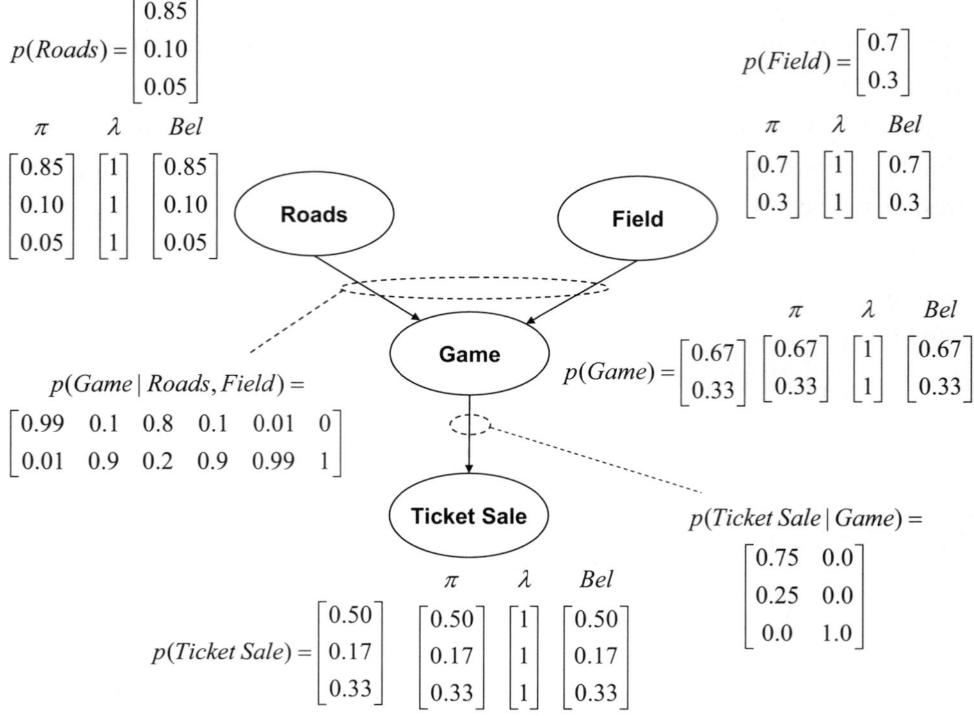

Figure 6-10: Initial probabilities, beliefs, and π and λ vectors

6.5   Belief Revision

In this section, we describe how a node revises its own belief upon receiving evidence on itself. Suppose a node X receives evidence e_X and the probability vector $p(X)$ is its current state certainties. Then its posterior probability is defined as:

$$p(X\,|\,e_X) = \frac{p(X)p(e_X\,|\,X)}{p(e_X)} = \alpha\, p(X)\lambda(X)$$

where the normalizing constant α is computed by summing over mutually exclusive and exhaustive states of the variable X (that is, $\sum_{X=x} p(X) = 1$):

$$\alpha = \frac{1}{p(e_X)} = \frac{1}{\sum_X p(X, e_X)} = \frac{1}{\sum_X p(e_X\,|\,X)p(X)}$$

Therefore, the belief of the node X after receiving evidence e_X becomes the normalized product of its prior probability vector $p(X)$ with the likelihood vector $\lambda(X)$.

Example

Consider the node *Weather* whose prior probability $p(Weather)$ and posted evidence $e_{Weather}$ is shown in Figure 6-11. A particular evidence e_X on a variable X in a network will be hypothetically considered as a binary child node of the node X, where the CPT $p(e_X \mid X) = \lambda(X)$.

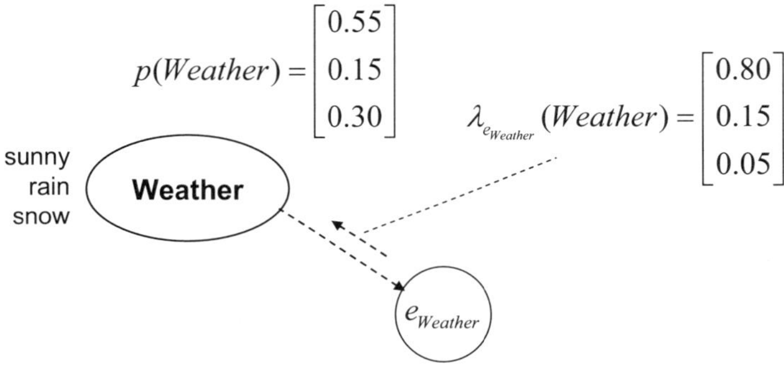

$$p(Weather) = \begin{bmatrix} 0.55 \\ 0.15 \\ 0.30 \end{bmatrix} \qquad \lambda_{e_{Weather}}(Weather) = \begin{bmatrix} 0.80 \\ 0.15 \\ 0.05 \end{bmatrix}$$

Figure 6-11: Posting evidence on a node

Posterior probability of the state *sunny*, for example, is computed as follows:

$$p(Weather \mid e_{Weather}) = \begin{bmatrix} \alpha \times 0.55 \times 0.80 \\ \alpha \times 0.15 \times 0.15 \\ \alpha \times 0.30 \times 0.05 \end{bmatrix} = \begin{bmatrix} 0.92 \\ 0.05 \\ 0.03 \end{bmatrix}$$

$$where \quad \alpha = \frac{1}{0.55 \times 0.80 + 0.15 \times 0.15 + 0.30 \times 0.05} = 2.09$$

Now consider belief revision in the case illustrated in Figure 6-9. The posterior probability of the node X upon receiving all the evidence is computed as follows:

$$Bel(X) = p(X \mid e_X^+, e_X^-)$$

$$= \frac{p(e_X^+, e_X^-, X)}{p(e_X^+, e_X^-)}$$

$$= \frac{p(e_X^- \mid e_X^+, X)\, p(X \mid e_X^+)\, p(e_X^+)}{p(e_X^+, e_X^-)}$$

$$= \alpha\; p(X \mid e_X^+)\, p(e_X^- \mid X)$$

$$= \alpha\; \pi(X)\lambda(X)$$

where

$$\alpha = \frac{p(e_X^+)}{p(e_X^-, e_X^+)} = \frac{1}{p(e_X^- \mid e_X^+)}$$

$$= \frac{1}{\sum_{X=x} p(e_X^- \mid e_X^+, X)\, p(X)}$$

$$= \frac{1}{\sum_{X=x} p(e_X^- \mid e_X^+, X)\, p(X \mid e_X^+)}$$

$$= \frac{1}{\sum_{X=x} \pi(X)\lambda(X)}$$

Note that $p(e_X^- \mid e_X^+, X) = p(e_X^- \mid X)$ because X separates e_X^+ and e_X^-. The node belief is therefore the normalized product of its λ and π vectors, which can be factorized as follows.

Consider a concrete instantiation of the above case as shown in Figure 6-12. The node X has parents U, V, and W, through which it received evidence e_U^+, e_V^+, and e_W^+ respectively. Node X has children Y, Z, and Q, through which it received evidence e_Y^-, e_Z^-, and e_Q^- respectively. Thus,

$$e_X^+ = \{e_U^+, e_V^+, e_W^+\}$$

$$e_X^- = \{e_Y^-, e_Z^-, e_Q^-\}$$

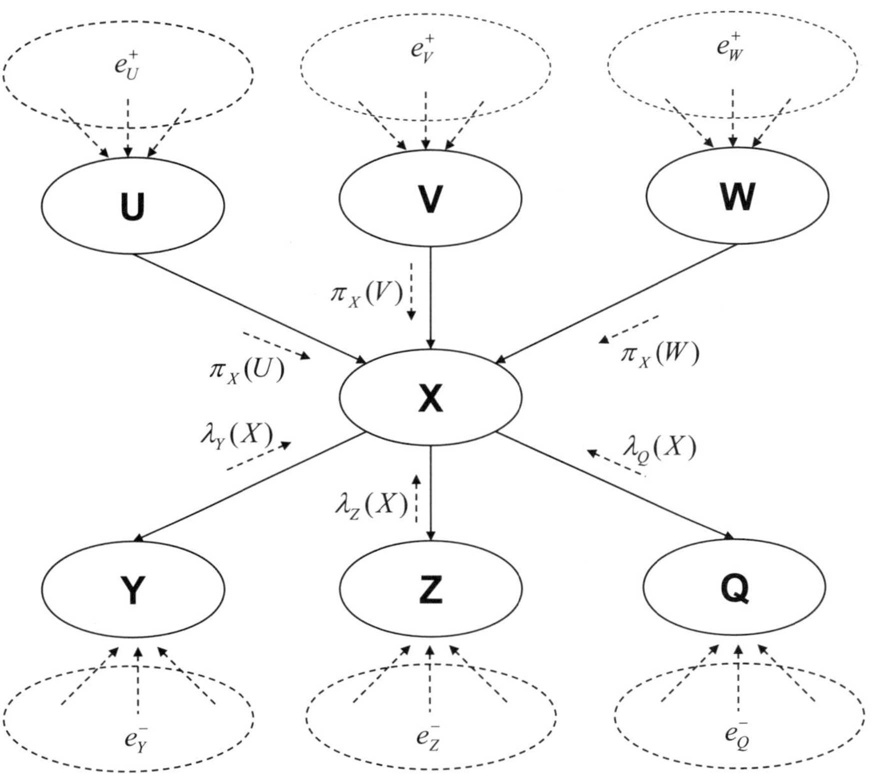

Figure 6-12: Node X has multiple parents and children

$$\begin{aligned}
\pi(X) \quad &= p(X \mid e_X^+) \\
&= p(X \mid e_U^+, e_V^+, e_W^+) \\
&= \sum_{U,V,W} p(X \mid U,V,W,e_U^+,e_V^+,e_W^+)\, p(U,V,W \mid e_U^+,e_V^+,e_W^+) \\
&= \sum_{U,V,W} p(X \mid U,V,W)\, p(U \mid V,W,e_U^+,e_V^+,e_W^+) \\
&\qquad\qquad p(V \mid W,e_U^+,e_V^+,e_W^+)\, p(W \mid e_U^+,e_V^+,e_W^+) \\
&= \sum_{U,V,W} p(X \mid U,V,W)\, p(U \mid e_U^+)\, p(V \mid e_V^+)\, p(W \mid e_W^+) \\
&= \sum_{U,V,W} p(X \mid U,V,W)\, \pi_X(U)\, \pi_X(V)\, \pi_X(W)
\end{aligned}$$

The above derivation uses the conditional assumption since U, V, and W separate X from e_U^+, e_V^+, and e_W^+, respectively. The derivation also uses independence relationships, such as U is independent of V, W, e_V^+, and e_W^+, V is independent of U, W, e_U^+, and e_W^+, and W is independent of U, V, e_U^+, and e_V^+. Similarly,

$$\lambda(X) = p(e_X^- \mid X)$$
$$= p(e_Y^-, e_Z^-, e_Q^- \mid X)$$
$$= p(e_Y^- \mid e_Z^-, e_Q^-, X) p(e_Z^- \mid e_Q^-, X) p(e_Q^- \mid X)$$
$$= p(e_Y^- \mid X) p(e_Z^- \mid X) p(e_Q^- \mid X)$$
$$= \lambda_Y(X) \lambda_Z(X) \lambda_Q(X)$$

This derivation uses the conditional independence assumptions that given X, e_Y^- is independent of e_Z^- and e_Q^-, and e_Z^- is independent of e_Q^-.

Now consider the case shown in Figure 6-13, when a given node X receives evidence e_X. In addition, suppose e_X^+ and e_X^- are the total evidence connected to X through its parents and children, respectively. Then the revised $\lambda(X)$ can be computed by using the conditional independence assumption derived from the fact that X separates e_X and e_X^-:

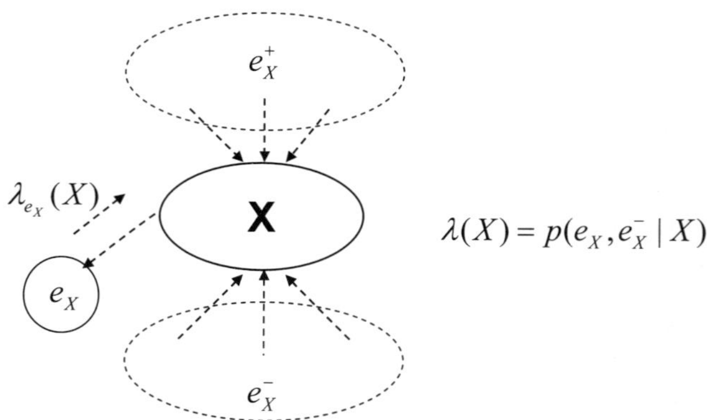

Figure 6-13: Node X receives evidence e_X

$$\lambda_{new}(X) = p(e_X\, e_X^- \mid X)$$
$$= p(e_X^- \mid e_X, X) p(e_X \mid X)$$
$$= p(e_X^- \mid X) p(e_X \mid X)$$
$$= \lambda(X) \lambda_{e_X}(X)$$

Thus, a node revises its λ vector by multiplying its λ vector with the likelihood vector for the evidence. Note that $\pi(X)$ remains unchanged as e_X^+ is unchanged.

The revised belief of X is computed as follows, using the necessary independence assumption derived from the fact that X separates e_X^+ from e_X^-, and e_X :

$$Bel_{new}(X) = p(X \mid e_X^+, e_X^-, e_X)$$

$$= \frac{p(e_X^+, e_X^-, e_X, X)}{p(e_X^+, e_X^-, e_X)}$$

$$= \frac{p(e_X^-, e_X \mid e_X^+, X)\, p(X \mid e_X^+)\, p(e_X^+)}{p(e_X^+, e_X^-, e_X)}$$

$$= \alpha\, p(X \mid e_X^+) p(e_X^-, e_X \mid X) \qquad \left[\text{where } \alpha \text{ is } \frac{1}{p(e_X^-, e_X \mid e_X^+)} \right]$$

$$= \alpha\, \pi(X) \lambda_{new}(X)$$

Therefore, the revised belief is simply the product of the revised $\lambda(X)$ with unchanged $\pi(X)$.

Example

Consider the network fragment at the top half of Figure 6-14 along with the π, λ, and belief vectors.

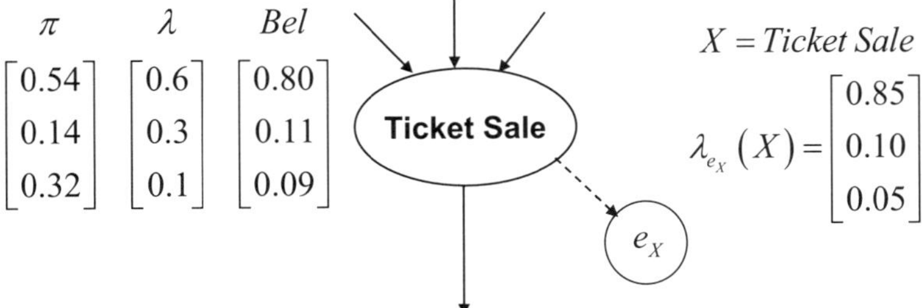

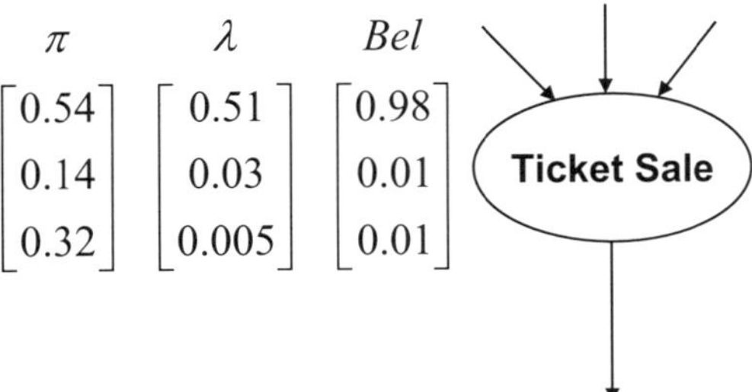

Figure 6-14: Example belief revision

As shown in the figure, evidence e_X has been posted into the node *Ticket Sale*. Then the revised λ and belief vectors are computed below (also shown at the bottom half of the figure).

$$\lambda_{new}(X) = \lambda(X)\lambda_{e_X}(X) = \begin{bmatrix} 0.6 \\ 0.3 \\ 0.1 \end{bmatrix}\begin{bmatrix} 0.85 \\ 0.10 \\ 0.05 \end{bmatrix} = \begin{bmatrix} 0.51 \\ 0.03 \\ 0.005 \end{bmatrix}$$

$$\pi(X) = \begin{bmatrix} 0.54 \\ 0.14 \\ 0.32 \end{bmatrix}$$

$$Bel(X) = \alpha\pi(X)\lambda_{new}(X) = \alpha\begin{bmatrix} 0.54 \\ 0.14 \\ 0.32 \end{bmatrix}\begin{bmatrix} 0.51 \\ 0.03 \\ 0.005 \end{bmatrix} = \begin{bmatrix} 0.98 \\ 0.015 \\ 0.005 \end{bmatrix}$$

6.6   Evidence Propagation in Polytrees

In the previous section, we detailed how a node updates its own belief upon receiving evidence on itself. This section discusses how a node X in a polytree updates its own beliefs when evidence is observed on one of its neighboring nodes and how it propagates the effects of that evidence to the neighboring nodes to help update their beliefs. We consider six different cases corresponding to possible polytree fragments around the node X that contain its immediate neighbors:

- *Upward* propagation in a *linear* fragment: X has only one parent U, and one child Y, and the child Y receives evidence

- *Downward* propagation in a *linear* fragment: X has only one parent U, and one child Y, and the parent U receives evidence

- *Upward* propagation in a *tree* fragment: X has only one parent U, and three children Y, Z, and Q, and one of the children, say Y, receives evidence

- *Downward* propagation in a *tree* fragment: X has only one parent U, and three children Y, Z, and Q, and the parent U receives evidence

- *Upward* propagation in a *polytree* fragment: X has parents U, V, and W, and three children Y, Z, and Q, and one of the children, say Y, receives evidence

- *Downward* propagation in a *polytree* fragment: X has parents U, V, and W, and three children Y, Z, and Q, and one of the parents, say U, receives evidence

6.6.1   Upward Propagation in a Linear Fragment

This case is illustrated in Figure 6-15, where the node X has only one parent U, and one child Y, and the child Y receives evidence e_Y. The node Y updates its belief and sends the message $\lambda_Y(X)$ to X. The node X updates its belief upon receiving the message from Y and, in turn, sends the message $\lambda_X(U)$ to U to help update its belief. All the π vectors remain unchanged, as there is no new causal evidence. Next, we compute the values of $\lambda_Y(X)$, $\lambda_X(U)$, and their relations to the new beliefs of X and U.

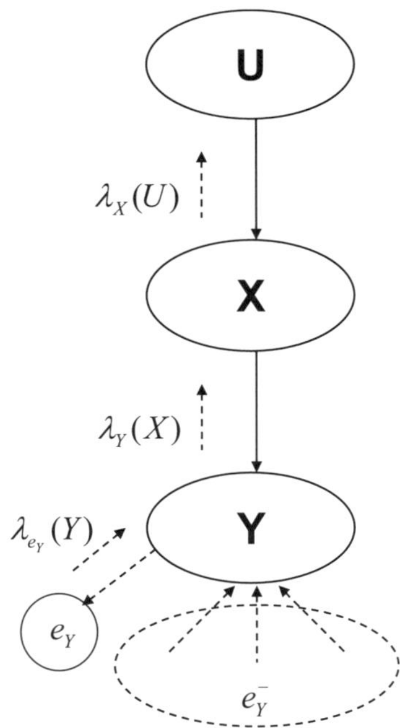

Figure 6-15: Upward propagation of evidence e_Y in a linear fragment

Let e_Y^- be the total evidence connected to Y, but not through its parent X. If Y now receives evidence e_Y, then

$$\lambda_{new}(Y) = p(e_Y, e_Y^- \mid Y)$$

Since Y separates X from e_Y and e_Y^-, the revised $\lambda(X)$ can now be computed as follows:

$$\begin{aligned}
\lambda_{new}(X) &= p(e_Y \, e_Y^- \mid X) \\
&= \sum_Y p(e_Y \, e_Y^- \mid Y, X) p(Y \mid X) \\
&= \sum_Y p(e_Y \, e_Y^- \mid Y) p(Y \mid X) \\
&= \sum_Y \lambda_{new}(Y) p(Y \mid X)
\end{aligned}$$

Therefore, the revised $\lambda(X)$ can be computed at the node Y by taking the product of the revised $\lambda(Y)$ and the CPT $p(Y \mid X)$. The revised value $\lambda_{new}(X)$ is then sent to the node X from Y as the message $\lambda_Y(X) = \lambda_{new}(X)$. Note that $\pi(X)$ remains unchanged as e_X^+ is unchanged. Since X separates e_X^+ from e_Y and e_Y^-, the node X revises its belief as follows:

$$\begin{aligned}
Bel_{new}(X) &= p(X \mid e_Y, e_Y^-, e_X^+) \\
&= \alpha \, p(X \mid e_X^+) p(e_Y, e_Y^- \mid X) \\
&= \alpha \, \pi(X) \lambda_{new}(X)
\end{aligned}$$

Therefore, X revises its belief by multiplying the revised $\lambda(X)$, sent as a message by Y, with its unchanged $\pi(X)$. The revised $\lambda(U)$ can now be computed as follows:

$$\lambda_{new}(U) = \sum_X \lambda_{new}(X) p(X \mid U)$$

X sends $\lambda_{new}(U)$ as a message $\lambda_X(U)$ to U.

Example

Consider the linear fragment shown in Figure 6-16 along with the π, λ, and belief vectors.

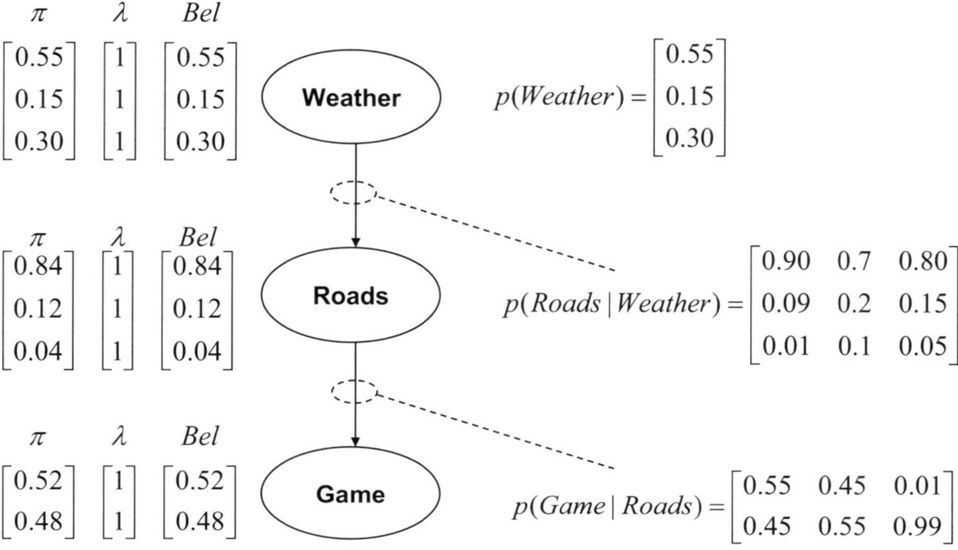

Figure 6-16: Example linear fragment

The upward propagation of evidence due to the posting of evidence e_{Game} at the node *Game* is shown in Figure 6-17.

The two λ-values $\lambda_{Game}(Roads)$ and $\lambda_{Roads}(Weather)$ in the figure are computed as follows:

$$\lambda_{Game}(Roads) = \lambda_{new}(Roads)$$

$$= \sum_{Game} \lambda_{new}(Game)\,p(Game \mid Roads)$$

$$= \begin{bmatrix} 0.9 \times 0.55 + 0.1 \times 0.45 \\ 0.9 \times 0.45 + 0.1 \times 0.55 \\ 0.9 \times 0.01 + 0.1 \times 0.99 \end{bmatrix} = \begin{bmatrix} 0.54 \\ 0.46 \\ 0.11 \end{bmatrix}$$

$$\lambda_{Rodas}(Weather) = \lambda_{new}(Weather)$$

$$= \sum_{Roads} \lambda_{new}(Roads)\,p(Roads \mid Weather)$$

$$= \begin{bmatrix} 0.54 \times 0.90 + 0.46 \times 0.09 + 0.11 \times 0.01 \\ 0.54 \times 0.7 + 0.46 \times 0.2 + 0.11 \times 0.1 \\ 0.54 \times 0.80 + 0.46 \times 0.50 + 0.11 \times 0.05 \end{bmatrix} = \begin{bmatrix} 0.53 \\ 0.48 \\ 0.51 \end{bmatrix}$$

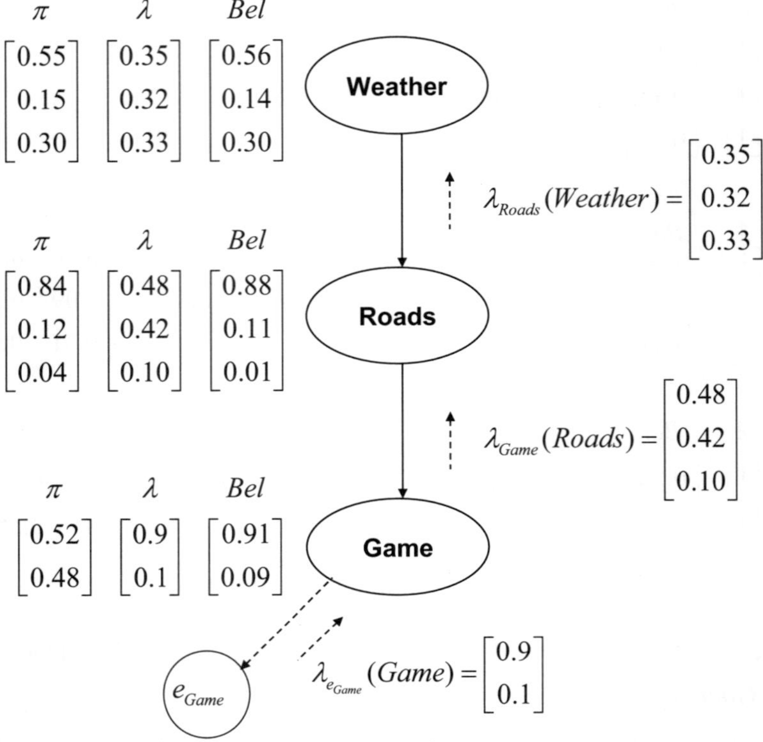

Figure 6-17: Example upward propagation in a linear fragment

6.6.2 Downward Propagation in a Linear Fragment

This case is illustrated in Figure 6-18, where the node X has only one parent U, and one child Y, and the parent node U receives evidence e_U. The node U updates its belief and sends the message $\pi_X(U)$ to X. The node X updates its belief upon receiving the message from U and, in turn, sends the message $\pi_Y(X)$ to X to help update its belief. Next, we compute the values of $\pi_X(U)$, $\pi_Y(X)$, and their relations to the new beliefs of X and Y respectively.

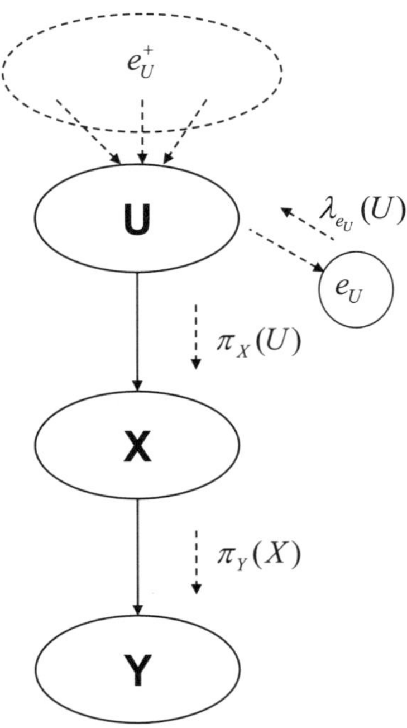

Figure 6-18: Downward propagation of evidence e_U in a linear fragment

Let e_U^+ be the total evidence connected to U but not through its child X. If U now receives evidence e_U, then $\lambda_{new}(U) = \lambda_{e_U}(U)\lambda(U)$. Note that $\pi(U)$ remains unchanged since there is no change in evidence connected through the parents of U. The revised $\pi(X)$ can now be computed as follows:

$$\pi_{new}(X) = p(X \mid e_U\, e_U^+)$$
$$= \sum_U p(X \mid U, e_U, e_U^+) p(U \mid e_U, e_U^+)$$
$$= \sum_U p(X \mid U) p(U \mid e_U\, e_U^+) \left[U \text{ separates } X \text{ from } e_U \text{ and } e_U^+ \right]$$
$$= \sum_U \pi_X(U)\, p(X \mid U)$$

where $\pi_X(U) = p(U \mid e_U\, e_U^+)$ is simplified as follows:

$$\pi_X(U) = p(U \mid e_U^-,\, e_U^+)$$

$$= \frac{p(e_U^-,\, e_U^+,\, U)}{p(e_U^-,\, e_U^+)}$$

$$= \frac{p(e_U^- \mid e_U^+,\, U)\, p(U \mid e_U^+)\, p(e_U^+)}{p(e_U^-,\, e_U^+)}$$

$$= \alpha\ p(U \mid e_U^+) p(e_U^- \mid U) \qquad \left[\text{since } U \text{ separates } e_U^- \text{ and } e_U^+\right]$$

$$= \alpha\ \pi(U)\lambda_{e^U}(U) \qquad \left[\alpha \text{ is } \frac{1}{p(e_U^- \mid e_U^+)}\right]$$

The node U can compute $\pi_X(U)$ by multiplying its likelihood vector for the evidence with its π vector. Therefore, the revised $\pi(X)$, $\pi_{new}(X)$, can be computed at the node X by taking the product of $\pi_X(U)$ and the CPT $p(X \mid U)$. The revised value $\pi_{new}(X)$ is then sent to the node Y from X as the message $\pi_Y(X)$. Note that $\lambda(X)$ remains unchanged since e_X^- is unchanged. The node X revises its belief as follows:

$$Bel_{new}(X) = p(X \mid e_U^-,\, e_U^+,\, e_U^-)$$

$$= \alpha\ p(e_U,\, e_U^+ \mid X)\, p(X \mid e_U^-) \left[\text{since } X \text{ separates } e_U^- \text{ from } e_U \text{ and } e_U^+\right]$$

$$= \alpha \pi_{new}(X)\lambda(X)$$

$$= \alpha\ \pi_X(U) p(X \mid U)\lambda(X)$$

Therefore, X revises its belief by multiplying message $\pi_X(U)$ sent by U, with its unchanged λ vector $\lambda(X)$ and the CPT $p(X \mid U)$. Similarly, X sends a message $\pi_Y(X)$ to Y to help revising its belief.

Example

Consider the linear fragment shown in Figure 6-19 along with the π, λ, and belief vectors.

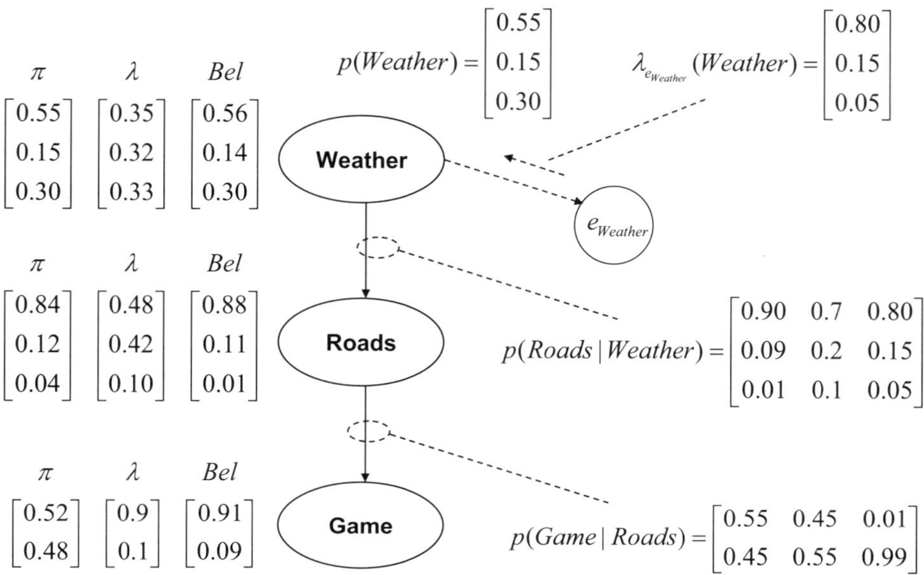

Figure 6-19: Example linear fragment

The downward propagation of evidence due to the posting of evidence $e_{Weather}$ at the node *Weather* is shown in Figure 6-20.

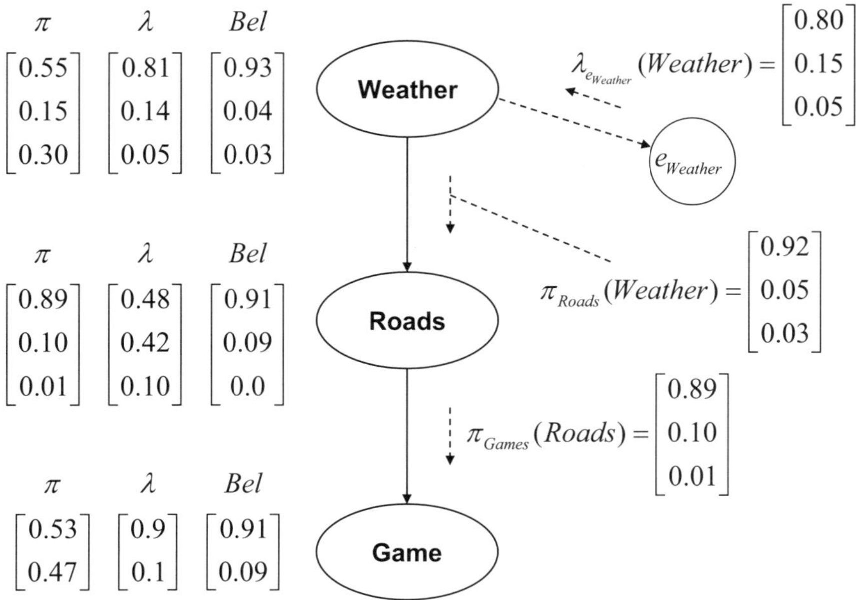

Figure 6-20: Example downward propagation in a linear fragment

6.6.3   *Upward Propagation in a Tree Fragment*

This case is illustrated in Figure 6-21, where the node X has only one parent U, and three children Y, Z, and Q, and the child Y receives evidence e_Y. The node Y updates its belief and sends the message $\lambda_Y(X)$ to X. The node X updates its belief upon receiving the message from Y and, in turn, sends the diagnostic message $\lambda_X(U)$ to U and the causal messages $\pi_Z(X)$ and $\pi_Q(X)$ to Z and Q, respectively, to help update their beliefs. The messages $\lambda_Y(X)$ and $\lambda_X(U)$ are computed as above in the case of upward propagation in a linear fragment. Next, we compute $\pi_Z(X)$, $\pi_Q(X)$, and their relations to the new beliefs of Z and Q.

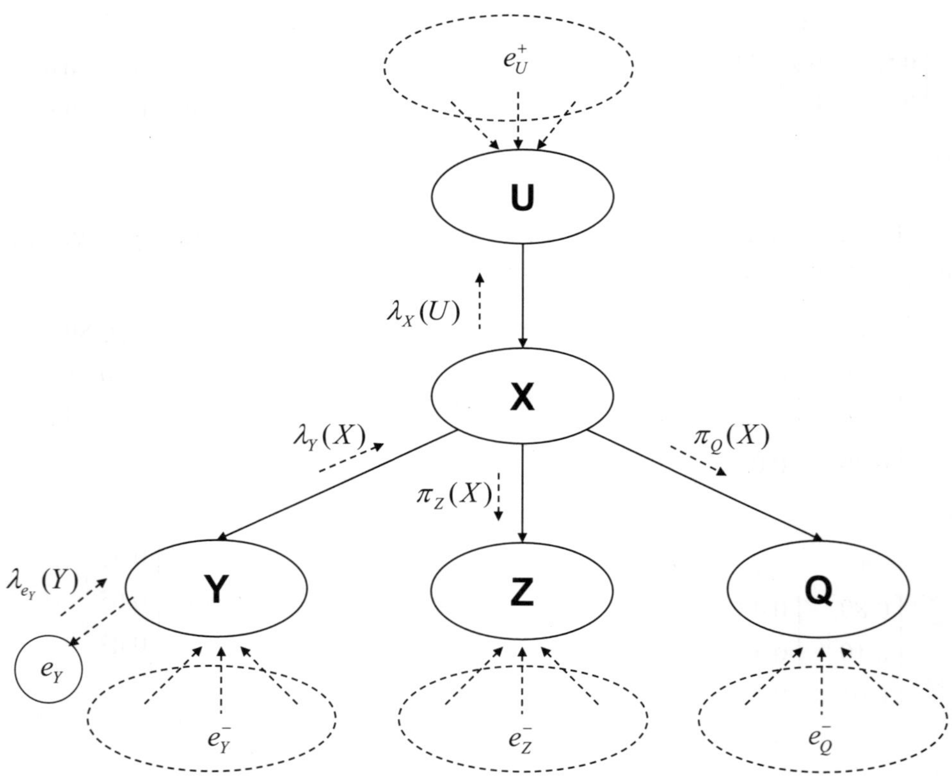

Figure 6-21: Upward propagation of evidence e_Y in a tree fragment

Let e_Y^-, e_Z^-, and e_Q^- be the total diagnostic evidence associated with Y, Z, and Q respectively that is not obtained through their parent X, and e_U^+ be the total causal evidence associated with U that is not obtained through its child X.

$$\pi_{new}(Z) = p(Z \mid e_Y, e_Y^-, e_Q^-, e_U^+)$$

$$= \sum_X p(Z \mid X, e_Y, e_Y^-, e_Q^-, e_U^+) p(X \mid e_Y, e_Y^-, e_Q^-, e_U^+)$$

$$= \sum_X p(Z \mid X, e_Y, e_Y^-, e_Q^-) p(X \mid e_Y, e_Y^-, e_Q^-, e_U^+)$$

$$\left[\text{given } X, Z \text{ is independent of } e_U^+ \right]$$

$$= \sum_X p(Z \mid X) \pi_Z(X) \left[\text{given } X, Z \text{ is independent of } e_Y, e_Y^-, e_Q^- \right]$$

where

$$\pi_Z(X) = p(X \mid e_Y, e_Y^-, e_Q^-, e_U^+) = \frac{p(e_Y, e_Y^-, e_Q^-, e_U^+, X)}{p(e_Y, e_Y^-, e_Q^-, e_U^+)}$$

$$= \frac{p(e_Y, e_Y^- \mid e_Q^-, e_U^+, X) p(e_Q^- \mid e_U^+, X) p(X \mid e_U^+) p(e_U^+)}{p(e_Y, e_Y^-, e_Q^-, e_U^+)}$$

$$= \alpha \, p(e_Y, e_Y^- \mid e_Q^-, e_U^+, X) p(e_Q^- \mid e_U^+, X) p(X \mid e_U^+) \left[\alpha = \frac{1}{p(e_Y, e_Y^-, e_Q^- \mid e_U^+)} \right]$$

$$= \alpha \, p(e_Y, e_Y^- \mid X) p(e_Q^- \mid e_U^+, X) p(X \mid e_U^+) \left[X \text{ separates } e_Y, e_Y^- \text{ from } e_Q^-, e_U^+ \right]$$

$$= \alpha \, p(e_Y, e_Y^- \mid X) p(e_Q^- \mid X) p(X \mid e_U^+) \left[X \text{ separates } e_Q^- \text{ from } e_U^+ \right]$$

$$= \alpha \, \lambda_Y(X) \lambda_Q(X) \pi(X)$$

$$= \alpha \, \frac{\lambda_Y(X) \lambda_Z(X) \lambda_Q(X) \pi(X)}{\lambda_Z(X)}$$

$$= \alpha \, \frac{Bel_{new}(X)}{\lambda_Z(X)}$$

Therefore, the revised $\pi(Z)$ can be computed at the node Z by taking the product of the message $\dfrac{Bel_{new}(X)}{\lambda_Z(X)}$, sent by its parent X, and the CPT $p(Z \mid X)$. Similarly,

X sends $\dfrac{Bel_{new}(X)}{\lambda_Q(X)}$ to Q to update its belief. Note that $\lambda(Z)$ remains unchanged

since e_Z^- is unchanged. The node Z revises its belief as follows:

$$Bel_{new}(Z) = p(Z \mid e_Y, e_Y^-, e_Q^-, e_U^+, e_Z^-)$$

$$= \alpha\, p(Z \mid e_Y, e_Y^-, e_Q^-, e_U^+)\, p(e_Z^- \mid Z) \left[\text{since } Z \text{ separates } e_Z^- \text{ from } e_Y, e_Y^-, e_Q^-, e_U^+ \right]$$

$$= \alpha\, \pi_{new}(Z)\, \lambda(Z)$$

$$= \alpha\, \pi_Z(X)\, p(Z \mid X)\, \lambda(Z)$$

Therefore, Z revises its belief by multiplying the message $\pi_Z(X)$, sent by its parent X, with its unchanged λ vector $\lambda(Z)$ and its CPT $p(Z \mid X)$.

6.6.4   *Downward Propagation in a Tree Fragment*

This case, which is illustrated in Figure 6-22, is similar to the case of downward propagation in a linear fragment presented earlier.

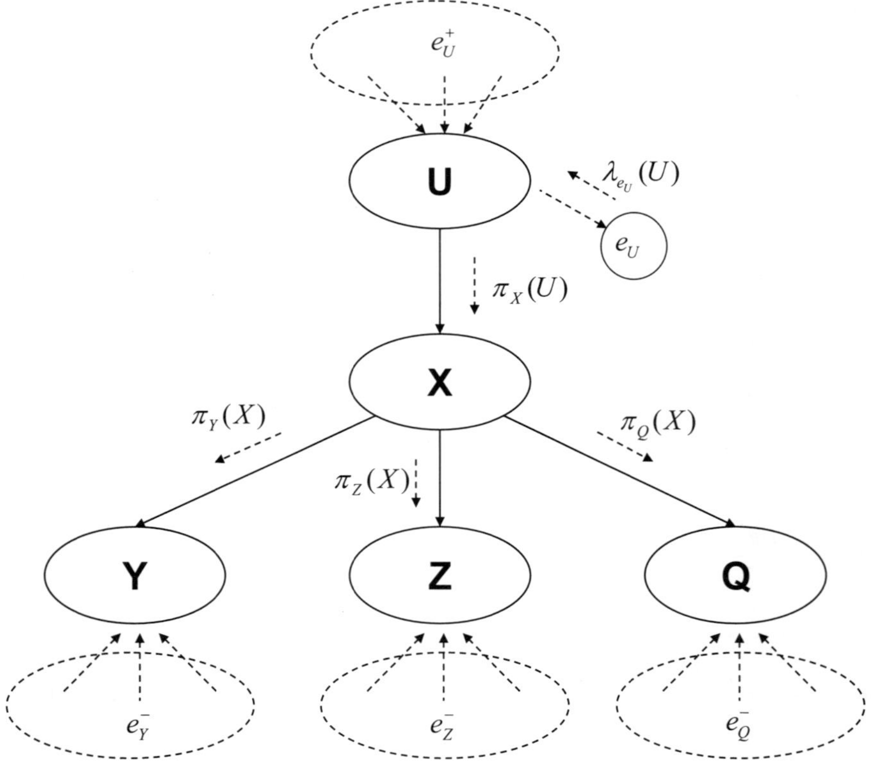

Figure 6-22: Downward propagation of evidence e_U in a tree fragment

6.6.5 *Upward Propagation in a Polytree Fragment*

This case is illustrated in Figure 6-23, where the node X has three parents U, V, and W, and three children Y, Z, and Q, and the child Y receives evidence e_Y. The node Y updates its belief and sends the message $\lambda_Y(X)$ to X. The node X updates its belief upon receiving the message from Y and, in turn, sends the causal messages $\pi_Z(X)$ and $\pi_Q(X)$ to Z and Q respectively to help update their beliefs.

The messages $\pi_Z(X)$ and $\pi_Q(X)$ are $\dfrac{Bel_{new}(X)}{\lambda(Z)}$ and $\dfrac{Bel_{new}(X)}{\lambda(Q)}$ computed as above in the case of upward propagation in a tree fragment. In the following, we show how U, V, and W update their beliefs upon receiving the messages $\lambda_X(U)$, $\lambda_X(V)$, and $\lambda_X(W)$ respectively from their common child X.

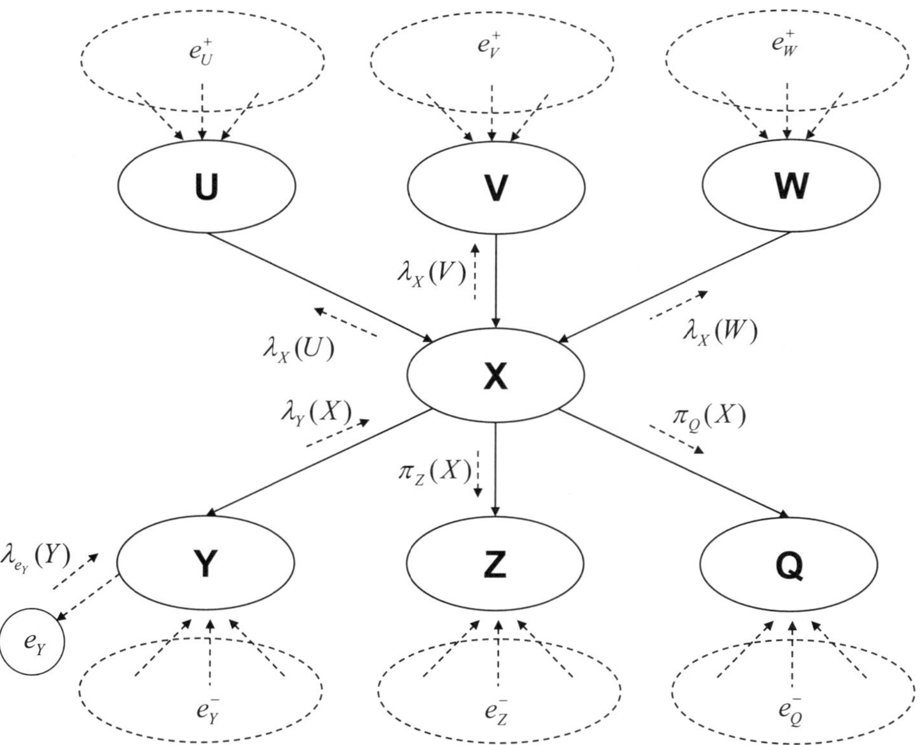

Figure 6-23: Upward propagation of evidence e_Y in a polytree fragment

The revised $\lambda(U)$ can now be computed as follows:

$$
\begin{aligned}
\lambda_{new}(U) &= p(e_Y, e_Y^-, e_Z^-, e_Q^-, e_V^+, e_W^+ \mid U) \\
&= \sum_X \sum_V \sum_W p(e_Y, e_Y^-, e_Z^-, e_Q^-, e_V^+, e_W^+ \mid U, V, W, X) p(V, W, X \mid U) \\[2ex]
&= \sum_X \sum_V \sum_W p(e_Y, e_Y^-, e_Z^-, e_Q^- \mid e_V^+, e_W^+, U, V, W, X) \\
&\qquad\qquad p(e_V^+, e_W^+ \mid U, V, W, X) p(V, W, X \mid U) \\[2ex]
&= \sum_X \sum_V \sum_W p(e_Y, e_Y^-, e_Z^-, e_Q^- \mid X) p(e_V^+ \mid e_W^+, U, V, W, X) \\
&\qquad\qquad p(e_W^+ \mid U, V, W, X) p(V, W, X \mid U) \\
&\qquad \left[\text{since } X \text{ seprates } e_Y, e_Y^-, e_Z^-, e_Q^- \text{ from } e_V^+, e_W^+, U, V, W \right] \\
&= \sum_X \sum_V \sum_W \lambda_{new}(X) p(e_V^+ \mid e_W^+, V, W) p(e_W^+ \mid V, W) p(V, W, X \mid U) \\
&\qquad \left[\begin{aligned} &\text{since } V \text{ seprates } e_V^+ \text{ from } U, X, \text{ and} \\ &W \text{ separates } e_W^+ \text{ from } U, X \end{aligned} \right] \\
&= \sum_X \lambda_{new}(X) \sum_V \sum_W p(e_V^+, e_W^+ \mid V, W) p(X \mid V, W, U) p(V, W \mid U)
\end{aligned}
$$

Since U, V, and W are marginally independent, $\lambda_{new}(U)$ can further be simplified as follows:

$$
\begin{aligned}
\lambda_{new}(U) &= \alpha \sum_X \lambda_{new}(X) \sum_V \sum_W \frac{p(V, W \mid e_V^+, e_W^+)}{p(V, W)} p(X \mid V, W, U) p(V, W) \\
&= \alpha \sum_X \lambda_{new}(X) \sum_V \sum_W p(V \mid W, e_V^+, e_W^+) p(W \mid e_V^+, e_W^+) p(X \mid V, W, U) \\
&= \alpha \sum_X \lambda_{new}(X) \sum_V \sum_W p(V \mid e_V^+) p(W \mid e_W^+) p(X \mid V, W, U) \\
&= \alpha \sum_X \lambda_{new}(X) \sum_V \sum_W \pi_X(V) \pi_X(W) p(X \mid V, W, U)
\end{aligned}
$$

Therefore, the message $\lambda_X(U)$, which will be sent to U from X as the new λ vector for U, is the above expression computed at X. Note that $p(X \mid V, W, U)$ is the CPT of X stored at the node X. The revised belief of U is obtained by multiplying its unchanged π vector with the above λ vector. The messages that are sent to V and W are the following:

$$\lambda_X(V) = \alpha \sum_X \lambda_{new}(X) \sum_U \sum_W \pi_X(U)\pi_X(W)p(X\,|\,U,W,V)$$

$$\lambda_X(W) = \alpha \sum_X \lambda_{new}(X) \sum_U \sum_V \pi_X(U)\pi_X(V)p(X\,|\,U,V,W)$$

Example

Consider the network fragment shown in Figure 6-24 along with the π, λ, and belief vectors, where the two CPTs for the nodes *Ticket Sale* and *Concession* are same as the two CPTs in Table 6-7 and Table 6-8, respectively, in our main belief network example.

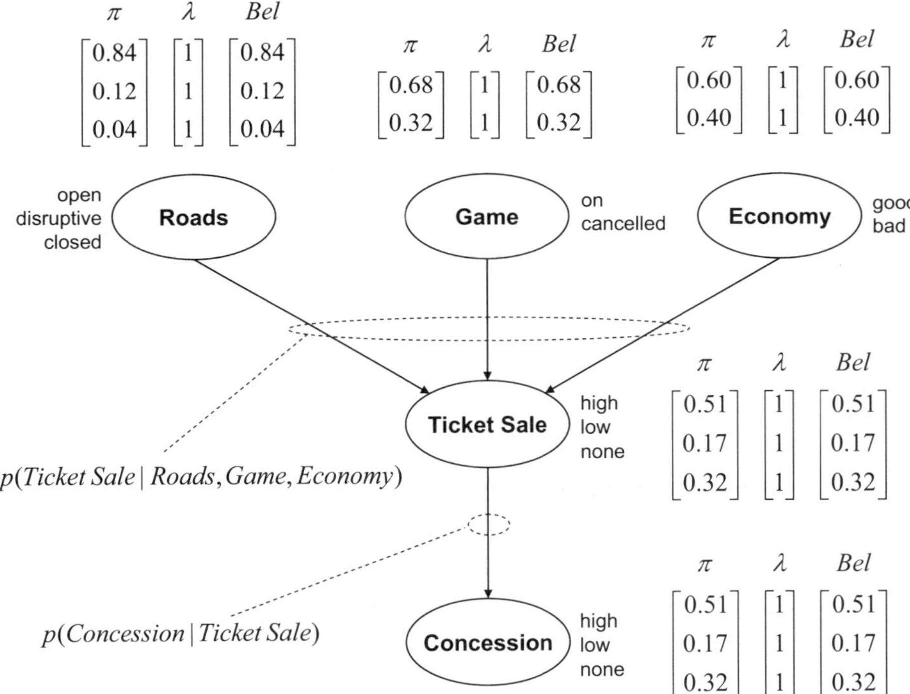

Figure 6-24: Example polytree fragment

The upward propagation of evidence due to the posting of evidence $e_{Concession}$ at the node *Concession* is shown in Figure 6-25.

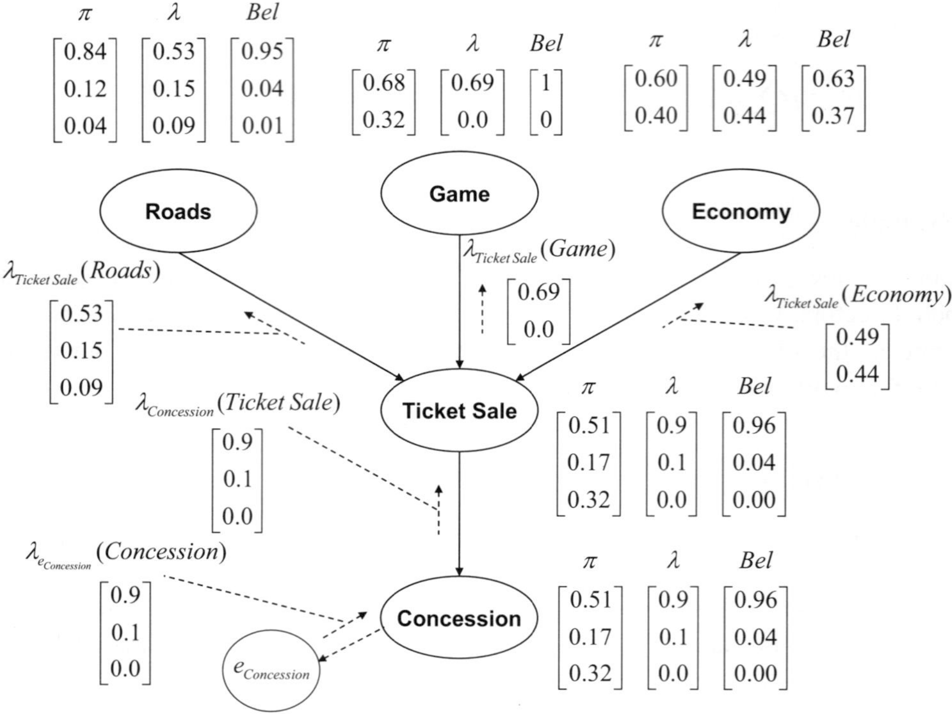

Figure 6-25: Example upward propagation in a polytree fragment

6.6.6   *Downward Propagation in a Polytree Fragment*

This case is illustrated in Figure 6-26, where the node X has three parents U, V, and W, and three children Y, Z, and Q, and the parent U receives evidence e_U. Node U updates its belief and sends the causal message $\pi_X(U)$ to X. Node X updates its belief upon receiving the message from Y and, in turn, sends the messages $\pi_Y(X)$, $\pi_Z(X)$ and $\pi_Q(X)$ to Y, Z and Q, respectively, to help update

their beliefs. The messages $\pi_Y(X)$, $\pi_Z(X)$, and $\pi_Q(X)$ are $\dfrac{Bel_{new}(X)}{\lambda(Y)}$,

$\dfrac{Bel_{new}(X)}{\lambda(Z)}$, and $\dfrac{Bel_{new}(X)}{\lambda(Q)}$, respectively, computed as above in the case of

upward propagation in a tree fragment. Next, we show how V and W update their beliefs upon receiving respectively the messages $\lambda_X(V)$ and $\lambda_X(W)$ from their common child X.

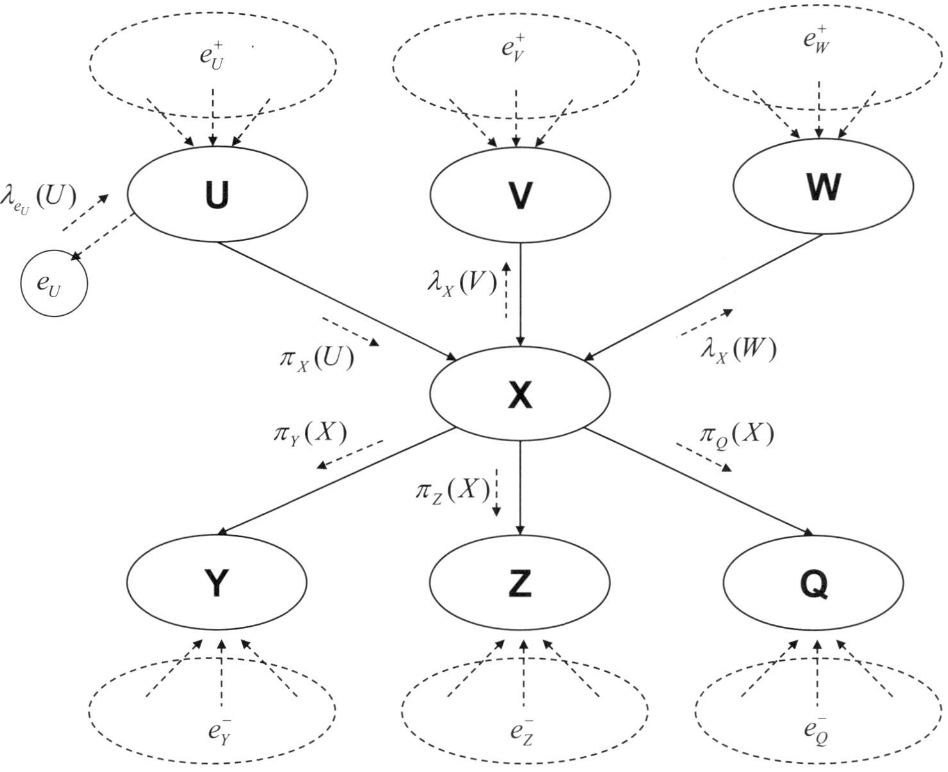

Figure 6-26: Downward propagation of evidence e_U in a polytree fragment

$$\lambda_{new}(V) = p(e_U, e_U^+, e_W^+, e_Y^-, e_Z^-, e_Q^- \mid V)$$

$$= \sum_X \sum_U \sum_W p(e_U, e_U^+, e_W^+, e_Y^-, e_Z^-, e_Q^- \mid V, U, W, X) p(U, W, X \mid V)$$

$$= \sum_X \sum_U \sum_W p(e_Y^-, e_Z^-, e_Q^- \mid e_U, e_U^+, e_W^+, V, U, W, X)$$
$$p(e_U, e_U^+, e_W^+ \mid V, U, W, X) p(U, W, X \mid V)$$

$$= \sum_X \sum_U \sum_W p(e_Y^-, e_Z^-, e_Q^- \mid X) p(e_U \mid e_U^+, e_W^+, V, U, W, X)$$
$$p(e_U^+, e_W^+ \mid V, U, W, X) p(U, W, X \mid V)$$
$$\left[\text{since } X \text{ seprates } e_Y^-, e_Z^-, e_Q^- \text{ from } e_U, e_U^+, e_W^+, U, V, W\right]$$

$$= \sum_X \sum_U \sum_W \lambda(X) p(e_U \mid U) p(e_U^+, e_W^+ \mid U, W, V) p(U, W, X \mid V)$$

$$\left[\begin{array}{l} \text{since } U \text{ seprates } e_U \text{ from } e_U^+, e_W^+, X, V, W, \text{and} \\ \text{since } U, W \text{ seprate } e_U^+, e_W^+ \text{ from } V, X \end{array} \right]$$

$$= \sum_X \lambda(X) \sum_U \sum_W \lambda_{e_U}(U) p(e_U^+ \mid U)$$

$$p(e_W^+ \mid W) p(U, W \mid V) p(X \mid U, W, V)$$

$$\left[\text{since } e_U^+, e_W^+ \text{ are independent of each other given } U \text{ or } W \right]$$

$$= \alpha \sum_X \lambda(X) \sum_U \sum_W \lambda_{e_U}(U) \pi_X(U) \pi_X(W) p(X \mid U, W, V)$$

$$\left[\text{since } U \text{ and } W \text{ are marginally independent} \right]$$

Note that if X did not receive any diagnostic evidence from its descendants Y, Z, and Q, then the λ vector for X would still be $(1,1,...,1)$. In this case, the above message sent to V from X due to the evidence on U would still be $(1,1,...,1)$, making no impact on the belief of V. This is consistent with the network marginal independence property, which says that U, V, and W are independent of each other if neither X nor any of its descendants received evidence. This is illustrated below.

Suppose $x_1,...,x_n$ (n states) are all possible instantiations of X. Then, from the CPT of X, U, V, W, we have for any u, v, w:

$$\sum_i p(x_i \mid u, v, w) = 1$$

If the variable X did not receive any evidence from its descendants then $\lambda(x_i) = 1$, for every i. If the variable V has m states and $v_1,...,v_m$ are all possible instantiations of V, then from the derivation of $\lambda_{new}(V)$ above,

$$\lambda_X(v_j) = \alpha \sum_i \lambda(x_i) \sum_U \sum_W \lambda_{e_U}(U) \pi_X(U) \pi_X(W) p(x_i \mid U, W, v_j)$$

$$= \alpha \sum_i \sum_U \sum_W \lambda_{e_U}(U) \pi_X(U) \pi_X(W) p(x_i \mid U, W, v_j)$$

$$= \alpha \sum_U \sum_W \lambda_{e_U}(U) \pi_X(U) \pi_X(W) \sum_i p(x_i \mid U, W, v_j)$$

$$= \alpha \sum_U \sum_W \lambda_{e_U}(U) \pi_X(U) \pi_X(W)$$

Therefore, each $\lambda_X(v_j)$ has the same value, making the vector $\lambda_X(V)$ a unit vector that does not change the belief of V.

Example

Consider the network fragment shown in Figure 6-27 along with the π, λ, and belief vectors, where the two CPTs for the nodes *Ticket Sale* and *Concession* are same as the two CPTs in Table 6-7 and Table 6-8, respectively, in our main belief network example.

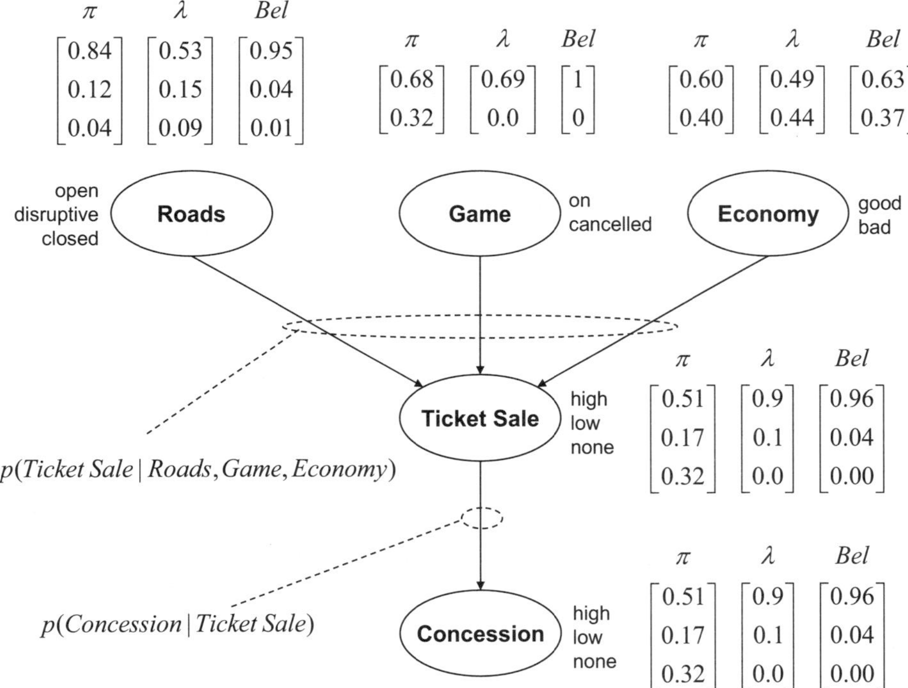

Figure 6-27: Example polytree fragment

The downward propagation of evidence due to the posting of evidence e_{Roads} at the node *Roads* is shown in Figure 6-28.

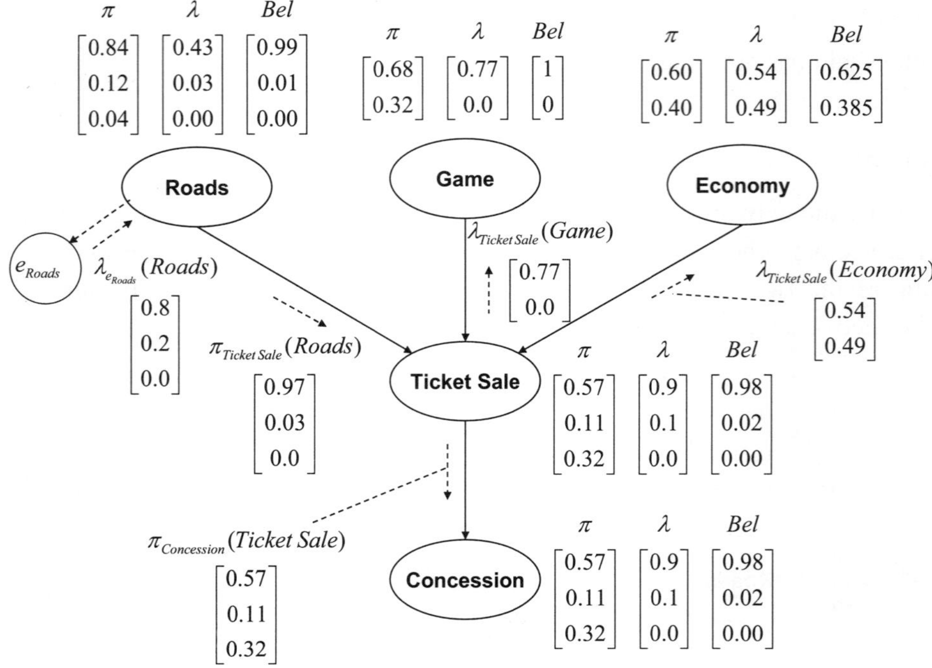

Figure 6-28: Example downward propagation in a polytree fragment

6.6.7 Propagation Algorithm

Now that we have illustrated different belief propagation with a series of examples, we generalize these steps to a concrete algorithm for belief propagation in polytrees.

Input

- Belief network $\mathcal{N}$ (causal structure with associated CPTs)
- Evidence e_X on the variable X of $\mathcal{N}$

Output

- Revised belief vector for each node, i.e., $p(Y \mid e_X)$, for each node Y

**Node Structure

- Each node X in $\mathcal{N}$ with p number of states stores the following information locally in a suitable data structure (as shown in Figure 6-29):

  - p-ary π vector $\pi(X)$

  - p-ary λ vector $\lambda(X)$

  - p-ary belief vector $Bel(X)$

  - p-ary evidence vector $\lambda_e(X)$

  - $p \times n$ CPT $p(X \mid U_1,...,U_n)$, if X has n parents $U_1, U_2,...,U_n$

  - q-ary parent π vector $\pi_X(U_i)$, for each parent U_i with q number of states

  - p-ary child λ vector $\lambda_{Y_i}(X)$, for each child Y_i

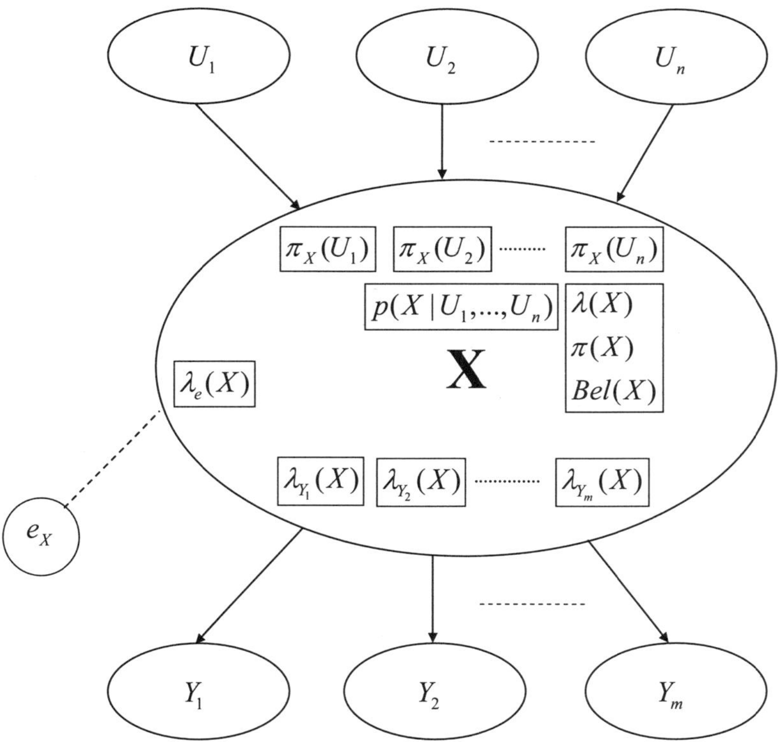

Figure 6-29: Node structure

Initial Probability Computation

- For each node X in $\mathcal{N}$ with p number of states, perform the following:

- If X is a root node, then set each of its π and belief vectors to its prior probability distribution
- Set the evidence vector $\lambda_e(X)$, λ vector $\lambda(X)$, and each child λ vector $\lambda_{Y_i}(X)$ to $(1,...,1)$

- Mark each root node and then recursively compute the π and belief vectors of the rest of the nodes through the following steps

- If the parents of a node X are already marked, then perform the following:
    - Set each parent π vector $\pi_X(U_i)$ to the π vector of the parent U_i

    - Set its π vector $\pi(X)$ to $\displaystyle\sum_{u_1,u_2,...,u_n} p(X\,|\,u_1,u_2,...,u_n)\prod_{i=1}^{n}\pi_X(u_i)$, if X has parents $U_1,U_2,...,U_n$

    - Set its belief vector $Bel(X)$ equal to its π vector

Evidence Propagation

- Set the evidence vector $\lambda_e(X)$ of X to the likelihood vector $p(e_X\,|\,X)$

- Revise the beliefs of X:
    - Compute the λ vector $\lambda(X)$ of X as the product of all its child λ vectors and the evidence vector
    - Compute the π vector $\pi(X)$ of X as the product of all its parent π vectors
    - Compute the belief vector of X as the product of its π and λ vectors

- Generate messages from X on the update of its evidence vector:
    - Send a message
    $$\alpha\sum_{x}\lambda(x)\sum_{u_1,...,u_{i-1},u_{i+1}...,u_n} p(x\,|\,u_1,...,u_{i-1},u_{i+1},...,u_n,U_i)\prod_{k\neq i}\pi_X(u_k)$$
    to each parent U_i of X

    - Send a message $\alpha\dfrac{Bel(X)}{\lambda_{Y_i}(X)}$ to each child Y_i of X

- Generate messages from X on the update of the λ_{Y_j} vector:

- Send a message

$$\alpha \sum_{x} \lambda(x) \sum_{u_1,\dots,u_{i-1},u_{i+1}\dots,u_n} p(x\,|\,u_1,\dots,u_{i-1},u_{i+1},\dots,u_n,U_i)\prod_{k\neq i} \pi_X(u_k)$$

to each parent U_i of X

- Send a message $\alpha\dfrac{Bel(X)}{\lambda_{Y_i}(X)}$ to each child Y_i of X other than the child Y_j

- Generate messages from X on the update of the $\pi_X(U_j)$ vector:

  - Send a message

$$\alpha \sum_{x} \lambda(x) \sum_{u_1,\dots,u_{i-1},u_{i+1}\dots,u_n} p(x\,|\,u_1,\dots,u_{i-1},u_{i+1},\dots,u_n,U_i)\prod_{k\neq i} \pi_X(u_k)$$

  to each parent U_i of X other than U_j

  - Send a message $\alpha\dfrac{Bel(X)}{\lambda_{Y_i}(X)}$ to each child Y_i of X

If a network already received evidence on some of its variables, then only the evidence propagation step is followed, bypassing the initial probability computation step.

6.7   Evidence Propagation in Directed Acyclic Graphs

The evidence propagation algorithm presented in the last section cannot usually handle DAGs (such as the one in Figure 6-1) because the evidence propagation on a variable in a DAG that is not a polytree may never terminate due to the recursive computation in the algorithm. This is explained in the context of the DAG shown in Figure 6-30.

Suppose no evidence has been posted to the variable *Weather*, making the two variables *Roads* and *Field* marginally independent. Assume further that the variable *Game* received some evidence, making *Roads* and *Field* conditionally dependent on each other. Now, if evidence is observed on *Roads*, then *Game* will receive causal evidence and will send diagnostic evidence to the *Field*. The node *Roads* then receives evidence from *Field* via *Weather*. This cyclic process of evidence propagation continues unless a stopping criterion is put in place based on repeated evidence and changes in node belief. A similar cyclic process continues in the other direction when *Roads* send diagnostic evidence to *Weather* upon receiving evidence on itself.

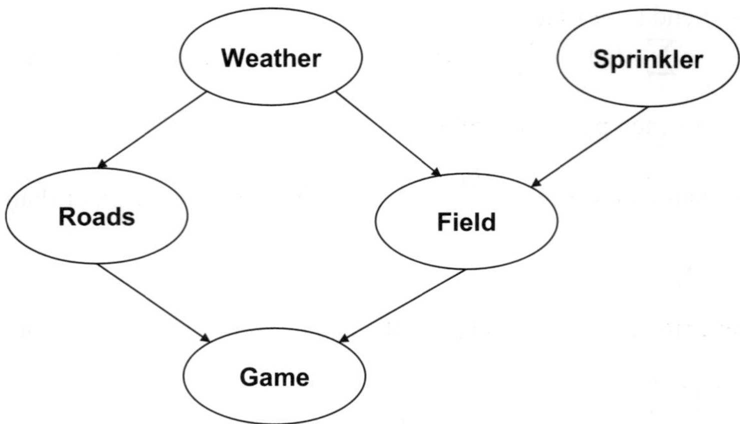

Figure 6-30: Example DAG which is not a polytree

A solution to this problem is to deal with the joint probability distribution of the variables in the network. Although belief network technology reduces the joint probability distribution of a network to merely the products of a few CPTs, we have ruled out this approach due to the large number of entries that need to be computed for a belief revision. Even for the small network in Figure 6-30, there are 72 entries in its joint probability distribution table that need to be computed using the following formula each time a node receives evidence. (Only the initial letter has been used from each variable name)

For example, *Roads* receives evidence e_{Roads} :

$$p(W,S,R,F,G \mid e_{Roads}) = p(W)p(S)p(R \mid W)p(F \mid W,S)p(G \mid R,F)p(e_{Roads} \mid R)$$

To make use of the simple and elegant message passing propagation algorithm while avoiding large joint probability distributions, one approach is to transform a DAG into a polytree by identifying and clustering loops into single nodes. Thus a node in a transformed cluster network is a set of nodes (also called a *clique*) instead of a single node, and joint probability distributions are computed in smaller chunks locally at the network nodes. For example, the network in Figure 6-30 is transformed into the network (undirected) in Figure 6-31 with two cliques.

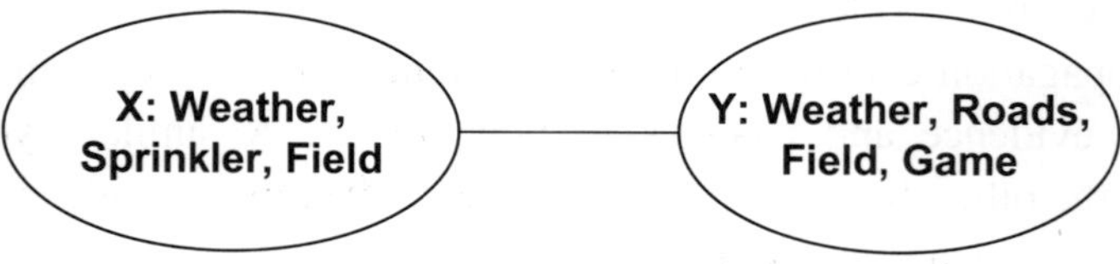

Figure 6-31: Transformed Network

Evidence on a variable in the original network is posted to the nodes in the transformed network containing the variable. The joint probability distributions are computed at the nodes, where evidence is posted and messages are passed to neighboring nodes. For example, if an evidence on the variable *Weather* is observed in the cluster network in Figure 6-31, then the joint probability distribution $p(W,R,F,G)$ is computed at the node Y using some of the CPTs. *Field* and *Weather* are the only variable in the intersection of X and Y. Thus Y sends the change in the joint probability distribution of *Field* and *Weather* as a message to the variable X, which then computes the joint probability distribution of its three variables and derives the new belief of the variable *Sprinkler* via marginalization.

But in a complex network, a cycle could be very large and therefore local computation of joint probability distribution is still impractical. A solution is to make smaller size clusters. Since the joint probability distribution of a network is derived from the CPTs and the CPT of a variable involves only its immediate neighbors, clusters around variables can be formed using only their neighboring nodes. One such cluster network constructed from the network in Figure 6-30 is shown in Figure 6-32.

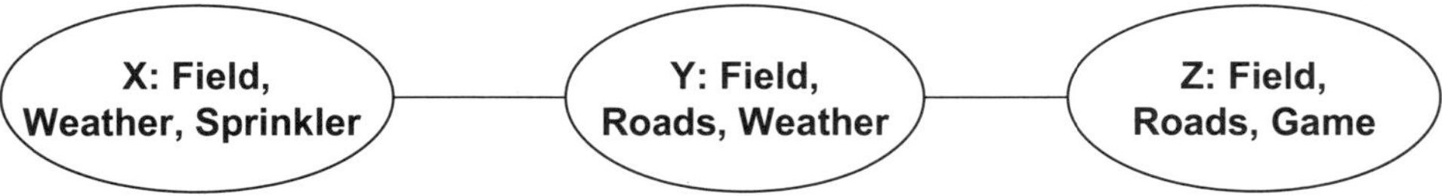

Figure 6-32: Cluster network

As before, if evidence on a variable, such as *Roads*, is observed, then it is posted to one of two nodes Y and Z (say, Y) containing the variable *Roads*. Each of these nodes then computes the joint distribution locally to find the beliefs of the individual variables. The node Y then passes the revised joint probability distributions of the variables *Field* and *Weather* to X (respectively, *Field* and *Roads* to Z) to help compute the revised belief for the node *Sprinkler* (respectively, *Game*) given the evidence on *Roads*.

In the following subsections, we present the *junction tree* algorithm for dealing with DAGs in general. This algorithm systematically constructs a cluster network from a belief network called a *junction tree* (or *clique tree* or *join tree*). The stages of the algorithm are shown on the left side of Figure 6-33.

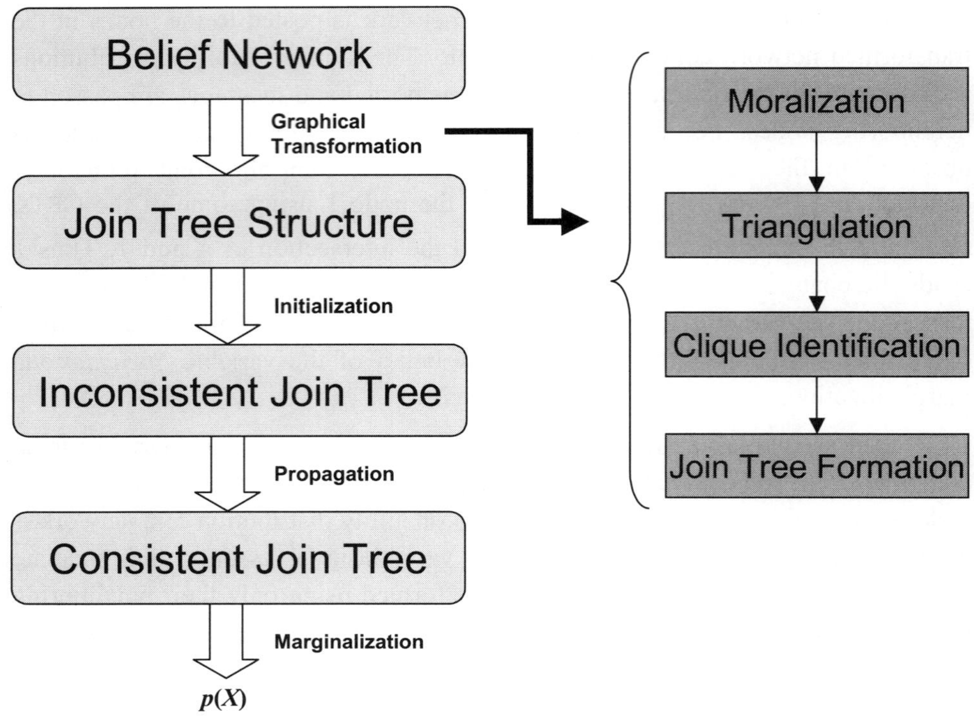

Figure 6-33: Steps for constructing join trees

The first stage, which is the graphical transformation stage, is expanded on the right side of the figure. This stage consists of four steps: moralization, triangulation, clique identification, and join tree formation. These steps construct a join tree from a belief network. The subsequent stages of the algorithm then compute prior beliefs of the variables in the network via an initialization of the join tree structure, followed by propagation and marginalization. The final subsection discusses how to handle evidence by computing the posterior beliefs of the variables in the network.

6.7.1 Graphical Transformation

The *moral graph* of a DAG is obtained by adding a link between any pair of variables with a common child, and dropping the directions of the original links in the DAG.

Example

The moral graph of the network in Figure 6-1 is shown in Figure 6-34. The dotted lines in the network in Figure 6-34 are the links added to the original network. For example, the nodes *Roads* and *Economy* have a common child *Ticket Sale*, and therefore are linked with a dotted line.

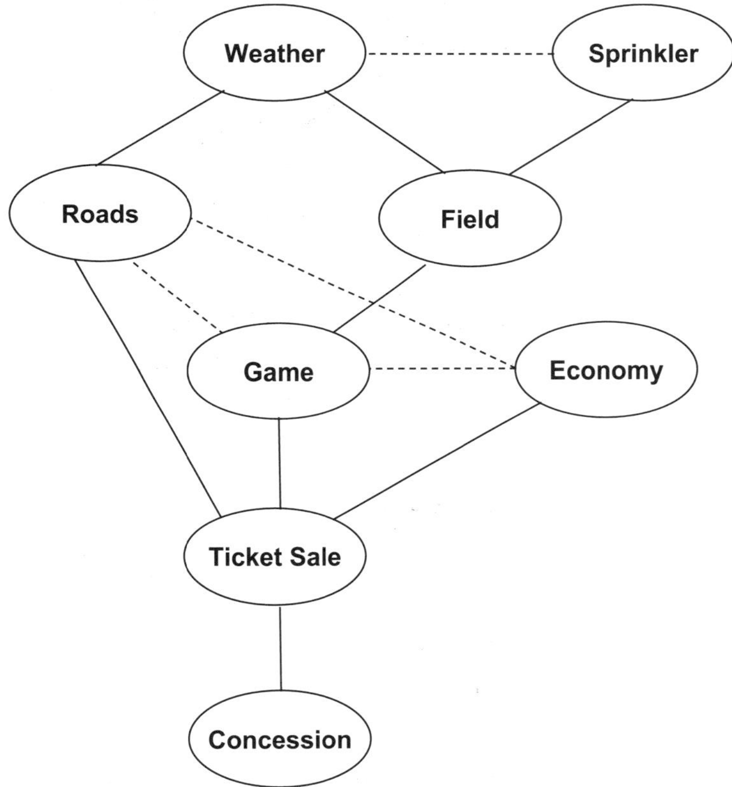

Figure 6-34: The moral graph of the network in Figure 6-1 (added links are indicated by dotted lines)

An undirected graph is *triangulated* if any cycle of length greater than 3 has a chord, that is, an edge joining two non-consecutive nodes along the cycle.

Example

The triangulated graph of the graph in Figure 6-34 is shown in Figure 6-35. The only link added by triangulation is the link between the nodes *Roads* and *Field*. These two nodes are two non-consecutive nodes along the cycle *Weather – Roads – Game – Field* of length 4.

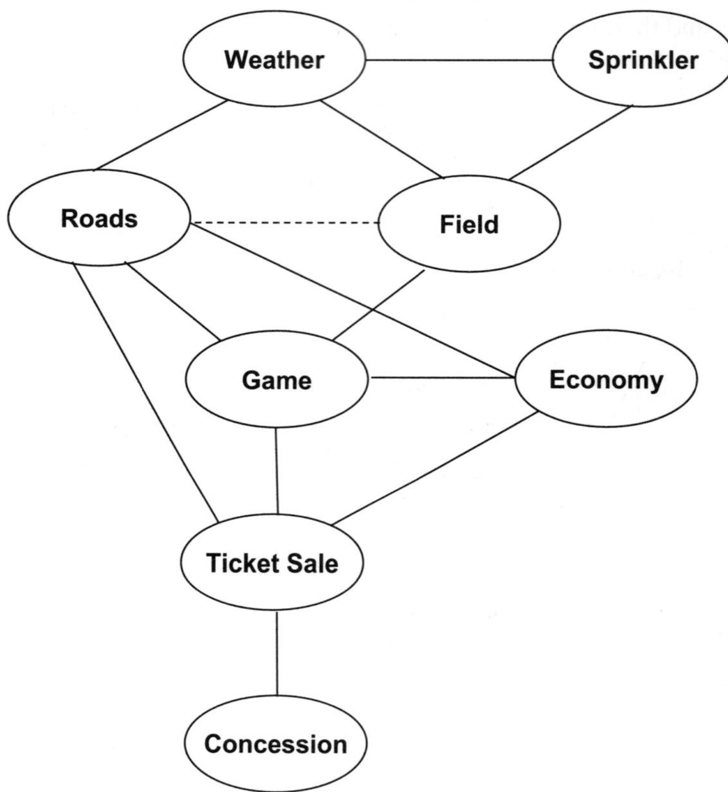

Figure 6-35: A triangulated graph of the network in Figure 6-34 (new link is indicated by dotted lines)

Note that there are, in general, more than one triangulation of a given graph. In the case of the graph in Figure 6-34, we could have added a link between the nodes *Weather* and *Game*, instead of between the nodes *Roads* and *Field* along the cycle *Weather – Roads – Game – Field* , yielding a different triangulation.

The nodes of a join tree for a graph are the *cliques* in the graph (maximal sets of variables that are all pairwise linked).

Example

The five cliques in the graph in Figure 6-35 are listed below:

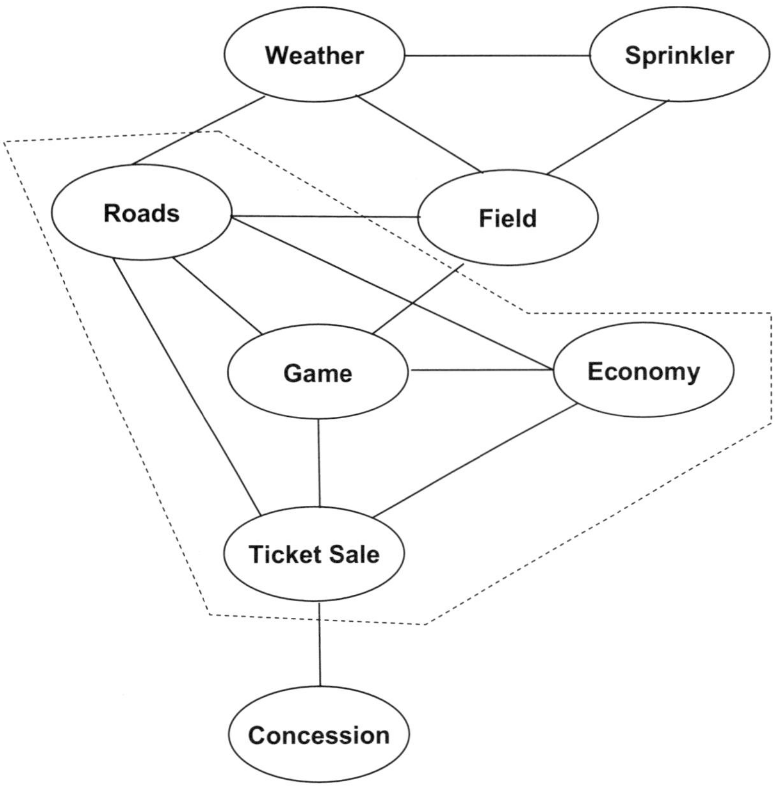

Figure 6-36: An example clique (surrounded by dotted lines) in the network in
Figure 6-35

C_1: {*Weather, Roads, Field*}

C_2: {*Weather, Sprinkler, Field*}

C_3: {*Roads, Field, Game*}

C_4: {*Roads, Game, Economy, Ticket Sale*}

C_5: {*Ticket Sale, Concession*}

The clique C_4 is shown in Figure 6-36, surrounded by dotted lines. Note that
the set {*Weather, Sprinkler, Roads, Field*} is not a clique because its nodes are
not all pairwise connected since *Roads* and *Sprinkler* are not connected. Though
the nodes in the set {*Roads, Game, Economy*} are pairwise connected, it is not a

clique because the set is contained in another clique {*Roads*, *Game*, *Economy*, *Ticket Sale*}, and it is therefore not maximal.

The triangulation and clique identification techniques described above are combined into a formal two-step recursive algorithm for systematically identifying cliques from an arbitrary undirected moral graph:

- Select a node X from the network $\mathcal{N}$ and make the cluster consisting of the node and its neighbor complete by adding the necessary edges. Choose the node that causes the least number of edges to be added. Break ties by choosing the node that induces the cluster with smallest weight, where the *weight* of a node is the number of states, and the weight of a cluster is the product of the weights of its constituent nodes.

- Remove X along with its edges (forming a clique) and repeat the previous step if there are still nodes left. Otherwise, if there are no nodes left, $\mathcal{N}$ is now triangulated. Note that a clique is formed in this stage only if it is not a subset of a subsequently induced clique.

Example

The graph in Figure 6-37 shows the triangulated graph obtained from the moral graph in Figure 6-34. The table in the figure shows the variable selection ordering in the first step of the algorithm. In each selection, the weight of the variable is computed and the induced cluster or clique is identified if it is bold and shaded.

The candidate variables to be selected first are *Sprinkler*, *Concession*, *Ticket Sale*, and *Economy* since, for each of these four variables, the cluster consisting of the variable and its neighbors is already complete without the requirements for adding any edge. For example, the cluster consisting of the variable *Sprinkler* and its two neighbors (*Weather* and *Field*) is complete, but the cluster consisting of the variable *Weather* and its two neighbors (*Roads* and *Field*) is not complete because an edge between *Roads* and *Field* needs to be added. The variable *Concession* is chosen among the four candidate variables because its weight 9 is the least among the four weights 12, 9, 108, and 36 for the four candidate variables respectively. The variable *Concession* is then removed along with the edge from it leading to *Ticket Sale*. The process then continues with the remainder of the network and the variable *Sprinkler* is chosen next. When the variable *Economy* is chosen, the corresponding induced cluster is not identified as a clique because it is a subset of a clique that is already identified.

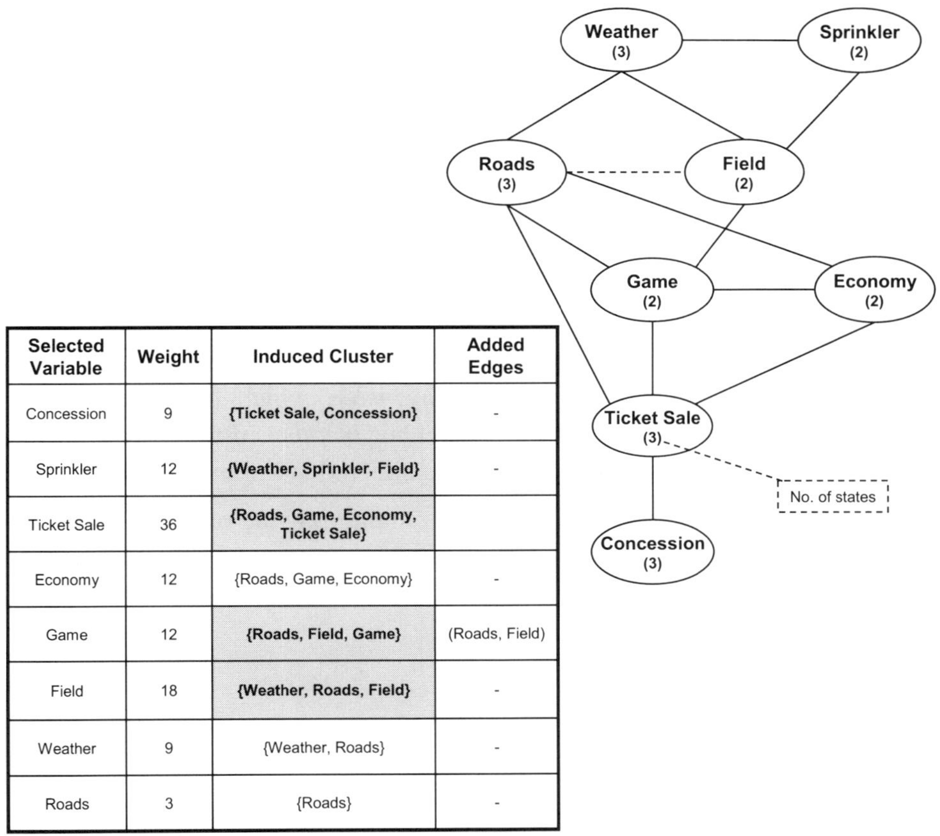

Selected Variable	Weight	Induced Cluster	Added Edges
Concession	9	{Ticket Sale, Concession}	-
Sprinkler	12	{Weather, Sprinkler, Field}	-
Ticket Sale	36	{Roads, Game, Economy, Ticket Sale}	
Economy	12	{Roads, Game, Economy}	-
Game	12	{Roads, Field, Game}	(Roads, Field)
Field	18	{Weather, Roads, Field}	-
Weather	9	{Weather, Roads}	-
Roads	3	{Roads}	-

Figure 6-37: Triangulation and clique identification

The network is now left with the four variables *Weather*, *Roads*, *Field*, and *Game*, and the selection of any of these will require an edge to be added. The two candidate variables are *Field* and *Game* as each of these has weight 12, and each of the remaining two variables *Weather* and *Roads* has weight 18. The selection of the variable *Field* needs the edge from *Weather* to *Game* to be added, and the selection of the variable *Game* needs the edge from *Roads* to *Field* to be added.

Once we have identified cliques, we rank the nodes by assigning numbers to systematically construct a join tree out of cliques. Nodes are numbered from 1 to *n* in increasing order by assigning the next number to the node with the largest set of previously numbered neighbors. For example, an ordering of the nodes of the graph in Figure 6-1 is given below:

1 – *Weather*

2 – *Sprinkler*

3 – *Field*

4 – *Roads*

5 – *Game*

6 – *Economy*

7 – *Ticket Sale*

8 – *Concession*

The choice between the two nodes *Weather* and *Sprinkler* as a starting node is arbitrary to break the tie. The node *Field* comes before the node *Roads* because *Field* has two neighbors, *Weather* and *Sprinkler*, which have already been numbered. On the other hand, only one neighbor *Weather* of the node *Roads* has been numbered. The rest of the sequence is numbered in a similar manner.

To form a join tree based on the above ranking scheme of graph nodes, first order the cliques of the graph by rank of the highest vertex of each clique. For example, the set of five cliques C_1, C_2, C_3, C_4, and C_5 in Figure 6-35 is as follows:

C_5: {*Ticket Sale, Concession*}

C_4: {*Roads, Game, Economy, Ticket Sale*}

C_3: {*Roads, Field, Game*}

C_1: {*Weather, Roads, Field*}

C_2: {*Weather, Sprinkler, Field*}

For example, the clique C_5 comes before the clique C_4 because C_5 contains the node *Ticket Sale* whose rank is higher than the rank of any node in C_4.

The join tree from an ordered set of cliques C_1, C_2, ..., C_n are formed by connecting each C_i to a predecessor C_j sharing the highest number of vertices with C_i. For example, a join tree for the above ordered set of cliques C_5, C_4, C_3, C_1, C_2 is shown in Figure 6-38. Each edge in the tree is labeled with the intersection of the adjacent cliques, which is called a *separator set* or a *sepset*.

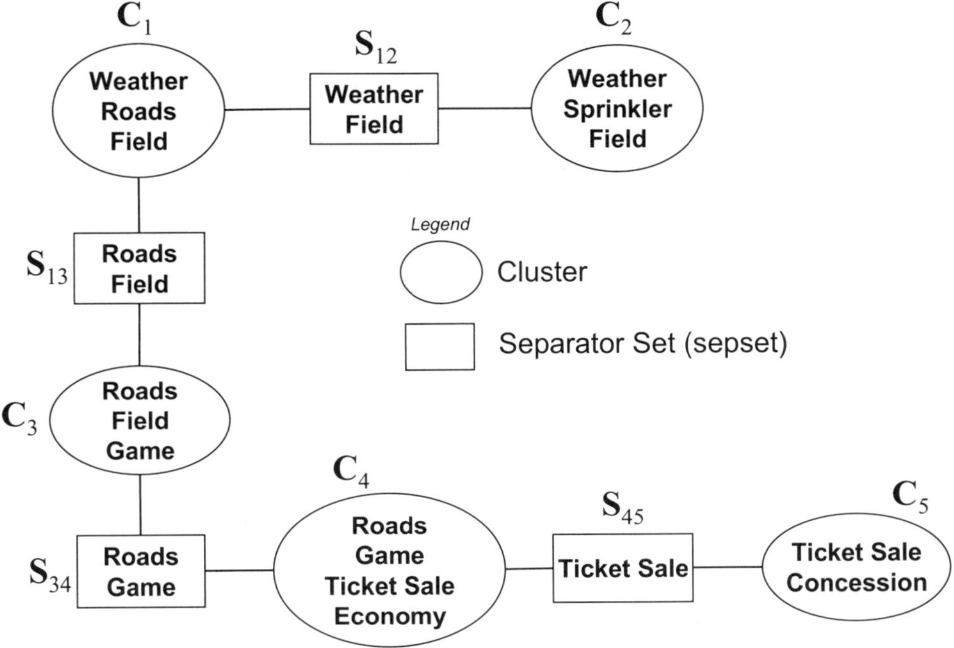

Figure 6-38: Join tree for the graph in Figure 6-35

The following algorithm helps systematically construct a join tree as shown in Figure 6-38. The algorithm is optimal with respect to the mass and cost criteria adopted during the selection of sepsets.

- Create a forest of n distinct trees, where each tree consists of only one node made out of the set of n cliques produced by the triangulation and clique identification procedure above. Also, create a set $\mathbf{S}$ of $n(n-1)$ distinct sepsets obtained by intersecting all possible pairs of distinct elements in the set of cliques.

- Repeat the following steps $n-1$ times:

  - Select a sepset $\mathbf{S}_{XY}$ (that is, $\mathbf{X} \cap \mathbf{Y}$) from $\mathbf{S}$ that has the largest mass, where mass of a sepset $\mathbf{S}_{XY}$ is the number of variables it contains. If two or more sepsets of equal mass can be chosen then choose the candidate sepset with the smallest cost, where the *cost of a sepset* $\mathbf{S}_{XY}$ is the sum of the product of the number of states of the variables in $\mathbf{X}$ and product of the number of states of the variables in $\mathbf{Y}$.

  - Insert the selected sepset $\mathbf{S}_{XY}$ between the cliques $\mathbf{X}$ and $\mathbf{Y}$ only if $\mathbf{X}$ and $\mathbf{Y}$ are on different trees in the forest.

In the construction of the join tree in Figure 6-38, first the forest is formed containing all five cliques C_1, C_2, C_3, C_4, and C_5. Each of the sepsets S_{12}, S_{13}, and S_{34} has a mass 2 and weight 6, are therefore inserted first into the join tree. Note that the sepset $\{Field\}$ was not inserted between C_2 and C_3 before S_{12} or S_{13} as $\{Field\}$ has lower mass than each of S_{12} or S_{13}.

6.7.2   *Join Tree Initialization*

A join tree maintains a joint probability distribution at each node, cluster, or sepset) in terms of a belief *potential*, which is a function that maps each instantiation of the set of variables in the node into a real number. The belief potential of a set $\mathbf{X}$ of variables will be denoted as $\varphi_{\mathbf{X}}$, and $\varphi_{\mathbf{X}}(x)$ is the number that the potential maps x onto. The probability distribution of a set $\mathbf{X}$ of variables is just the special case of a potential whose elements add up to 1. In other words,

$$\sum_{x \in \mathbf{X}} \varphi_{\mathbf{X}}(x) = \sum_{x \in \mathbf{X}} p(x) = 1$$

The marginalization and multiplication operations on potentials are defined in a manner similar to the same operations on probability distributions.

Belief potentials encode the joint distribution $p(\mathbf{X})$ of the belief network according to the following:

$$p(\mathbf{X}) = \frac{\prod_i \phi_{C_i}}{\prod_j \phi_{S_j}}$$

where φ_{C_i} and φ_{S_j} are the cluster and sepset potentials respectively. We have the following joint distribution for the join tree in Figure 6-38:

$$p(W,S,R,F,G,E,T,C) = \frac{\phi_{C_1} \phi_{C_2} \phi_{C_3} \phi_{C_4} \phi_{C_5}}{\phi_{S_{12}} \phi_{S_{13}} \phi_{S_{34}} \phi_{S_{45}}}$$

$$= \frac{\phi_{WRF} \phi_{WSF} \phi_{RFG} \phi_{RGTE} \phi_{TC}}{\phi_{WF} \phi_{RF} \phi_{RG} \phi_{T}}$$

It is imperative that a cluster potential agrees on the variables in common with its neighboring sepsets up to marginalization. This imperative is formalized by the concept of local consistency. A join tree is *locally consistent* if for each cluster $\mathbf{C}$ and neighboring sepset $\mathbf{S}$, the following holds:

$$\sum_{C \setminus S} \phi_C = \phi_S$$

To start initialization, for each cluster **C** and sepset **S**, set the following:

$$\phi_C \leftarrow 1, \quad \phi_S \leftarrow 1$$

Then assign each variable X to a cluster **C** that contains X and its parents $pa(X)$. Then set the following:

$$\phi_C \leftarrow \phi_C \, p(X \mid pa(X))$$

Example

To illustrate the initialization process, consider the join tree in Figure 6-38. The allocation of prior and conditional probability tables are shown in Figure 6-39. The variable *Field* has been assigned to cluster C_2 as it contains the variable and its parents *Weather* and *Sprinkler*. The probability table $p(Weather)$ could have been assigned to any of C_1 and C_2, but C_1 is chosen arbitrarily.

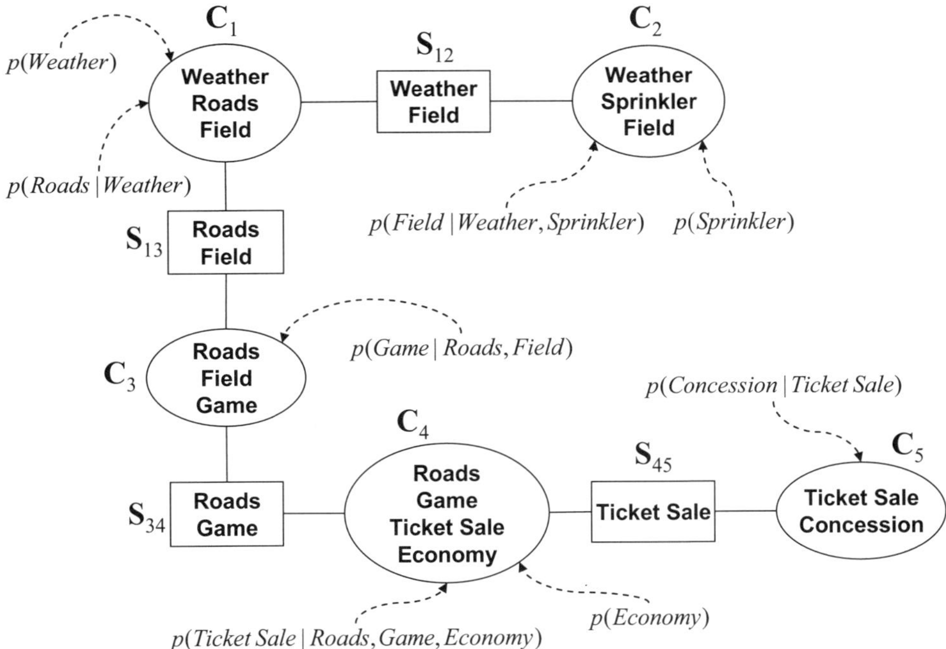

Figure 6-39: Allocation of prior and conditional probability tables for join tree initialization

As an example, Figure 6-40 shows the computation of potential for the clique C_1 by multiplying $p(Weather)$ and $p(Roads \mid Weather)$.

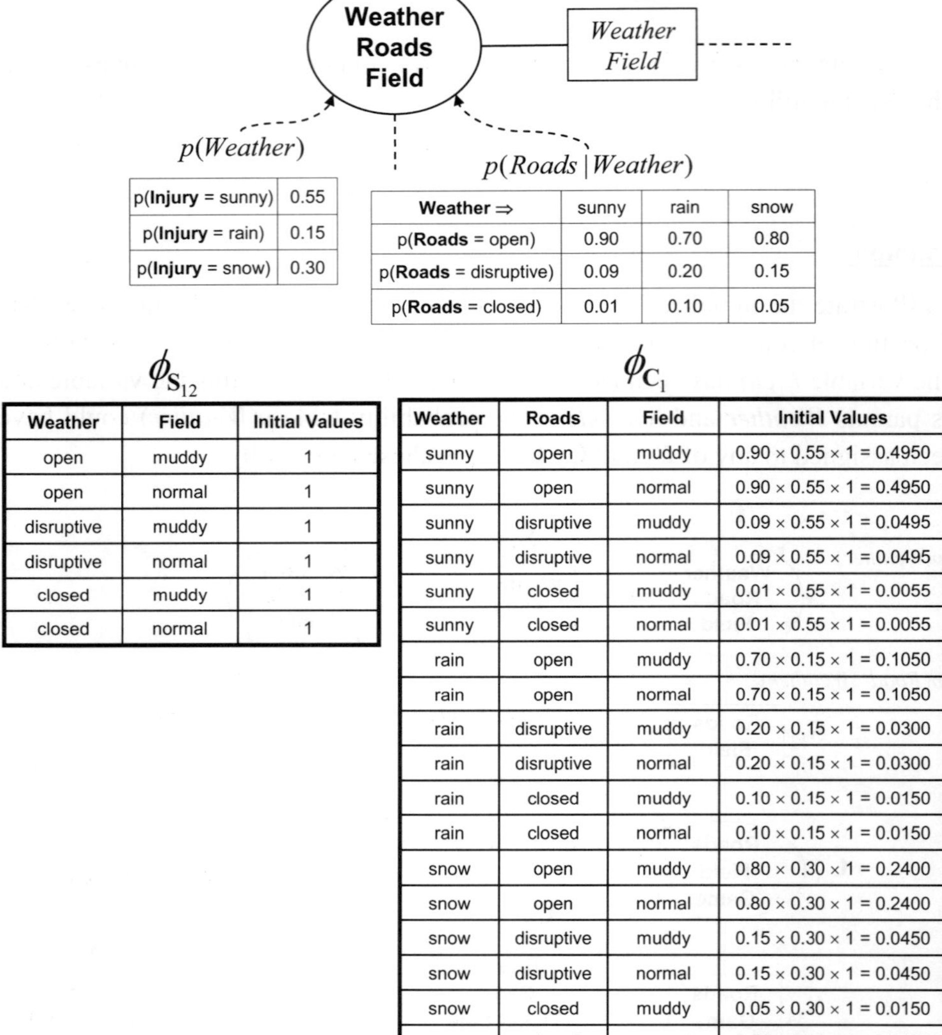

$p(Weather)$

p(**Injury** = sunny)	0.55
p(**Injury** = rain)	0.15
p(**Injury** = snow)	0.30

$p(Roads \mid Weather)$

Weather ⇒	sunny	rain	snow
p(**Roads** = open)	0.90	0.70	0.80
p(**Roads** = disruptive)	0.09	0.20	0.15
p(**Roads** = closed)	0.01	0.10	0.05

$\phi_{S_{12}}$

Weather	Field	Initial Values
open	muddy	1
open	normal	1
disruptive	muddy	1
disruptive	normal	1
closed	muddy	1
closed	normal	1

ϕ_{C_1}

Weather	Roads	Field	Initial Values
sunny	open	muddy	$0.90 \times 0.55 \times 1 = 0.4950$
sunny	open	normal	$0.90 \times 0.55 \times 1 = 0.4950$
sunny	disruptive	muddy	$0.09 \times 0.55 \times 1 = 0.0495$
sunny	disruptive	normal	$0.09 \times 0.55 \times 1 = 0.0495$
sunny	closed	muddy	$0.01 \times 0.55 \times 1 = 0.0055$
sunny	closed	normal	$0.01 \times 0.55 \times 1 = 0.0055$
rain	open	muddy	$0.70 \times 0.15 \times 1 = 0.1050$
rain	open	normal	$0.70 \times 0.15 \times 1 = 0.1050$
rain	disruptive	muddy	$0.20 \times 0.15 \times 1 = 0.0300$
rain	disruptive	normal	$0.20 \times 0.15 \times 1 = 0.0300$
rain	closed	muddy	$0.10 \times 0.15 \times 1 = 0.0150$
rain	closed	normal	$0.10 \times 0.15 \times 1 = 0.0150$
snow	open	muddy	$0.80 \times 0.30 \times 1 = 0.2400$
snow	open	normal	$0.80 \times 0.30 \times 1 = 0.2400$
snow	disruptive	muddy	$0.15 \times 0.30 \times 1 = 0.0450$
snow	disruptive	normal	$0.15 \times 0.30 \times 1 = 0.0450$
snow	closed	muddy	$0.05 \times 0.30 \times 1 = 0.0150$
snow	closed	normal	$0.05 \times 0.30 \times 1 = 0.0150$

Figure 6-40: Computation of potential

6.7.3　Propagation in Join Tree and Marginalization

The join tree thus formed is not locally consistent as, for example, $\sum_{C_1 \backslash S_{12}} \phi_{C_1} \neq \phi_{S_{12}}$.

An inconsistent join tree can be made consistent using a *global propagation*. The message passing mechanism is at the heart of global propagation.

Consider two adjacent clusters $\mathbf{C}_1$ and $\mathbf{C}_2$ with sepset $\mathbf{S}$. A message pass from $\mathbf{C}_1$ to $\mathbf{C}_2$ consists of the following two steps:

- Projection: $\phi_\mathbf{S}^{old} \leftarrow \phi_\mathbf{S}$ $\qquad \phi_\mathbf{S} \leftarrow \sum_{\mathbf{C}_1 \backslash \mathbf{S}} \phi_{\mathbf{C}_1}$

- Absorption: $\phi_{\mathbf{C}_2} \leftarrow \phi_{\mathbf{C}_2} \dfrac{\phi_\mathbf{S}}{\phi_\mathbf{S}^{old}}$

It can be easily verified that any number of messages passing as shown above encodes the joint distribution $p(\mathbf{X})$ of the belief network. Global propagation is a systematic collection of message passing via the following two recursive procedures: *Collect Evidence* and *Distribute Evidence*.

- Choose an arbitrary cluster $\mathbf{C}$.

- Unmark all clusters and call the following three steps of *Collect Evidence*($\mathbf{C}$):

  - Mark $\mathbf{C}$.

  - Recursively call *Collect Evidence* on each unmarked neighboring cluster of $\mathbf{C}$.

  - Pass a message from $\mathbf{C}$ to the cluster which invoked *Collect Evidence*($\mathbf{C}$) .

- Unmark all clusters and call the following three steps of *Distribute Evidence*($\mathbf{C}$):

  - Mark $\mathbf{C}$.

  - Pass a message from $\mathbf{C}$ to each of its unmarked neighboring cluster.

  - Recursively call *Distribute Evidence* on each unmarked neighboring cluster of $\mathbf{C}$.

Example

Figure 6-41 shows the message flow order when the cluster $\mathbf{C}_3$ is chosen as the starting cluster in the above algorithm. First, *Collect Evidence* is called on cluster $\mathbf{C}_3$, which causes two calls of *Collect Evidence* on each of $\mathbf{C}_1$ and $\mathbf{C}_4$. The first call of these two calls on $\mathbf{C}_1$ triggers a call of *Collect Evidence* on $\mathbf{C}_2$. The node then passes message 1 to $\mathbf{C}_1$. The process continues, yielding a total of eight messages as shown in Figure 6-41.

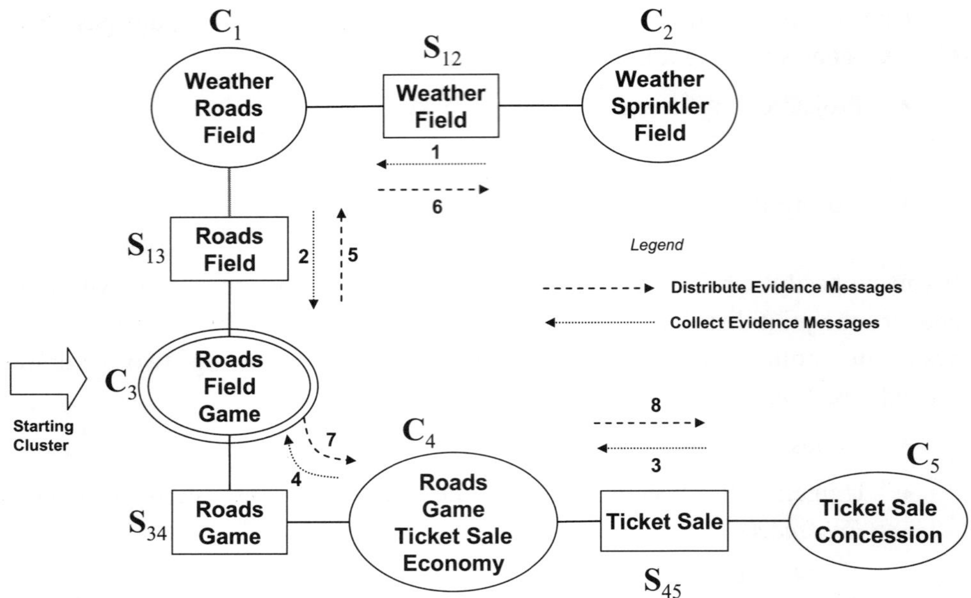

Figure 6-41: Message flow in global propagation

Once the join tree has been made consistent, prior probabilities of variables can be obtained using marginalization. First, identify a cluster $\mathbf{C}$ that contains the variable X of interest. Then compute $p(X)$ by marginalizing $\phi_{\mathbf{C}}$ as $p(X) = \sum_{\mathbf{C} \backslash \{X\}} \phi_{\mathbf{C}}$. An example of this computation for the variable *Field* from cluster $\mathbf{C}_2$ is shown in Figure 6-42.

Clique $\mathbf{C}_2 = \{Weather, Sprinkler, Field\}$

Weather	Sprinkler	Field	Potential ϕ_{C_2}
sunny	burst pipe	muddy	0.8
sunny	normal	normal	0.2
sunny	burst pipe	muddy	0.99
sunny	normal	normal	0.01
rain	burst pipe	muddy	0.1
rain	normal	normal	0.9
rain	burst pipe	muddy	0.6
rain	normal	normal	0.4
snow	burst pipe	muddy	0.01
snow	normal	normal	0.99
snow	burst pipe	muddy	0.15
snow	normal	normal	0.85

Sum (Shaded Rows): 2.65
Sum (Clear Rows): 3.35

p(**Field** = muddy)	0.44 = 2.65 / (2.65 + 3.35)
p(**Field** = normal)	0.56 = 3.35 / (2.65 + 3.35)

Figure 6-42: Marginalization from potential and normalization

6.7.4 Handling Evidence

Figure 6-43 shows the overall flow for using evidence in join trees to compute the variables' posterior probabilities. Compare this figure with Figure 6-33, which shows the flow for computing only prior probabilities. When new evidence on a variable is entered into the tree, it becomes inconsistent and requires a global propagation to make it consistent. The posterior probabilities can be computed via marginalization and normalization from the global propagation. If evidence on a variable is updated, then the tree requires initialization. Next, we present initialization, normalization, and marginalization procedures for handling evidence.

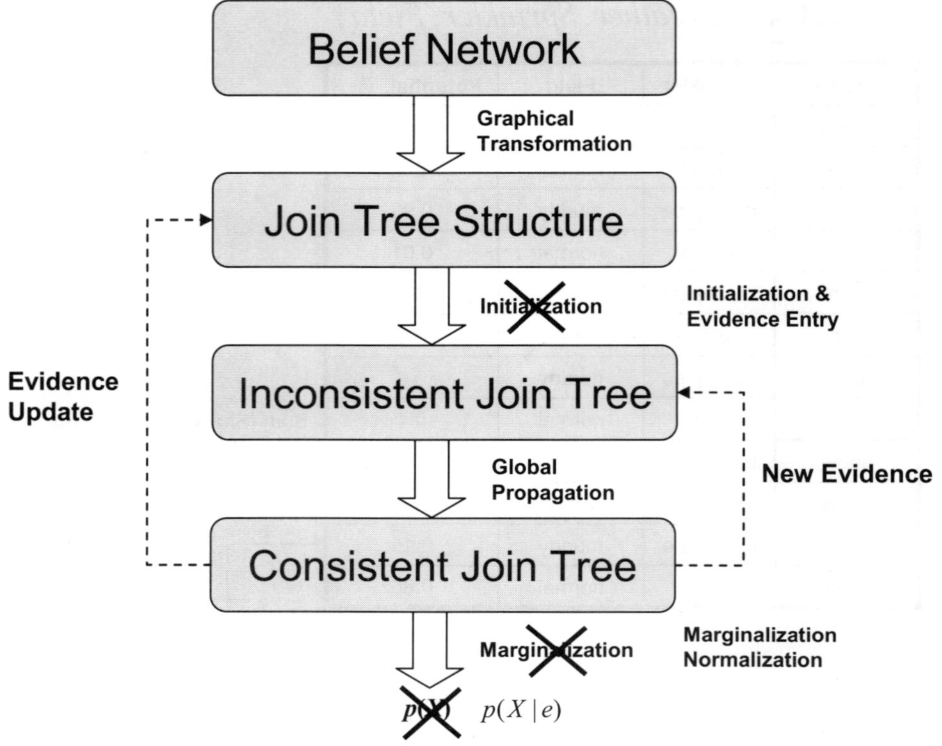

Figure 6-43: Steps for handling evidence in join trees

As before, to start initialization, for each cluster **C** and sepset **S**, set the following:

$$\phi_C \leftarrow 1, \ \phi_S \leftarrow 1$$

Then assign each variable X to a cluster **C** that contains X and its parents $pa(X)$, and then set the following:

$$\phi_C \leftarrow \phi_C p(X \mid pa(X))$$
$$\lambda_X \leftarrow 1$$

where λ_X is the likelihood vector for the variable X. Now, perform the following steps for each piece of evidence on a variable X:

- Encode the evidence on the variable as a likelihood λ_X^{new}.

- Identify a cluster **C** that contains X (e.g. one containing the variable and its parents).

- Update as follows:

$$\phi_C \leftarrow \phi_C \frac{\lambda_X^{new}}{\lambda_X}$$

$$\lambda_X \leftarrow \lambda_X^{new}$$

Now perform a global propagation using the *Collect Evidence* and *Distribute Evidence* procedures. Note that if the belief potential of one cluster **C** is modified, then it is sufficient to unmark all clusters and call only *Distribute Evidence*(**C**).

The potential φ_C for each cluster **C** is now $p(\mathbf{C}, e)$, where e denotes evidence incorporated into the tree. Now marginalize C into the variable as

$$p(X, e) = \sum_{C \backslash \{X\}} \phi_C$$

Compute posterior $p(X \mid e)$ as follows:

$$p(X \mid e) = \frac{p(X, e)}{p(e)} = \frac{p(X, e)}{\sum_X p(X, e)}$$

To update evidence, for each variable X on which evidence has been obtained, update its likelihood vector. Then initialize the join tree by incorporating the observations. Finally, perform global propagation, marginalization, etc.

6.8   Complexity of Inference Algorithms

Probabilistic inference using belief networks is computationally intractable, that is, NP-hard (Cooper 1990). Informally, this means if there exists an algorithm that solves our problems in polynomial time, then the polynomial-time algorithm would exist for practically all discrete problems, such as the propositional satisfiability problem. The fact that inferencing in belief networks is not tractable does not mean it can never be applied; it simply means that there are cases when its inferencing time will take too long for this algorithm to be practical.

The computational complexity of Pearl's message passing algorithm for acyclic networks can be shown to be $O(n \times d \times 2^d)$, where n is the number of vertices in the network and d is the network's maximal in-degree. The computational complexity of Lauritzen and Spiegelhalter's junction tree algorithm equals $O(n \times 2^c)$, where n is the number of vertices in the network and c is the number of vertices in the largest clique in the clique tree that is constructed from the network. Note that the algorithm's complexity is exponential in the size of the largest clique.

If the clique sizes in the junction tree algorithm are bounded by a constant, then the algorithm takes linear time. Since the computational complexity of the junction tree algorithm relates exponentially to clique size, the best clique tree to use in practical applications is a tree inducing smallest state space. The problem of finding such a clique tree is known to be NP-hard (Wen, 1990). Various efficient heuristic algorithms are available for finding a clique tree for an acyclic network. However, these algorithms do not exhibit any optimality properties.

Therefore, it seems unlikely that an exact algorithm can be developed to perform probabilistic inference efficiently over all classes of belief networks. This result suggests that research should be directed away from the search for a general, efficient probabilistic inference algorithm, and towards the design of efficient special-case (for example, tree structure or inherently modular network), average-case, and approximation algorithms.

6.9   Acquisition of Probabilities

The acquisition of probabilities for belief network structures involves eliciting conditional probabilities from subject matter experts along causal directions. These probabilities are "causal" conditional probabilities of the form $p(Report \mid Rain, Sensor)$, indicating the chance of obtaining a sensor report from the appropriate sensor when it is raining. This chance is related to the sensor functionality in the presence of rain, which may be best estimated by the sensor designers. Similarly, a causal probability of the form $p(Game \mid Field, Roads)$ indicates the chance of a game given the field and road conditions. The game referees, together with the people managing the game, can best estimate this probability. On the other hand, the "diagnostic" conditional probabilities in the belief context are probabilities of the form $p(Field \mid Game)$, indicating of the chance that various field conditions will affect the status of the game. An experienced groundskeeper or a regular game observer in the field may best estimate this probability. Both causal and diagnostic probabilities can be used to compute joint probability distributions.

The major issue related to eliciting probabilities from subject matter experts is how to phrase questions to experts so as to accurately and efficiently determine relevant prior and conditional probabilities (Druzdel and van der Gaag, 1995).

Example

Consider, for example, a revised version of the fragment of the network in Figure 6-1 shown in Figure 6-44, which says that sunny weather (*SW*), good economy (*GE*), and holiday season (*HS*) together cause tickets to be sold out (*TS*). Each variable X (*SW*, *GE*, *HS*, and *TS*) in the network is binary with two states X and $\neg X$. If we are eliciting causal probabilities, then experts will be asked questions for determining the priors $p(SW)$, $p(GE)$, and $p(HS)$, and the conditional $p(TS \mid SW, GE, HS)$.

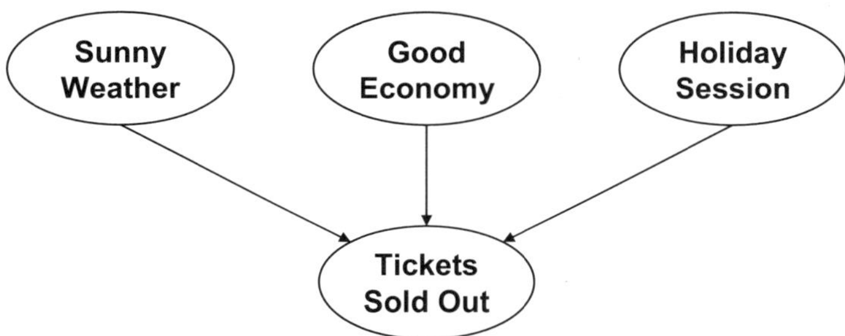

Figure 6-44: BN illustrating probability acquisition

If prior probability $p(TS)$ is available, then diagnostic conditional probabilities are elicited by employing the arc reversal approach (Shachter and Heckerman, 1987). But even if the diagnostic conditional probabilities $p(SW \mid TS)$, $p(GE \mid TS)$, and $p(HS \mid TS)$ are known, it is not possible to uniquely determine the causal probability $p(TS \mid SW, GE, HS)$.

Our experience suggests that subject matter experts, scientists, and knowledge engineers often comfortable drawing arrows in the causal direction once the term "causality" is explained. The directionality has great impact on the resultant ease of knowledge elicitation. For example, if all three arrows in the network in Figure 6-44 are reversed, then the resultant network is equally effective in determining whether the tickets are going to be sold out or not, but now the a priori probability $p(TS)$ is required, as are the conditional probabilities $p(SW \mid TS)$, $p(GE \mid TS)$, and $p(HS \mid TS)$.

The number of conditional probabilities required for a node to complete its conditional probability table, like $p(TS \mid SW, GE, HS)$ for *TS*, grows exponentially with the number of parents. But we can employ the *noisy-or* technique to avoid building large probability tables, provided certain conditions are met (Pearl, 1988). Informally, the noisy-or technique states that any members of a set of independent conditions are likely to cause a certain event and that this chance is not reduced when several of these conditions occur simultaneously. In the case of the node *TS*, chances for the tickets to be sold out can only increase when several of the three conditions, sunny weather, good economy, and holiday season, occur simultaneously. Given this knowledge, we can generate the CPT $p(TS \,|\, SW, GE, HS)$ of 8 independent entries from only three values. Formally, the noisy-or is a belief network node representing an event (e.g. *Tickets Sold Out*) at which the following assumptions are made:

- The *accountability* condition requires that we explicitly list as parent nodes all the conditions likely to influence the event. In the example, this assumption requires that we explicitly list all the conditions likely to cause the tickets to be sold out. In other words, the condition states that an event is presumed false if all conditions listed as causes of the event are false.

- The *exception independence* condition requires that whatever inhibits each parent node from implying the event is independent of whatever inhibits other parent nodes from implying the event. For example, assume that the only factor inhibiting the tickets to be sold out when the weather is sunny is high temperature, and the only factor inhibiting the tickets to be sold out when the economy is good is very high ticket price. The exception independence condition holds since the two inhibitory factors can be assumed independent of each other. On the other hand, going to the beach inhibits the tickets to be sold out when the weather is sunny, and the same factor inhibits the tickets to be sold when it is holiday season, thus violating the exception independence condition.

Example

An incomplete CPT, like the one in Table 6-9, can be completed by deriving the missing probabilities through the noisy-or technique. Suppose, only the following entries in the CPT are known:

$$p(TS \,|\, SW, \neg GE, \neg HS) = 0.7$$
$$p(TS \,|\, \neg SW, GE, \neg HS) = 0.6$$
$$p(TS \,|\, \neg SW, \neg GE, HS) = 0.9$$

Sunny Weather (SW)		SW				¬SW			
Good Economy (GE)		GE		¬GE		GE		¬GE	
Holiday Session (HS)		HS	¬HS	HS	¬HS	HS	¬HS	HS	¬HS
Tickets Sold Out (TS)	TS	?	?	?	0.7	?	0.6	0.9	?
	¬TS	?	?	?	0.3	?	0.4	0.1	?

Table 6-9: An incomplete CPT to be completed by the noisy-or technique

Therefore, we have the following:

$$p(\neg TS \mid SW, \neg GE, \neg HS) = 0.3$$
$$p(\neg TS \mid \neg SW, GE, \neg HS) = 0.4$$
$$p(\neg TS \mid \neg SW, \neg GE, HS) = 0.1$$

Now,

$$p(TS \mid SW, GE, \neg HS) = 1 - p(\neg TS \mid SW, GE, \neg HS).$$

The accountability condition states that *TS* is false if all conditions listed as causes of *TS* are false. Therefore,

$$p(\neg TS \mid SW, GE, \neg HS) = p(\neg TS \mid SW, GE)$$

Thus,

$$p(TS \mid SW, GE, \neg HS) = 1 - p(\neg TS \mid SW, GE)$$

Now, the exception independence condition states that the inhibitory conditions *SW* and *GE* for *TS* are independent of each other. Therefore,

$$p(\neg TS \mid SW, GE) = p(\neg TS \mid SW)p(\neg TS \mid GE)$$

Thus, $p(TS \mid SW, GE, \neg HS)$

$$= 1 - p(\neg TS \mid SW)p(\neg TS \mid GE)$$
$$= 1 - (1 - p(TS \mid SW))(1 - p(TS \mid GE))$$
$$= 1 - (1 - 0.7)(1 - 0.6) = 0.88$$

The rest of the entries in the CPT can be computed in a similar manner.

6.10   Advantages and Disadvantages of Belief Networks

Like any other computational formalism, belief network technology offers certain advantages and disadvantages. Advantages of belief networks include:

- Sound theoretical foundation: The computation of beliefs using probability estimates is guaranteed to be consistent with probability theory. This advantage stems from the Bayesian update procedure's strict derivation from the axioms of probability.

- Graphical models: Belief networks graphically depict the interdependencies that exist between related pieces of domain knowledge, enhancing understanding of the domain. The structure of a belief network captures the cause-effect relationships that exist amongst the variables of the domain. The ease of causal interpretation in belief network models typically makes them easier to construct than other models, minimizing the knowledge engineering costs and making them easier to modify.

- Predictive and diagnostic reasoning: Belief networks combine both deductive/predictive and abductive/diagnostic reasoning. Interdependencies among variables in a network are accurately captured and speculative if-then type computation can be performed.

- Computational tractability: Belief networks are computationally tractable for most practical applications. This efficiency stems principally from the exploitation of conditional independence relationships over the domain. We have presented an efficient single-pass evidence propagation algorithm for networks without loops.

- Evidence handling: Evidence can be posted to any node in a belief network. This means that subjective evidence can be posted at an intermediate node representing an abstract concept.

A major disadvantage of belief network technology is the high level of effort required to build network models. Although it is relatively easy to build a belief network structure with the help of subject matter experts, the model will require a significant amount of probability data as the number of nodes and links in the structure increase. The size of a CPT corresponding to a node with multiple parents can potentially be huge. For example, the number of independent entries in the CPT of a binary node (a node with two states) with 8 binary parent variables is 128.

Belief networks are also poor at handling continuous variables. Current software handles continuous variables in a very restrictive manner (for example, they must be Gaussian and can only be children). Lener et al. (2001) developed an inference algorithm for static hybrid belief networks, which are Conditional Linear Gaussian models, where the conditional distribution of the continuous variables assigned to the discrete variables is a multivariate Gaussian. Cob and Shenoy (2004) developed an inference algorithm in hybrid belief networks using Mixtures of Truncated Potentials. But these techniques are yet to be incorporated in commercial software.

6.11 Belief Network Tools

Various free and commercial software tools and packages are currently available for manipulating Bayesian belief networks incorporating some of the above functionalities. Of these, HUGIN (www.hugin.com) is the most widely commercial package that contains a flexible, user friendly and comprehensive graphical user interface. The package allows modeling decision-making problems via influence diagrams and handles continuous variables with some restrictions. Other popular tools and packages include Netica (http://www.norsys.com/), BayesiaLab (http://www.bayesia.com/), and BNet^{TM} (http://www.cra.com/).

6.12 Further Readings

The book by Pearl (1988) is still the most comprehensive account on belief networks, and more generally on using probabilistic reasoning to handle uncertainty. Various cases of the evidence propagation algorithms in polytrees presented here closely follow Pearl's book. Though Pearl himself developed an exact inference algorithm for DAGs, called loop cutset conditioning (Pearl, 1986), the junction tree algorithm of Lauritzen and Spiegelhalter (1988), as refined by Jensen et al (1990) in HUGIN, is more general and the most popular inference algorithm for general belief networks. A good comprehensive procedural account of the algorithm can be found in (Huang and Darwiche, 1996). Jensen's books (1996, 2002) are also useful guides in this field.

Chapter 7

Influence Diagrams for Making Decisions

This chapter describes how to use Bayesian belief networks to make decisions, and therefore provides a foundation for building probabilistic epistemic models for decision making agents. Belief networks do not explicitly incorporate the concepts of action and utility, which are ubiquitous in the decision-making context. By incorporating the concepts of action and utility, belief networks are converted to influence diagrams, subsuming the functionality of the normative theory of decision making under expected utility theory and decision trees. For inferencing in influence diagrams, we extend the junction tree algorithm for belief networks presented in the last chapter. The extended algorithm presented in this chapter compiles an influence diagram into a strong junction tree in which the computation of maximum expected utility can be done by local message passing in the tree.

7.1   Expected Utility Theory and Decision Trees

A *decision* is a choice between several alternative courses of risky or uncertain action. Expected Utility Theory (EUT) states that the decision-maker chooses between alternative courses of action by comparing their *expected utility* values, which is the weighted sum obtained by adding the utility values of possible outcomes multiplied by their respective probabilities. Formally, if $A = \{a_1,...,a_n\}$ is the set of all possible actions as decision alternatives and $W = \{w_1,...,w_n\}$ is the corresponding set of possible outcomes or world states, then the *expected utility* (EU) for action a_k is the following:

$$EU(a_k) = \sum_i U(w_i)p(w_i \mid a_k)$$

where $U(w_i)$ is the utility of the outcome w_i and $p(w_i \mid a_k)$ is the probability that the outcome is w_i if the action a_k is taken. The Maximum Expected Utility (MEU) operation is used to choose the best alternative course of action:

$$MEU(A) = \max_k EU(a_k)$$

In cases where an outcome w_i does not depend on the decision-maker's action but on other context-dependent factors which the decision maker has no control over, then the expected utility is simply defined without actions, that is,

$$EU = \sum_i U(w_i)p(w_i)$$

The computation of MEU is not relevant in this case.

A *decision tree* poses an ordered list of systematic questions that leads the decision-maker through a series of decisions to a logical endpoint that results in a solution. These likely outcomes of possible solutions to the problem are projected as utilities. The systematic exploration of likely outcomes is organized in a tree in which each branch node represents a choice between a number of alternatives, and each leaf node represents a solution. A branch node is either a *chance node* or an *action node*. MEUs are computed at action nodes and EUs are computed at chance nodes.

Example

Figure 7-1 shows an example decision tree in the context of our ball-game example in which ovals, rectangles, and diamonds represent chance nodes, decision nodes, and terminal nodes respectively. The root node is a chance node representing the status of rain, and branches are its uncertain outcomes. It is 70% likely that rain will be present and 30% likely that rain will be absent. These two possible outcomes are represented as two branches coming out of the chance node for rain. Given either of these two alternatives of rain status, the next step is to decide whether to proceed with the game or abandon it. This decision requires exploring four possible combinations of options between the rain status and the game action. For each of these combinations, the likely ticket sale volumes need to be considered to compute profits that are projected as utilities.

For example, if the action taken is to proceed with the game, and rain is present, then there is a 10% chance that the ticket sale volume will be high and 90% chance that the sale volume will be low. Each of these two possibilities yields different profit as utility and are attached to the terminal node. So the top branch of the tree in the figure encodes a possible outcome, which states that if the rain is present, action to proceed with the game is taken, and the ticket sale volume is high, then the utility is 900. Eight possible outcomes are explored in this manner in the context of the example.

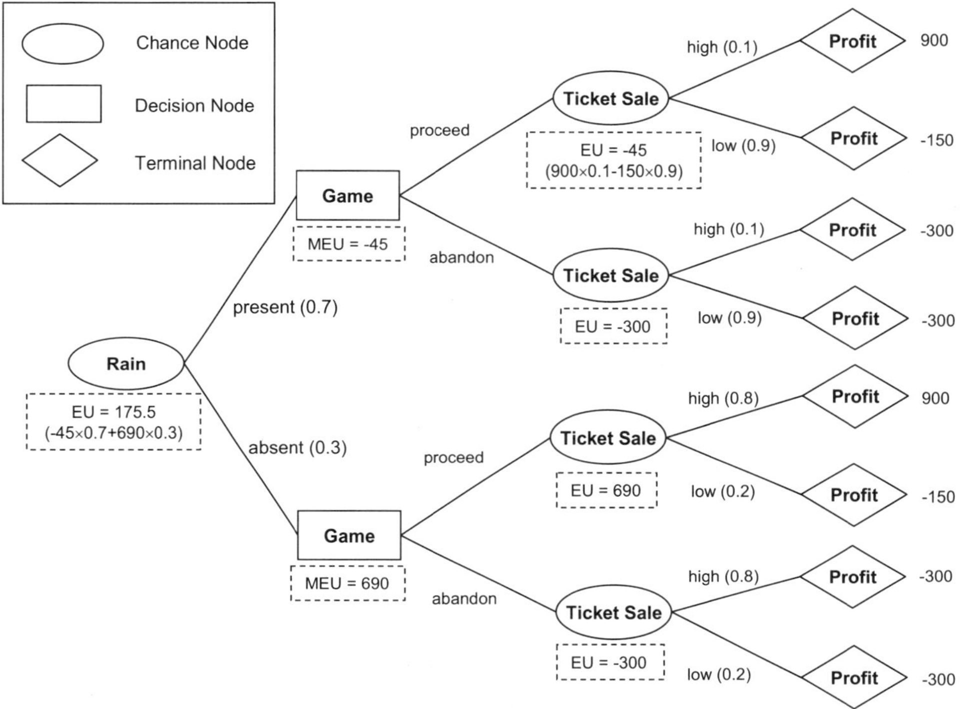

Figure 7-1: An example decision tree

EUs and MEUs at branch nodes are computed bottom up along the tree (right to left in the figure). For example, the EU at the *Ticket Sale* node of the topmost branch is computed using the formula $\sum_i U(w_i)p(w_i)$, yielding $900 \times 0.1 + (-150) \times 0.9$, that is, –45. To compute the MEU at the node *Game* in the upper branch of the tree, EUs of the two actions *Proceed* and *Abandon*, *EU(Proceed)* and *EU(Abandon)*, need to be computed using the formula $\sum_i U(w_i)p(w_i \mid a_k)$. Now, the EUs of the two *Ticket Sale* nodes of the upper branch are –45 and –300. Therefore, the MEU of the node *Game* is the maximum of –45 and –300, that is, –45. Finally, the EU of the node *Rain* is computed using the formula $\sum_i U(w_i)p(w_i)$, yielding 175.5.

The procedure for computing utilities using decision trees, as explained above, is a simple statistical procedure. Decision trees also provide easy-to-understand graphic representations. But laying out all the possible options is not feasible when there are several factors, each with multiple outcomes. Moreover, the procedure provides no scope for incorporating subjective knowledge. Influence diagrams address some of these issues.

7.2   Influence Diagrams

Influence diagrams are belief networks augmented with decision variables and utility functions, and used to solve decision problems. There are three types of nodes in an influence diagram:

- *Chance nodes* (that is, belief network nodes), represented by ovals
- *Decision nodes*, represented by rectangles
- *Value* or *utility nodes*, represented by diamonds

As opposed to chance nodes representing probabilistic variables, decision nodes represent actions that are under the full control of decision-makers and hence no CPT is attached to a decision node.

Example

Figure 7-2 shows an example influence diagram from our game example. There are two chance nodes (*Rain* and *Ticket Sale*), one decision node (*Game*), and one value node (*Profit*). The arrow from the node *Rain* to the node *Ticket Sale* represents the causal influence of the status of rain on the volume of ticket sales. The CPT quantifies this causality as in usual belief networks. As we mentioned before, there is no CPT attached to a decision node, but the arrow from the chance node *Rain* to the decision node *Game* represents the fact that the knowledge of the status of rain should be known before making the decision (whether to *Proceed* or *Abandon*) about the game. The value node *Profit* has two parents, representing the causal influence of the ticket sale volume and the game status on the profit amount. The table represents a *utility function* whose definition quantifies the utilities for all possible combinations of the parent values. For example, if the decision-maker proceeded with the game and the ticket sale is low then the profit is negative 150.

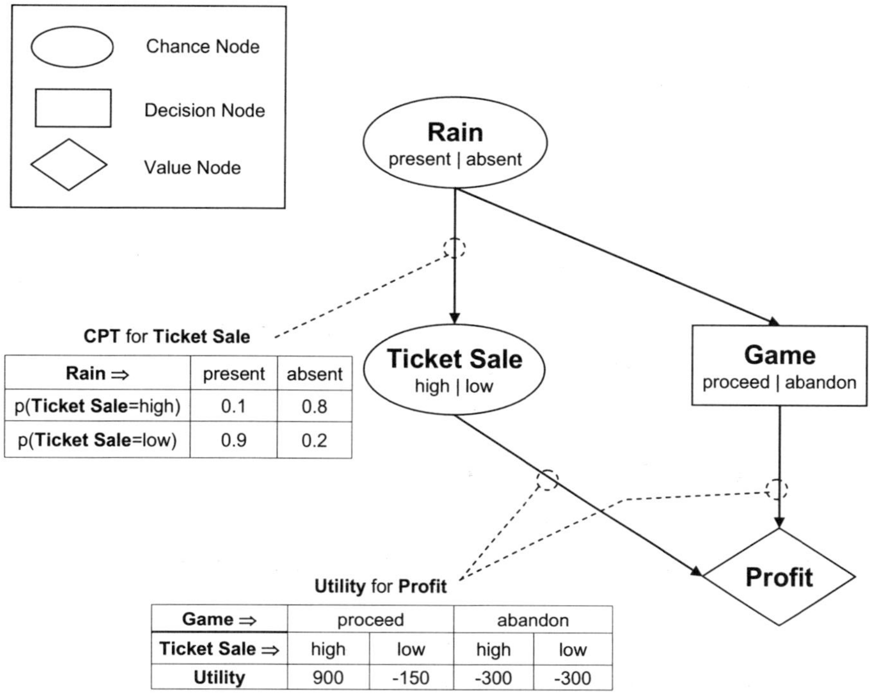

Rain $\Rightarrow$	present	absent
p(**Ticket Sale**=high)	0.1	0.8
p(**Ticket Sale**=low)	0.9	0.2

Game $\Rightarrow$	proceed		abandon	
Ticket Sale $\Rightarrow$	high	low	high	low
Utility	900	-150	-300	-300

Figure 7-2: An example influence diagram

Inferences in influence diagrams involve computing the EU for each of the action alternatives. This kind of computation is sensitive to the temporal order in which the decisions are made. Therefore, influence diagrams require a directed path connecting all the decision nodes sequentially. EUs are evaluated on the condition that an optimal choice has been made in all previous steps.

Suppose, $A_1,...,A_m$ is the ordered set of actions to be decided (A_m is the last action), where each A_i is a set of mutually exclusive action options; $V_0,...,V_m$ is a partition of the chance variables $X_1,...,X_n$ such that V_{i-1} is the set of variables instantiated before deciding on the action A_i. Therefore we have the following ordering:

$$V_0 \prec A_1 \prec V_1 \prec ... \prec A_m \prec V_m$$

Then MEU for a chosen set of alternatives for $A_1,...,A_m$ is given below:

$$MEU(A_1,...,A_m) =$$

$$\sum_{V_0} \max_{A_1} \sum_{V_1} ... \sum_{V_{m-1}} \max_{A_m} \sum_{V_m} \sum_{U} U \times p(V_0,V_1,...,V_m \mid A_1,...,A_m)$$

The MEU for action A_k is given as

$$MEU(A_k) =$$

$$\max_{A_k} \sum_{V_k} ... \sum_{V_{m-1}} \max_{A_m} \sum_{V_m} \sum_{U} U \times p(V_k,...,V_m \mid V_0,...,V_{k-1},A_1,...,A_m)$$

Thus the computation of MEU in an influence diagram involves a series of alternating sum-marginalization and max-marginalization to eliminate the variables.

7.3   Inferencing in Influence Diagrams

In this section, we illustrate inferencing in influence diagrams considering the following two special cases of interactions between decisions and other types of variables:

- Non-intervening interactions: Actions which have no impact on variables (or probability distribution) in the network

- Intervening interactions: Actions which have an impact on variables in the network that then affect the beliefs of other variables, or affect the utilities of value nodes

Next, we present a junction-tree-based algorithm for inferencing in influence diagrams containing arbitrary sequences of interactions.

<u>Example</u>

Figure 7-3 shows an influence diagram with one node of each type: chance, decision, and value. The decision node represents a non-intervening action *Game*, which has no impact on $p(Rain)$, that is, $p(Rain \mid Game) = p(Rain)$.

Given the likelihood e as shown in the figure, the posterior probability for *Rain* and expected utilities for each of the actions are computed as follows:

$$p(Rain \mid e) = \begin{bmatrix} p(Rain = Present \mid e) \\ p(Rain = Absent \mid e) \end{bmatrix} = \alpha \begin{bmatrix} 0.9 \\ 0.1 \end{bmatrix} \begin{bmatrix} 0.2 \\ 0.8 \end{bmatrix} = \begin{bmatrix} 0.692 \\ 0.308 \end{bmatrix}$$

EUs for each of the actions based on the posterior probabilities are computed as follows:

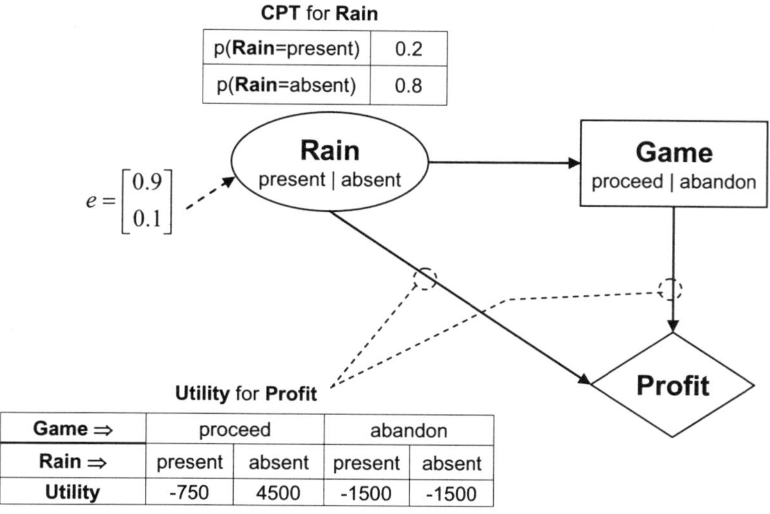

Figure 7-3: Influence diagram with non-intervening action

$EU(Game = Proceed \,|\, e)$

$\quad = U(Game = Proceed, Rain = Present) \times p(Rain = Present \,|\, e) +$
$\quad\quad U(Game = Proceed, Rain = Absent) \times p(Rain = Absent \,|\, e)$
$\quad = (-750) \times 0.692 + 4500 \times 0.308 \approx 865$

$EU(Game = Abandon \,|\, e)$

$\quad = (-1500) \times 0.692 + (-1500) \times 0.308 = -1500$

The MEU operation to choose the best alternative course of action is computed below:

$MEU(Game \,|\, e)$

$\quad = max \,\{EU(Game = Proceed \,|\, e), EU(Game = Abandon \,|\, e)\}$
$\quad = max \,\{865, -1500\} = 865$

This MEU computation can be executed in a decision tree equivalent to the influence diagram in Figure 7-3. One such tree is shown in Figure 7-4, where the decision and chance variables are systematically laid out (parent variables before their children) to explore all possible combinations of states. Therefore, there will be as many leaf nodes in the tree as the number of states in the joint distribution of the set of all decision and chance variables. Each branch yields some utility in terms of positive or negative profit obtained from the utility table in Figure 7-3. The computation of EU in the tree is performed in the usual manner starting at the leaf nodes and moving backwards.

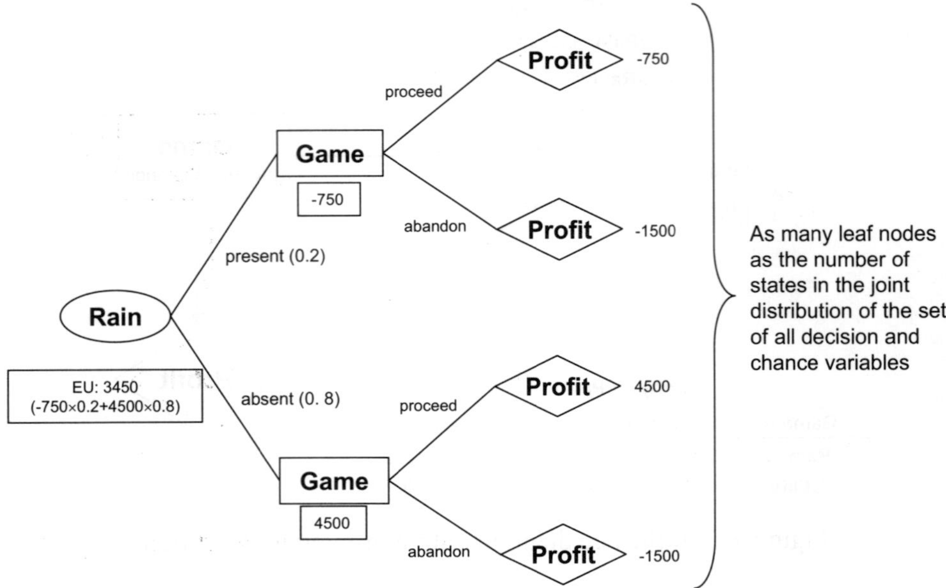

Figure 7-4: Decision tree equivalent to the influence diagram in Figure 7-3

When evidence on the variable *Rain* is obtained, its posterior probability is computed by component-wise multiplication followed by normalization. The EU is then computed based on the posterior probability of *Rain* as shown in Figure 7-5.

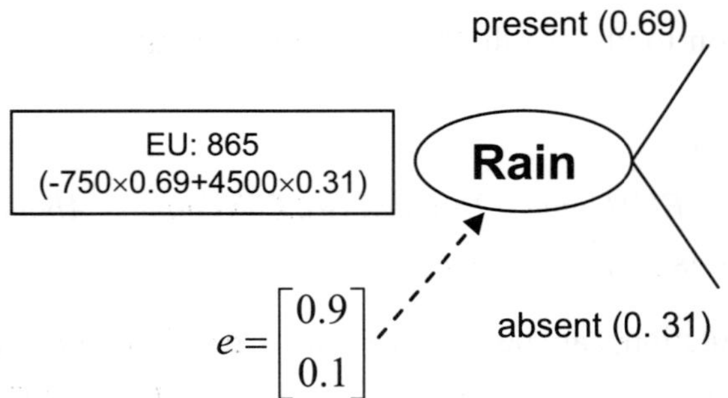

Figure 7-5: Decision tree of Figure 7-4 with evidence entered

Example

Figure 7-6 shows an influence diagram with a set of non-intervening actions. The EU is computed by summing up the MEU of each of the constituent actions. In other words,

$$MEU(Clean\,Field, Sale\,Ticket \mid e) = MEU(Clean\,Field \mid e) + MEU(Sale\,Ticket \mid e)$$

MEU for the variable *Sale Ticket* can be computed as follows:

$$EU(Sale\,Ticket = Proceed \mid e) = (-750) \times 0.692 + 4500 \times 0.308 \approx 865$$

$$EU(Sale\,Ticket = Abandon \mid e) = (-1500) \times 0.692 + (-1500) \times 0.308 = -1500$$

$$MEU(Sale\,Ticket \mid e) \approx 865$$

Similarly, MEU of *Clean Field* can be computed, giving $MEU(Clean\,Field \mid e) = -200$.

Therefore, $MEU(Clean\,Field, Sale\,Ticket \mid e) \approx 665$

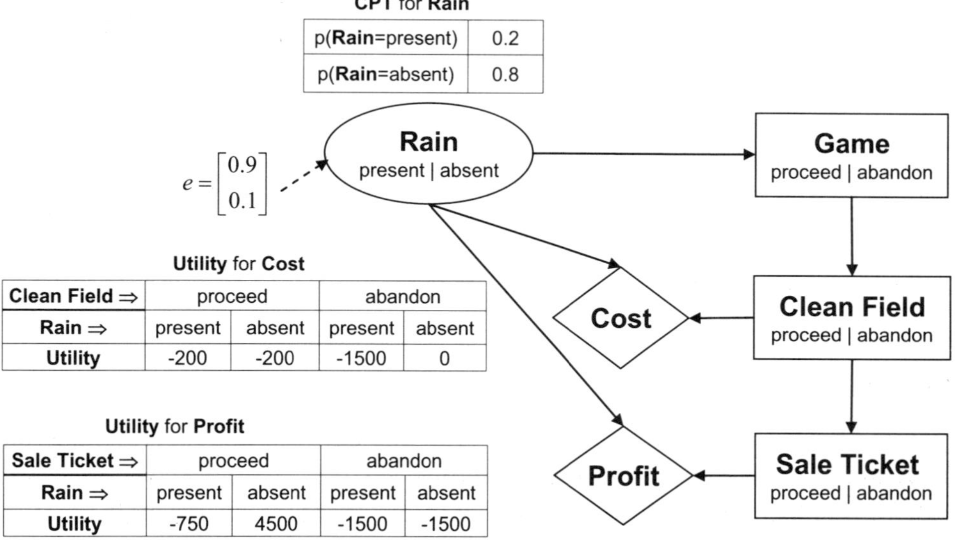

Figure 7-6: Influence diagram with a set of non-intervening actions

A decision tree equivalent to the influence diagram of Figure 7-6 is shown in Figure 7-7. The utility along a branch of the tree is computed by summing up the cost and profit corresponding to the actions specified along the branch. For example, the topmost branch states that if rain is present and the option *Proceed* is chosen for each of the three action variables *Game*, *Clean Field*, and *Sale Ticket*, then the utility is −950. This utility is obtained by adding the cost of proceeding with the *Clean Field* action in the presence of rain (-200) to the profit for proceeding with *Sale Ticket* action in the presence of rain (-750).

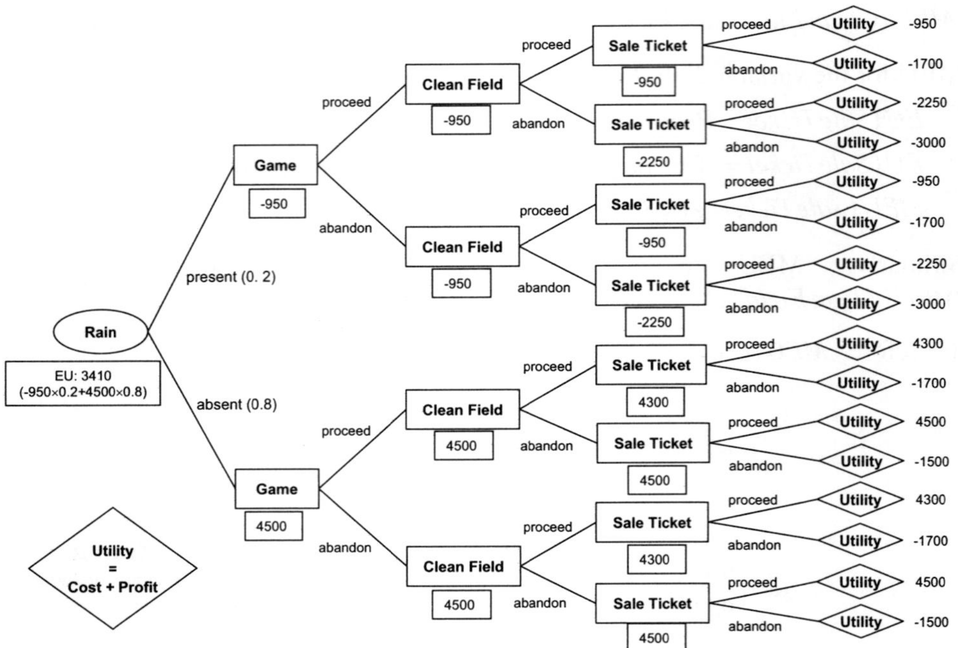

Figure 7-7: Decision tree equivalence to the influence diagram in Figure 7-6

Example

Figure 7-8 shows an influence diagram with one decision node representing the intervening action *Game*, that directly influences the variable *Ticket Sale*. The posterior probability and EU for the variable *Ticket Sale* when the option *Proceed* is chosen is computed as follows:

$$p(Ticket\ Sale \mid e, Game = Proceed) = \begin{bmatrix} 0.42 \\ 0.51 \\ 0.07 \end{bmatrix}$$

$$EU(Game = Proceed \mid e)$$
$$= 4500 \times 0.42 + (-500) \times 0.51 + (-1500) \times 0.07 \approx 1504$$

Similarly, EU for the option *Abandon* is computed similarly, yielding the following:

$$EU(Game = Abandon \mid e) = -1500$$

Therefore, $MEU(Game \mid e) = -1500$

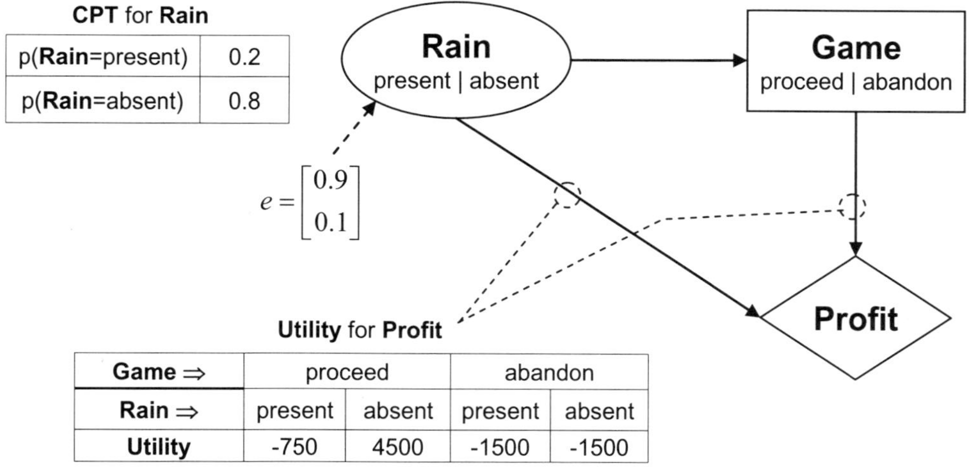

$$e = \begin{bmatrix} 0.9 \\ 0.1 \end{bmatrix}$$

Game ⇒	proceed		abandon	
Rain ⇒	present	absent	present	absent
Utility	-750	4500	-1500	-1500

Figure 7-8: Influence diagram with intervening action

A decision tree equivalent to the influence diagram in Figure 7-8 is shown in Figure 7-9.

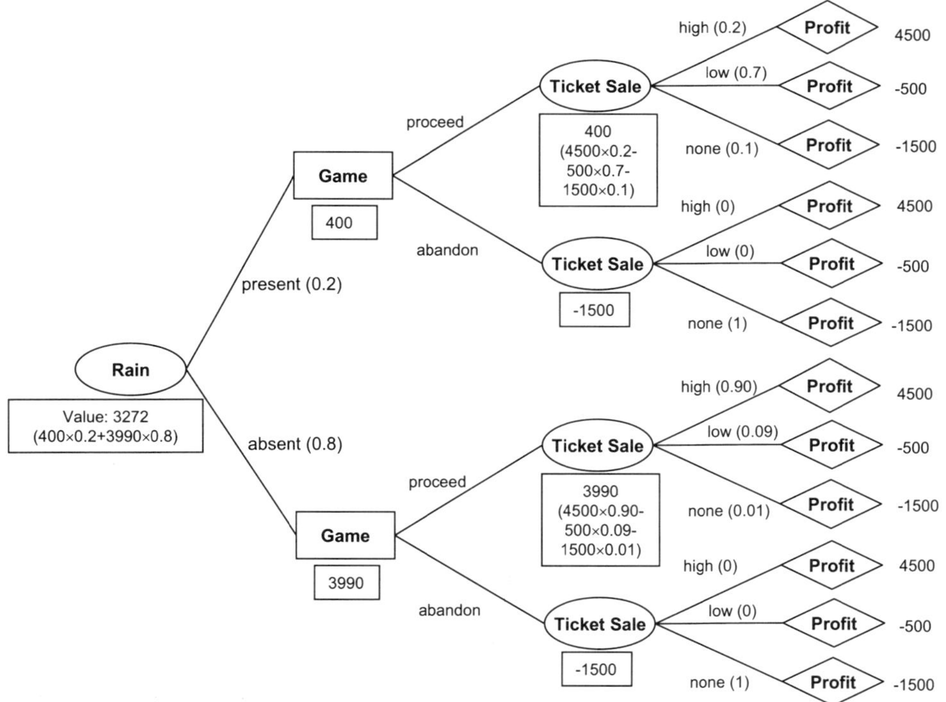

Figure 7-9: Decision tree equivalent to the influence diagram in Figure 7-8

7.4 Compilation of Influence Diagrams

Compiling an influence diagram involves transforming it to a *strong junction tree* (or rooted junction tree), that maintains a special ordering relative to the root clique. Just to recall, a *junction tree* is a tree of cliques if for each pair $\mathbf{C}_1$ and $\mathbf{C}_2$ of cliques, $\mathbf{C}_1 \cap \mathbf{C}_2$ is contained in every clique on the path connecting $\mathbf{C}_1$ and $\mathbf{C}_2$. A *separator* of the two adjacent cliques $\mathbf{C}_1$ and $\mathbf{C}_2$ is the intersection $\mathbf{C}_1 \cap \mathbf{C}_2$.

The compilation steps for influence diagrams are shown on the right side of Figure 7-10. Steps to compile belief networks into junction trees are shown on the left. Details of each step are explained below.

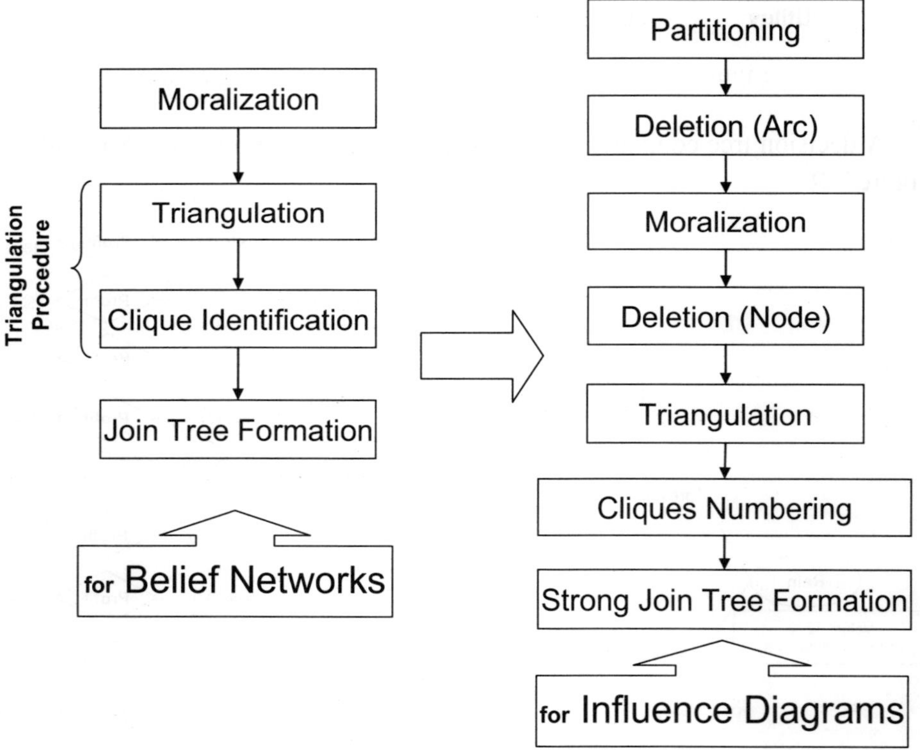

Figure 7-10: Steps in the compilation process of influence diagrams

- *Partitioning*: Partition the chance variables $X_1,...,X_n$ into $V_0,...,V_p$ such that V_{i-1} is the set of variables instantiated before deciding on the action A_i, yielding the ordering $V_0 \prec A_1 \prec V_1 \prec ... \prec A_m \prec V_m$.

- *Deletion (Arc)*: Delete edges pointing into decision nodes.

- *Moralization*: Marry parents with common children including parents of utility nodes and drop the directions of the original links.

- *Deletion (Node)*: Delete utility nodes along with their edges.

- *Triangulation*: Triangulate in such as a way that it facilitates the computation of MEU. Do this by adopting a special elimination order based on the ordering found during partitioning. Start eliminating variables using the triangulation procedure in the following order: eliminate variables from the set V_m, then the decision variable A_m, then variables from the set V_{m-1}, then the decision variable A_{m-1}, and so on. The elimination sequence constitutes a numbering where the first eliminated variable is assigned the highest number.

- *Clique Numbering*: A clique is numbered after its variable with the highest number k (otherwise 1) such that the rest of the variables in the clique have a common neighbor outside the clique with a number lower than k.

- *Strong Join Tree Formation*: A *strong root* of a junction tree is a distinguished clique $\mathbf{R}$ such that for each pair $\mathbf{C}_1$ and $\mathbf{C}_2$ of adjacent cliques with $\mathbf{C}_1$ closer to $\mathbf{R}$ than $\mathbf{C}_2$, there is an ordering in $\mathbf{C}_2$ that respects $\prec$ and with the vertices of the separator $\mathbf{C}_1 \cap \mathbf{C}_2$ precede the vertices of $\mathbf{C}_2 - \mathbf{C}_1$. A junction tree is *strong* if it has a strong root. The ordering within a junction tree ensures that the computation of the MEU can be done by local message passing in the tree. The following steps are followed to form a strong junction tree from the sequence $\mathbf{C}_1, ..., \mathbf{C}_n$ of cliques arranged in increasing order according to their indices:

  - Start with $\mathbf{C}_1$ as the root clique with number 1.

  - Connect the cliques as follows in the increasing order of their numbers: First, compute the following for each clique $C_k\ (k > 1)$:

$$S_k = C_k \cap \bigcup_{i=1}^{k-1} C_i$$

    Connect the clique C_k with another clique which contains S_k.

  - Variables that were eliminated first are placed farthest from the root in the strong junction tree.

The following example illustrates these steps to construct a strong junction tree.

Example

Figure 7-11 is an example influence diagram modified from our original belief network for the game example. The chance variable *Field* is the only parent of the decision variable *Game*, indicating that information about field condition is required before the action for canceling the game or going ahead with the game. If the game is on, there will be a cost for organizing the game, but profits can be made from both ticket sales and concession.

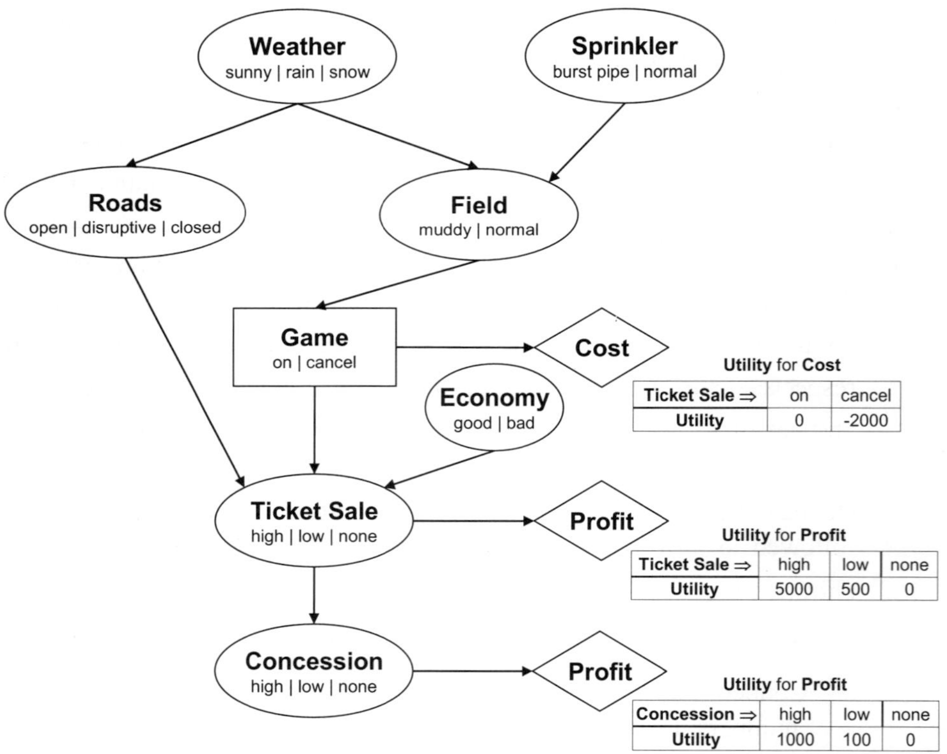

Utility for **Cost**

Ticket Sale ⇒	on	cancel
Utility	0	-2000

Utility for **Profit**

Ticket Sale ⇒	high	low	none
Utility	5000	500	0

Utility for **Profit**

Concession ⇒	high	low	none
Utility	1000	100	0

Figure 7-11: Example influence diagram

Consider the following ordering of the chance and decision variables of the belief network:

$$\{Sprinkler, Weather, Field, Roads\} \prec \{\textbf{\textit{Game}}\} \prec$$
$$\{Economy, Ticket\ Sale, Concession\}$$

The sequence in which the variables are eliminated during the triangulation step is shown in Figure 7-12, along with the assigned numbers.

Variable	Number
Concession	8
Ticket Sale	7
Economy	6
Game	5
Roads	4
Field	3
Weather	2
Sprinkler	1

Figure 7-12: Numbering of variables during triangulation

The five cliques identified during the triangulation step and their numbers are given below:

C_1: {*Weather, Sprinkler, Field*} → 3

C_2: {*Weather, Roads, Field*} → 4

C_3: {*Roads, Field, Game*} → 5

C_4: {*Roads, Game, Economy, Ticket Sale*} → 7

C_5: {*Ticket Sale, Concession*} → 8

To construct a strong junction tree, we start with C_1 as the root. Figure 7-13 shows the strong junction tree based on these cliques. The variables that were eliminated first (such as *Concession* and *Ticket Sale*) are placed farthest from the root (clique number 1) in the strong junction tree.

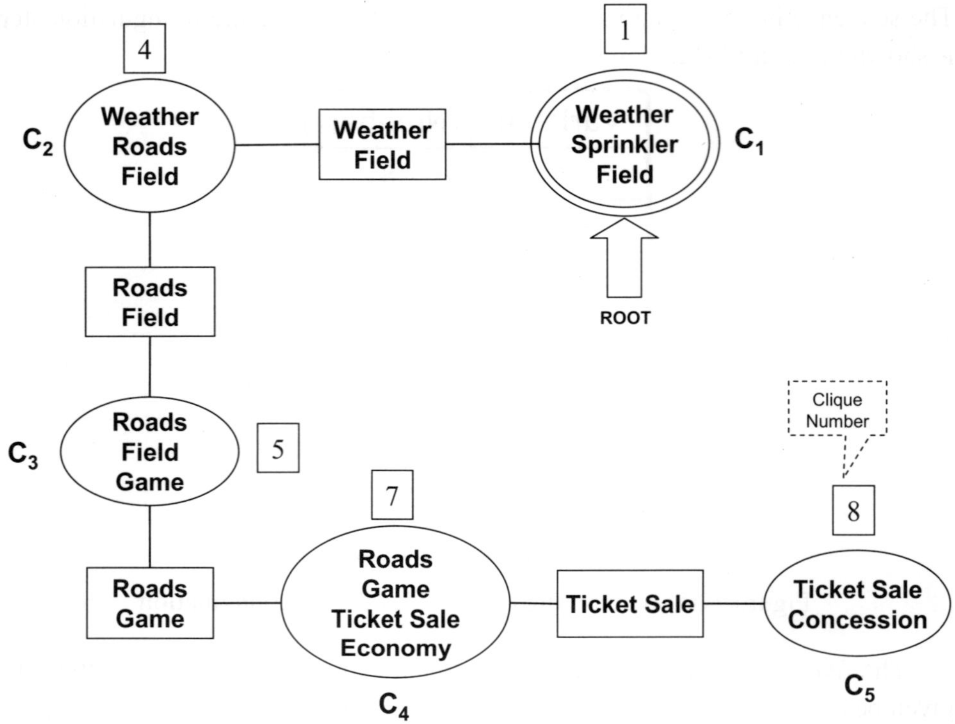

Figure 7-13: Strong junction tree

7.5 Inferencing in Strong Junction Tress

Inferencing in a strong junction tree employs a special collect operation from the leaves to the root of the tree. This operation is similar to the one for computing EU at the root node of a decision tree as shown in Figure 7-1. The difference is that a node here takes the form of a clique rather than a single variable. In addition to associating a probability potential with each clique of the tree, a utility potential is also associated with each clique. The utility potential for a clique is the sum of the utility functions assigned to it. The utility potential for a clique is a null function if no utility functions are assigned to it.

<u>Example</u>

Continuing with our earlier example, Figure 7-14 shows an assignment of probability distributions and utility functions. The two profit utility functions are assigned to the clique $\mathbf{C}_5$ since the parents of both these two utility nodes belong to $\mathbf{C}_5$. Similarly, the cost utility function could have been assigned to any of the two cliques $\mathbf{C}_3$ and $\mathbf{C}_4$ containing the variable *Game*, and $\mathbf{C}_3$ was chosen arbitrarily.

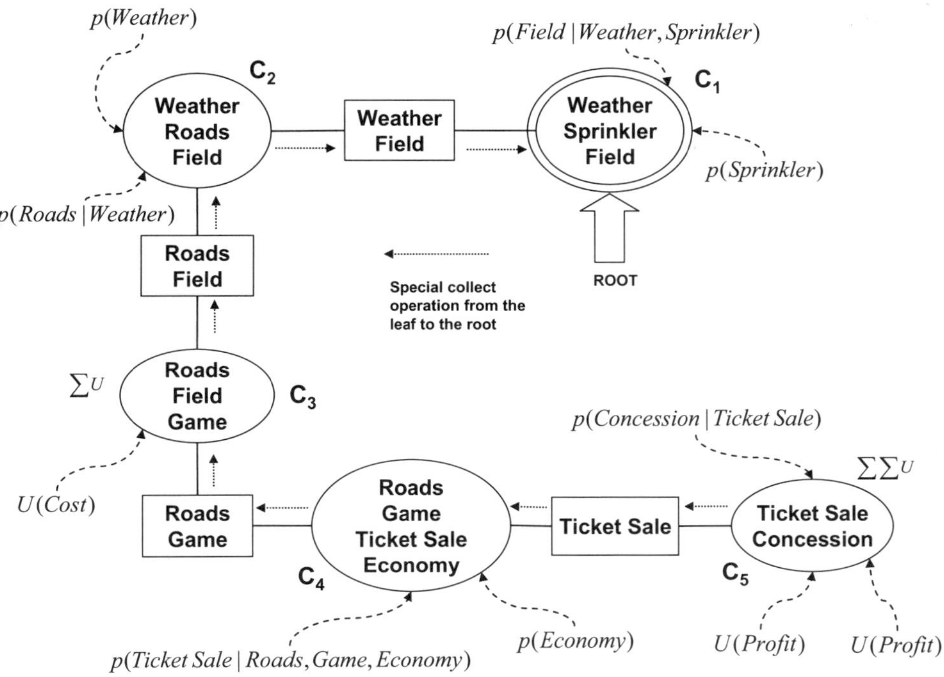

Figure 7-14: Assignment of probability distributions and utility functions to cliques

The computation of the utility potential for the node $\mathbf{C}_5$ is given below in Figure 7-15 based on the two utility functions for the two profit nodes.

Consider the following ordering of the chance and decision variables as before:

$$\{Sprinkler, Weather, Field, Roads\} \prec \{\textbf{Game}\} \prec$$

$$\{Economy, Ticket\ Sale, Concession\}$$

Then the MEU for *Game* based on this ordering is:

$$MEU(Game) = \max_{G} \sum_{\{E,T,C\}} \sum_{U} U \times p(E,T,C \mid S,W,F,R,G)$$

Utility for Profit

Ticket Sale ⇒	high	low	none
Utility	5000	500	0

Utility for Profit

Concession ⇒	high	low	none
Utility	1000	100	0

Ticket Sale	Concession	Utility
high	high	6000
high	low	5100
high	none	5000
low	high	1500
low	low	600
low	none	500
none	high	1000
none	low	100
none	none	0

Figure 7-15: Utility potential

But the special collect operation on the strong junction tree in Figure 7-14 allows the MEU to be computed at node 5 (clique $\mathbf{C}_3$) as follows:

$$MEU(Game) =$$

$$\max_G \sum_{\{R,F\}} U(Cost) \times p(R,F,G) \sum_{\{E,T\}} p(R,G,T,E) \sum_C \sum_U U(P) \times p(T,C)$$

where $U(P)$ is the sum utility as shown in Figure 7-15. In this example, node 5 could also have been considered as the root node.

7.6   Further Readings

More details on the strong junction tree algorithm presented in this chapter can be found in (Jensen et al., 1994) and (Shenoy, 1992). From the historical perspective, Howard and Matheson (1981) first introduced influence diagrams, and their ability to model decision problems with uncertainty by transforming them into decision trees, for a single decision-maker. Shachter (1986) described a reduction-based approach for evaluating influence diagrams.

Chapter 8

Modal Logics for the Possible
World Epistemic Model

This chapter presents modal and epistemic logics for the possible world model that provide a foundation for decision-making agents. Unlike our earlier chapter on the logical (classical) and probabilistic foundations of epistemic models, in this chapter we start with a brief presentation of the history of modal logics since the field is not as widely known. We then introduce some relevant systems of modal logics, including $\mathcal{K}$, $\mathcal{D}$, $\mathcal{T}$, $\mathcal{S4}$, $\mathcal{B}$, and $\mathcal{S5}$, and their interrelationships. We develop possible world semantics of modal logics using accessibility relations and establish the associated soundness and completeness results. We then define a modal resolution scheme to efficiently reason within various systems of modal logics. We present the epistemic instantiation of modal logics as a way of reasoning with various mentalistic constructs of agents such as "belief" and "knowledge."

Modal logics are often criticized for their coarse-grained representation of knowledge of assertion possibilities. That is to say, if two assertions in an application are possible in the current world, their further properties are indistinguishable in the modal formalism even if an agent knows that one of them is true in twice as many of its possible worlds as compared to the other one. Epistemic logic, that is, the logic of knowledge and belief, cannot avoid this shortcoming because it inherits the syntax and semantics of modal logics. In the penultimate section of this chapter, we develop an extended formalism of modal epistemic logic that allows an agent to represent its *degrees of support* about an assertion. The degrees are drawn from qualitative and quantitative dictionaries that are accumulated from an agent's a priori knowledge about the application domain. Using the accessibility *hyperelation* concept, possible-world semantics of the extended logic and its rational extension, is developed, and the soundness and completeness results are established.

255

8.1   Historical Development of Modal Logics

Propositions in classical logics are categorized as either true or false. In modal logics, true propositions can be divided into necessary and contingently true propositions in a given context. No context allows straightforward inconsistency (that is, *A* and *not A* are true simultaneously). Within the context where mathematical and physical theories are accepted, propositions like "there are infinitely many prime numbers" and "the sum of two sides of a triangle is greater than the remaining third side" are true by the necessity of mathematical and physical theories. On the other hand, propositions like "Tajmahal is in India" and "Paris is the capital of France" are contingently true, that is, true only by their existence or specific conditions. Similarly, propositions like "Pyramids are in France" are contingently false propositions, and "there is only a finite number of prime numbers" or "Maria loves me and I am not loved" is false by the necessity of mathematical or logical properties and therefore termed as impossible. Figure 8-1 presents this categorization of propositions in modal logic.

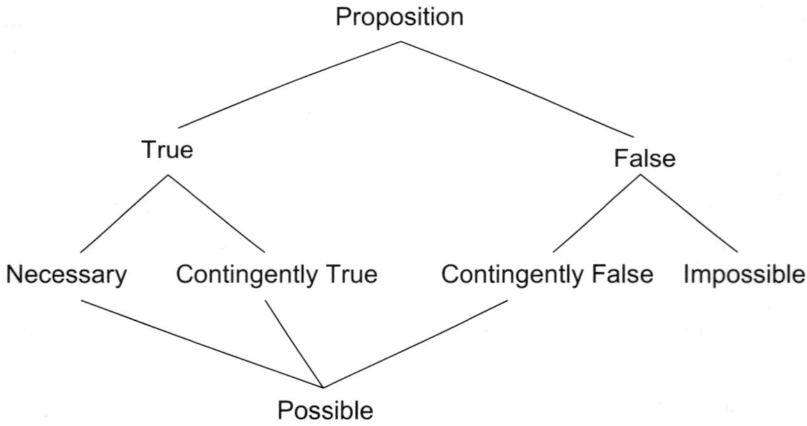

Figure 8-1: Categorization of propositions

Traditionally, the modality is restricted to the two basic modal concepts, possibility and necessity, and the ones that are definable in terms of these two. A logic based on these concepts is called a *classical modal logic*, and a *classical modal statement* "is a statement containing the words "necessary", "possible", or words definable in terms of these two words. However, the scope of modal logic has been widened by extensive work throughout the last century to include: 1) deontic modals such as obligations, permission, may, must, ought; 2) temporal modals such as since, after, next; and 3) epistemic modals such as knowledge and belief. Modal logic is a study of valid arguments involving modal concepts.

The origin of modal logics goes back to the work of Aristotle. In his work of *Organon, Prior Analytics of De Interpretation,* he developed a theory of modal statements. He called statements of the form "It is necessary ..." and "It is possible ..." apodeictic and problematic, respectively, and he called ordinary propositions assertoric statements. Aristotle developed a theory of a modal logic, called *modal syllogism* (a syllogism is an inference with two premises, major and minor, and one conclusion) by generalizing his theory of pure assertoric or categorical syllogism. An example modal syllogism from the 72 different types that Aristotle studied is as follows:

It is necessary that all M is P

All S is M

It is necessary that all S is P

Unlike categorical syllogism, Aristotle's modal syllogism contains a number of controversial ideas which were discussed by several authors (see Lukasiewicz's book (1951)). Aristotle also showed that the terms "necessary" and "possible" can be defined in terms of each other using negation.

Aristotle's modal syllogism remained almost neglected until Hugh MacColl published a series of papers in *Mind* during the late 19th century and early 20th century under the title *Symbolic Reasoning.* MacColl's work on symbolic reasoning is based on implication (:), which is different from the classical implication, and is defined in terms of classical implication using the necessity operator as follows:

$$A : B = (A \rightarrow B)^{\varepsilon}$$

But unlike Aristotle's modal syllogism, MacColl's study did not produce a complete axiomatic system.

The root of modern modal logics can be found in Lewis's book *Survey of Symbolic Logic* (Lewis, 1918; later also included in (Lewis and Langford, 1930)). Following MacColl, Lewis adopted a version of implication called *strict implication,* different from the "material implication" of Whitehead and Russel's *Principia Mathematica* (Whiteheadn and Russell, 1925-27). In Lewis's strict implication, "*P* implies *Q*" is synonymous with "*P* is deducible from *Q*." Lewis pointed out that this synonymy does not hold in the case of material implication. Using material implication leads to paradoxes like "a false proposition implies every proposition" and "a true proposition is implied by any proposition". Lewis also pointed out that if "P is consistent with Q" means "P does not imply the falsity of Q" and "Q is independent of P" means "P does not imply Q" then, in terms of material implication, no two propositions can be at once consistent and independent.

Let us consider the following example from Lewis' *Survey of Symbolic Logic*. Consider P as "roses are red" and Q as "sugar is sweet". P is consistent with Q, and P is independent of Q. Now, it is not the case that "roses are red" is true and "sugar is sweet" is false, since "sugar is sweet" is true. Thus, $\neg(P \wedge \neg Q)$, that is, $P \rightarrow Q$. But P does not strictly imply Q as "sugar is sweet" cannot be derived from "roses are red".

Lewis introduced self-consistency, or the possibility of a proposition P as $\Diamond P$, which is read as "P is self-consistent" or "P is possible". Then he defined strict implication in terms of negative possibility as follows (Lewis used conjunction $P \wedge Q$ as simply PQ and the symbol ' $\neg$ ' for negation):

$$P \propto Q \equiv \neg \Diamond (P \wedge \neg Q)$$

Thus, "P strictly implies Q" means "It is false that it is possible that P should be true and Q false." Consider an example where P is "he doesn't do his homework" and Q is "he will fail in the examination". Therefore, the negation of $P \propto Q$ is $\Diamond(P \wedge \neg Q)$, which means that it is possible he doesn't do his homework and he will not fail in the examination. The possibility of "he doesn't do his homework and he will not fail in the examination" suggests that $P \propto Q$ does not hold in this case as opposed to $P \rightarrow Q$. Lewis also distinguished between intensional and extensional disjunctions. Lewis argued that the meanings of either-or in these disjunctions are different from each other.

Unlike MacColl, Lewis developed several complete axiomatic systems of modal logic. The axiomatic system set out by him first in 1918, known as the *Survey*, takes logical impossibility as a primitive symbol. Later in his book, the impossibility concept was eliminated and the primitive idea of possibility, using the symbol $\Diamond$, was introduced. Impossibility was then defined as $\neg \Diamond$. This definition gave Lewis's system of strict implication the following set of axioms (with changes suggested by E. L. Post), where the strict implication $\propto$ is defined as $P \propto Q \equiv \neg \Diamond (P \wedge \neg Q)$:

A1 : $P \wedge Q \propto Q \wedge P$

A2 : $Q \wedge P \propto P$

A3 : $P \propto P \wedge P$

A4 : $P \wedge (Q \wedge R) \propto Q \wedge (P \wedge R)$

A5 : $P \propto \neg(\neg P)$

A6 : $((P \propto Q) \wedge (Q \propto R)) \propto (P \propto R)$

A7 : $\neg \Diamond P \propto \neg P$

A8 : $(P \propto Q) \propto (\neg \Diamond Q \propto \neg \Diamond P)$

Lewis called the axiom A8 as the *axiom of syllogism*. The system included multiple inference rules, including substitution, detachment, and adjunction, as follows:

- Substitution: The result of replacing at one or more places an expression in a theorem by its equivalent expression is itself a theorem. The result of replacing any variable in a theorem by any wff (well-formed formula) is itself a theorem.

- Detachment: If P is a theorem and $P \propto Q$ is a theorem, then so is Q. This is, of course, the modal logic equivalent of classical modus ponens

- Adjunction: If P and Q are theorems, then so is $P \wedge Q$. In classical logic adjunction is redundant because $P \wedge Q$ can be derived from P and Q.

McKinsey (1934) showed that A5 can be deduced from the rest of the axioms and inference rules. In his book, Lewis proposed a new set of axioms for various modal systems as follows:

$$B1\text{-}B6 : Same\, as\, A1\text{-}A6$$

$$B7 : (P \wedge (P \propto Q)) \propto Q$$

$$B8 : \Diamond(P \wedge Q) \propto \Diamond Q$$

$$B9 : (\exists P \exists Q) \neg (P \propto Q) \wedge \neg (P \propto \neg Q)$$

The first seven axioms (B1-B7) are deductively equivalent to the axioms A1-A7, and constitute Lewis' most primitive system *S1* of modal logics. The system *S2* is obtained by adding to *S1* the axiom B8, which Lewis called the *consistency postulate*. The axioms B1-B8 can be derived from his Survey system axioms A1-A8. The axiom B9 states that there are at least two propositions, say P and Q, such that P implies nothing about the truth or falsity of Q. Lewis introduced this axiom to prevent interpretation of the postulates B1-B8 of strict implication as a system for material implication by interpreting $\Diamond P$ as equivalent to P and the symbol '$\propto$' is syntactically equivalent to '$\rightarrow$'. Lewis argued that the formula $(P \rightarrow Q) \vee (P \rightarrow \neg Q)$ is a theorem of the system of material implication, but $(P \propto Q) \vee (P \propto \neg Q)$ contradicts B9.

One aspect of the study of modal systems is to determine properties of their modalities and modal functions, for example, whether $\Diamond P$ and $\neg\Diamond\neg\Diamond P$ are equivalent. A *modal function* is recursively defined as follows:

- Each propositional variable is a modal function of one variable.

- If F and G are modal functions, then each of $\neg F$, $\Diamond F$, $F \wedge G$ is a modal function of the number of variables it contains.

- Any function equivalent by definition to a modal function is a modal function.

A *modality* is a modal function of one variable in whose construction no operator other that $\Diamond$ and $\neg$ are used. For example, $\neg\Diamond\neg\Diamond P$ is a modality, whereas $\Diamond P \wedge \Diamond\neg P$ is a modal function but not a modality.

With a view that some modalities in Lewis's system have no intuitive interpretation, Becker (1930) proposed the following set of axioms, any one or more of which can be added to the Survey system (A1-8) to reduce the number of modalities:

$$C10 : \neg\Diamond\neg P \propto \neg\Diamond\neg\neg\Diamond\neg P$$
$$C11 : \Diamond P \propto \neg\Diamond\neg\Diamond P$$
$$C12 : P \propto \neg\Diamond\neg\Diamond P$$

Becker calls C12 as *Brouwerian Axiom*. C11 is deductively equivalent to C10 and C12. Becker also introduces an inference rule (we will refer to it as Becker's rule) as "if $P \propto Q$ is established as a theorem then so is $\Diamond P \propto \Diamond Q$."

The addition of C11 to the Survey system allows us to reduce all complex modalities to six, namely, P, $\neg P$, $\Diamond P$, $\neg\Diamond P$, $\neg\Diamond\neg P$, and $\Diamond\neg P$. Parry (1939) proved that there are only 42 non-reducible complex modalities in Lewis's Survey system. McKinsey (1940) demonstrated that *S2*, and hence *S1*, has an infinite number of complex modalities. In particular, he proved that all modalities of the form $\Diamond\Diamond...\Diamond$ or $\Diamond^n$ are not reducible, which makes Lewis's systems *open*.

Lewis in his book defined a necessarily true proposition P as "it is not logically inconceivable that P should be false" (pp.161), that is, $\neg\Diamond\neg P$ symbolically. Axioms C10-C12 can therefore be rewritten as follows with $\Box$ as the symbol of *necessary*:

$$C10 : \Box P \propto \Box\Box P$$
$$C11 : \Diamond P \propto \Box\Diamond P$$
$$C12 : P \propto \Box\Diamond P$$

Lewis's system *S4* is determined by B1-B7 (which is equivalent to A1-A7) and C10. Axiom A8 can be derived from *S4*, thus making *S4* stronger than the Survey system (*S3*). Lewis's system *S5* is determined by B1-B7 plus C11. The system *S5* is stronger than *S4*.

In the early 1930s, Gödel reviewed Becker's work in connection with intuitionistic logic, and formalized assertions of *provability* by a connective B, reading $B\alpha$ as "α is provable." Gödel defined an axiomatic system consisting of

$$B\alpha \to \alpha$$
$$B\alpha \to (B(\alpha \to \beta) \to B\beta)$$
$$B\alpha \to BB\alpha$$

and the inference rule "from α infer $B\alpha$." Gödel showed that the system is equivalent to Lewis's *S4* when $B\alpha$ is translated as $\Box\alpha$. Modal logics are currently presented in the above axiomatic style of Gödel.

The *Decision problem* of a logical system is to find whether an arbitrary given sentence of the system is provable. McKinsey (1941) was the first to offer a solution to the decision problem of *S2* and *S4* by means of algebraic interpretation via finite matrices. He first defined a matrix $\mathcal{M}$ as a quintuple $\langle \mathbf{K}, \mathbf{D}, -, *, \times \rangle$, where $\mathbf{K}$ is a set and $\mathbf{D}$ is a non-empty proper subset of $\mathbf{K}$, called the designated elements, $-$ and $*$ are unary functions defined over $\mathbf{K}$, and $\times$ is a binary function defined over $\mathbf{K}$. The symbols $-$, $*$, and $\times$ correspond to the negation, possibility, and conjunction of a modal system, respectively. A normal matrix is one that satisfies conditions corresponding to Lewis's rules of inference, adjunction, and replacement. McKinsey proved that a normal *S2* matrix is a Boolean algebra with certain characteristic properties, and provided a set of necessary and sufficient conditions for a matrix to be a normal *S2* matrix. He developed similar results for *S4*. McKinsey transformed the decision problem of a modal system to the decision problem in the equivalent Boolean algebra. McKinsey's proposal is interesting from theoretical point of view, but highly impractical as it requires a huge number of matrices be considered to decide on a formula.

During the late 1950s and early 1960s, Kripke published several papers on model theoretic semantics of modal logics. Kripke defined (1963) a normal model structure for the systems *S4*, *S5*, *T*, and *B* in terms of *possible worlds* and accessibility relation. Semantic tableaux methods were used to prove completeness results.

During the latter half of the last century, the traditional concept of modality (that is, "necessary" and "possible") was interpreted in a variety of ways, including deontic notions of obligation and permissibility, epistemic notions of knowledge and belief, and tense notions of now, since, until, etc. Our specific interest here is the epistemic interpretation that includes reasoning with an agent's knowledge. The formal study of epistemic logic was initiated by von Wright (1951), and his insights were extended by Hintikka (1962).

8.2   Systems of Modal Logic

Many scientists, logicians, and philosophers have proposed hundreds of systems of modal logics since the revival of the area by Lewis in the beginning of the last century. The aim here is not to cover each and every system, but introduce a few important ones that are useful and practical, and can potentially form a basis for representing and reasoning with agents' knowledge. From this perspective, we first lay down some needed requirements from candidate modal systems, including Lewis's distinction between strict and material implications and their relations to the modal notions of possibility and necessity. We then introduce some important modal systems that satisfy those requirements, including some of Lewis' with different axiomatic bases.

- Any modal system should contain "properly" the propositional logic or the system of material implication as termed by Lewis. This means that the axioms and inference rules of a candidate modal system should contain all the axioms and inference rules of propositional logic. In addition, the possibility operator $\Diamond$ should not be definable in terms of a combination of the logical connectives.

- The definition of the necessity of a proposition P, denoted as $\Box P$, is that it is not possible that P should be false. Symbolically, $\Box P \equiv \neg \Diamond \neg P$.

- P strictly implies Q means that it is not possible that P should be true and Q false. Symbolically, $P \propto Q \equiv \neg \Diamond (P \wedge \neg Q)$, that is, $P \propto Q \equiv \Box (P \rightarrow Q)$ since $\Box P \equiv \neg \Diamond \neg P$ and $P \rightarrow Q \equiv \neg (P \wedge \neg Q)$.

- Whatever follows logically from a necessarily true proposition is necessarily true. Symbolically, $\Box P \wedge (P \propto Q) \rightarrow \Box Q$.

- Valid propositions are not merely true but necessarily true. Symbolically, if $\vdash P$ then $\vdash \Box P$.

The second and third properties come from Lewis's own definitions of necessity and strict implication. The last two somewhat correspond to the inference rule in

Lewis's systems. The set of requirements listed above yield the most basic system $\mathcal{K}$ of modal logic. Next, we introduce this system formally and then bring additional modal properties to yield various other modal systems.

A formal axiomatic theory $\mathcal{K}$ for the modal logic is defined as follows:

- The symbols of $\mathcal{K}$ are a modal propositional alphabet:

    - Parentheses (and)

    - Logical connectives $\neg$ and $\rightarrow$

    - Modality $\Diamond$

    - Propositional variables $P, Q, R, \ldots$

- Formulae of $\mathcal{K}$ are inductively defined as follows:

    - All propositional variables are formulae.

    - If F is a formula then so are $\neg F$ and $\Diamond F$.

    - If F and G are formulae the so is $F \rightarrow G$.

    - An expression is a formula if and only if it can be shown to be a formula on the basis of the above three conditions.

- If F, G, and H are any formula of $\mathcal{K}$, then the following are axioms of $\mathcal{K}$:

    - Propositional axioms:

        PC1: $F \rightarrow (G \rightarrow F)$

        PC2: $F \rightarrow (G \rightarrow H) \rightarrow ((F \rightarrow G) \rightarrow (F \rightarrow H))$

        PC3: $(\neg G \rightarrow \neg F) \rightarrow (F \rightarrow G)$

    - Modal axiom K:

        K: $\Box(F \rightarrow G) \rightarrow (\Box F \rightarrow \Box G)$

- The following are the rules of inference of $\mathcal{K}$ ($\vdash$ means deducibility in $\mathcal{K}$):

    - Modus Ponens (MP): If $\vdash F$ and $\vdash F \rightarrow G$ then $\vdash G$

    - Rule of Necessitation (RN): If $\vdash F$ then $\vdash \Box F$

- The following are the definitions in $\mathcal{K}$:

$$F \wedge G \equiv \neg(F \rightarrow \neg G)$$
$$F \vee G \equiv \neg F \rightarrow G$$
$$F \leftrightarrow G \equiv (F \rightarrow G) \wedge (G \rightarrow F)$$
$$\Box F \equiv \neg \Diamond \neg F$$

It is clear that $\mathcal{K}$ includes the classical propositional calculus $\mathcal{PC}$.

An intuitively sound principle that can be considered as part of this very basic system of modal logic is that whatever is necessarily true is true. This is called the *axiom of necessity* or T axiom. Symbolically

$$T : \Box F \rightarrow F$$

But, for the purpose of reasoning with, for example, an agent's permissible and obligatory actions via deontic logic, the symbol $\Box$ will be interpreted "it is obligatory that." The axiom T in this context is intuitively unsatisfactory because whatever is obligatory is not necessarily true. The following weaker deontic axiom D is considered instead:

$$D : \Box F \rightarrow \Diamond F$$

For the purpose of reasoning with an agent's knowledge and belief via epistemic logic, the symbols $\Box$ and $\Diamond$ will be taken as knowledge and belief operators respectively, and interpreted as "agent knows that" and "agent believes that," respectively. In this context, an intuitively sound assumption is that an agent knows what it knows and knows its beliefs. These two principles yield the following two potential axioms 4 and E (C10 and C11, respectively, presented above) to be considered for modal systems-based epistemic logics:

$$4 : \Box F \rightarrow \Box \Box F$$
$$E : \Diamond F \rightarrow \Box \Diamond F$$

Finally, an intuitively sound principle is that if something is true then it is necessarily possible, which can be axiomatized as the following Brouwerian axiom B (axiom C12 presented in the historical development section):

$$B : F \rightarrow \Box \Diamond F$$

We now introduce a few renowned modal systems with $\mathcal{K}$ as their common foundation. Later we will study their properties and interrelationships. The system $\mathcal{D}$ is obtained from $\mathcal{K}$ by adding the axiom D into $\mathcal{K}$. The system $\mathcal{T}$ is obtained from $\mathcal{K}$ by adding T into $\mathcal{K}$. The systems that are obtained from $\mathcal{T}$ by adding separately the two axioms 4 and E are equivalent to Lewis's $\mathcal{S}4$ and $\mathcal{S}5$, respectively. The system $\mathcal{B}$ is obtained from $\mathcal{T}$ by adding the axiom B. The system $S5'$ is obtained from $\mathcal{T}$ by adding axioms 4 and E.

8.3 Deductions in Modal Systems

This section presents some deduction principles and techniques, and then derives some well-known results within the modal systems $\mathcal{K}$, $\mathcal{D}$, $\mathcal{T}$, $\mathcal{B}$, $S4$, and $S5$ to be used later for establishing the soundness and completeness results. The bracketed information alongside a derivation step lists any axioms, steps, or inference rules used to derive the step (PL stands for Propositional Logic).

8.3.1 Principle of Duality

Proof by duality eliminates tedious repetitions. The dual of a formula F, written as F^*, is obtained from F by repeatedly applying the following set of rules:

DR1: $P^* = P$, where P is a propositional variable

DR2: $(\neg F)^* = F^*$

DR3: $(F \wedge G)^* = F^* \vee G^*$

DR4: $(F \vee G)^* = F^* \wedge G^*$

DR5: $(F \rightarrow G)^* = \neg(G^* \rightarrow F^*)$

DR6: $(F \leftrightarrow G)^* = \neg(F^* \leftrightarrow G^*)$

DR7: $(\Box F)^* = \Diamond(F^*)$

DR8: $(\Diamond F)^* = \Box(F^*)$

Note that if a formula only contains $\neg$, $\wedge$, $\vee$, $\Diamond$, and $\Box$ symbols, then its dual can be obtained by replacing each propositional variable by its negation and interchanging all occurrences of $\wedge$ with $\vee$, and $\Diamond$ with $\Box$.

Example

Steps for computing the dual of $\Box(P \rightarrow Q) \rightarrow (\Box P \rightarrow \Box Q)$ are given below:

Step 1: $(\Box(P \rightarrow Q) \rightarrow (\Box P \rightarrow \Box Q))^*$ [Dual]

Step 2: $\neg((\Box P \rightarrow \Box Q)^* \rightarrow (\Box(P \rightarrow Q))^*)$ [DR 5]

Step 3: $\neg(\neg((\Box Q)^* \rightarrow (\Box P)^*) \rightarrow \Diamond(P \rightarrow Q)^*)$ [DR 5 & DR 7]

Step 4: $\neg(\neg(\Diamond(Q^*) \to \Diamond(P^*)) \to \Diamond\neg(Q^* \to P^*))$ [DR 5 & DR 7]

Step 5: $\neg(\neg(\Diamond Q \to \Diamond P) \to \Diamond\neg(Q \to P))$ [DR 1]

Step 6: $\neg(\neg\Diamond\neg(Q \to P) \to (\Diamond Q \to \Diamond P))$ [PL]

Result

If $\vdash F \to G$ then $\vdash G^* \to F^*$

If $\vdash F \leftrightarrow G$ then $\vdash F^* \leftrightarrow G^*$

8.3.2   Theorems of $\mathcal{K}$

TK 1: **If $\vdash F \to G$ then $\vdash \Box F \to \Box G$**

Step 1: $\vdash F \to G$

Step 2: $\vdash \Box(F \to G)$ RN]

Step 3: $\vdash \Box(F \to G) \to (\Box F \to \Box G)$ [Axiom K]

Step 4: $\vdash \Box F \to \Box G$ [Step 2, Step 3, & MP]

TK 2: **If $\vdash F \to G$ then $\vdash \Diamond F \to \Diamond G$**

Step 1: $\vdash F \to G$

Step 2: $\vdash \neg G \to \neg F$ [PL]

Step 3: $\vdash \Box(\neg G \to \neg F)$ [RN]

Step 4: $\vdash \Box\neg G \to \Box\neg F$ [TK 1]

Step 5: $\vdash \neg\Box\neg F \to \neg\Box\neg G$ [PL]

Step 6: $\vdash \Diamond F \to \Diamond G$ [Definition]

TK 3: $\Box(F \wedge G) \leftrightarrow (\Box F \wedge \Box G)$

Step 1: $F \wedge G \to F$	[PL]
Step 2: $\Box(F \wedge G \to F)$	[RN]
Step 3: $\Box(F \wedge G) \to \Box F$	[TK 1]
Step 4: $\Box(F \wedge G) \to \Box G$	[Similar to Step 3]
Step 5: $\Box(F \wedge G) \to \Box F \wedge \Box G$	[PL, Step 3, & Step 4]
Step 6: $F \to (G \to F \wedge G)$	[PL]
Step 7: $\Box(F \to (G \to F \wedge G))$	[Step 6 & RN]
Step 8: $\Box F \to \Box(G \to F \wedge G)$	[Step 7 & TK 1]
Step 9: $\Box(G \to F \wedge G) \to (\Box G \to \Box(F \wedge G))$	[Axiom K & RS]
Step 10: $\Box F \to (\Box G \to \Box(F \wedge G))$	[Step 8, Step 9, & PL]
Step 11: $\Box F \wedge \Box G \to \Box(F \wedge G)$	[Step 10 & PL]
Step 12: $\Box(F \wedge G) \leftrightarrow (\Box F \wedge \Box G)$	[Step 5 & Step 11]

TK 4: **If** $\vdash F \wedge G \to H$ **then** $\vdash \Box F \wedge \Box G \to \Box H$

Step 1: $\vdash F \wedge G \to H$	
Step 2: $\vdash \Box(F \wedge G) \to \Box H$	[TK 1 & RS]
Step 3: $\vdash \Box F \wedge \Box G \to \Box H$	[TK 3 & RS]

TK 5: $\vdash_{\mathcal{K}} \Box F \vee \Box G \to \Box(F \vee G)$

Step 1: $\vdash_{\mathcal{K}} F \to F \vee G$	[PL]
Step 2: $\vdash_{\mathcal{K}} \Box F \to \Box(F \vee G)$	[TK 1]
Step 3: $\vdash_{\mathcal{K}} \Box G \to \Box(F \vee G)$	[Similar to Step 2]
Step 4: $\vdash_{\mathcal{K}} \Box F \vee \Box G \to \Box(F \vee G)$	[Step 2, Step 3, & PL]

TK 6: $\Diamond(F \vee G) \leftrightarrow (\Diamond F \vee \Diamond G)$

TK6 follows from TK 3 by applying the principle of duality.

TK 7: $\Diamond(F \wedge G) \rightarrow (\Diamond F \wedge \Diamond G)$

Step 1: $F \wedge G \rightarrow F$	[PC]
Step 2: $\Diamond(F \wedge G) \rightarrow \Diamond F$	[TK 2]
Step 3: $\Diamond(F \wedge G) \rightarrow \Diamond G$	[Similar to Step 2]
Step 4: $\Diamond(F \wedge G) \rightarrow \Diamond F \wedge \Diamond G$	[Step2, Step 3 & PL]

8.3.3   Theorems of $\mathcal{D}$

TD 1: $\vdash_{\mathcal{D}} \Diamond F \vee \Diamond \neg F$

Step 1: $\vdash_{\mathcal{D}} \Box F \rightarrow \Diamond F$	[Axiom D]
Step 2: $\vdash_{\mathcal{D}} \neg \Diamond \neg F \rightarrow \Diamond F$	[Definition of $\Box$]
Step 3: $\vdash_{\mathcal{D}} \Diamond F \vee \Diamond \neg F$	[PL]

TD 2: $\vdash_{\mathcal{D}} \Diamond \top$ or $\vdash_{\mathcal{D}} \neg \Box \bot$

Step 1: $\vdash_{\mathcal{D}} \Diamond(\neg \top \vee \top) \leftrightarrow (\Diamond \neg \top \vee \Diamond \top)$	[TK 6]
Step 2: $\vdash_{\mathcal{D}} \Diamond \top \leftrightarrow (\Box \top \rightarrow \Diamond \top)$	[PL & Defn. of $\Box$]
Step 3: $\vdash_{\mathcal{D}} \Box \top \rightarrow \Diamond \top$	[Axiom D]
Step 4: $\vdash_{\mathcal{D}} \Diamond \top$	[Step 2, Step 3, & PL]

8.3.4   Theorems of $\mathcal{T}$

TT 1: $\vdash_{\mathcal{T}} F \rightarrow \Diamond F$

    Step 1: $\vdash_{\mathcal{T}} \Box\neg F \rightarrow \neg F$ [Axiom T]

    Step 2: $\vdash_{\mathcal{T}} \neg\neg F \rightarrow \neg\Box\neg F$ [PL]

    Step 3: $\vdash_{\mathcal{T}} F \leftrightarrow \neg\neg F$ [PL]

    Step 4: $\vdash_{\mathcal{T}} F \rightarrow \Diamond F$ [Step 2, Step 3, PL, and Defn. of $\Diamond$]

TT 2: $\vdash_{\mathcal{T}} \Box F \rightarrow \Diamond F$

Follows immediately from Axiom T and TT 1.

8.3.5   Theorems of *S4*

TS4 1: $\vdash_{S4} \Box F \leftrightarrow \Box\Box F$

    Step 1: $\vdash_{S4} \Box F \rightarrow F$ [Axiom T]

    Step 2: $\vdash_{S4} \Box\Box F \rightarrow \Box F$ [TK 1]

    Step 3: $\vdash_{S4} \Box F \rightarrow \Box\Box F$ [Axiom 4]

    Step 4: $\vdash_{S4} \Box F \leftrightarrow \Box\Box F$ [Step 2, Step 3, & PL]

TS4 2: $\vdash_{S4} \Diamond F \leftrightarrow \Diamond\Diamond F$

TS4 2 follows by applying the duality principle on TS4 1.

TS4 3: $\vdash_{S4} \Box F \rightarrow \Box\Diamond\Box F$

    Step 1: $\vdash_{S4} \Box\Box F \rightarrow \Diamond\Box F$ [Axiom D]

Step 2: $\vdash_{S4} \Box\Box\Box F \to \Box\Diamond\Box F$ [TK 1]

Step 3: $\vdash_{S4} \Box F \leftrightarrow \Box\Box\Box F$ [From TS4 1 & PL]

Step 4: $\vdash_{S4} \Box F \to \Box\Diamond\Box F$ [Step 2, Step 3 & PL]

TS4 4: $\vdash_{S4} \Diamond\Box\Diamond F \to \Diamond F$

TS4 4 follows by applying the duality principle on TS4 3.

TS4 5: $\vdash_{S4} \Box\Diamond\Box\Diamond F \leftrightarrow \Box\Diamond F$

Step 1: $\vdash_{S4} \Diamond\Box\Diamond F \to \Diamond F$ [TS4 4]

Step 2: $\vdash_{S4} \Box\Diamond\Box\Diamond F \to \Box\Diamond F$ [TK 1]

Step 3: $\vdash_{S4} \Box\Diamond F \to \Box\Diamond\Box\Diamond F$ [TS4 3 & RS]

Step 4: $\vdash_{S4} \Box\Diamond\Box\Diamond F \leftrightarrow \Box\Diamond F$ [Step 2, Step 3 & PL]

TS4 6: $\vdash_{S4} \Diamond\Box\Diamond\Box F \leftrightarrow \Diamond\Box F$

TS4 6 follows by applying the duality principle on TS4 5.

8.3.6 *Theorems of $\mathcal{B}$*

TB 1: $\vdash_{\mathcal{B}} \Diamond\Box F \to F$

Step 1: $\vdash_{\mathcal{B}} \neg F \to \Box\Diamond\neg F$ [Axiom B & RS]

Step 2: $\vdash_{\mathcal{B}} \neg\Box\Diamond\neg F \to F$ [PL]

Step 3: $\vdash_{\mathcal{B}} \Diamond\Box F \to F$ [Definition of $\Box$ and $\Diamond$]

TB 2: If $\vdash_\mathcal{B} \Diamond F \to G$ then $\vdash_\mathcal{B} F \to \Box G$

Step 1: $\vdash_\mathcal{B} \Diamond F \to G$

Step 2: $\vdash_\mathcal{B} \Box \Diamond F \to \Box G$ [TK 1]

Step 3: $\vdash_\mathcal{B} F \to \Box \Diamond F$ [Axiom B]

Step 4: $\vdash_\mathcal{B} F \to \Box G$ [Step 2, Step 3, & PL]

8.3.7   Theorems of S5

TS5 1: $\vdash_{S5} \Diamond \Box F \to \Box F$

Step 1: $\vdash_{S5} \Diamond \neg F \to \Box \Diamond \neg F$ [Axiom E and RS]

Step 2: $\vdash_{S5} \neg \Box \Diamond \neg F \to \neg \Diamond \neg F$ [PL]

Step 3: $\vdash_{S5} \Diamond \Box F \to \Box F$ [Definition of $\Box$ and $\Diamond$]

TS5 2: $\vdash_{S5} F \to \Box \Diamond F$

Step 1: $\vdash_{S5} \Diamond F \to \Box \Diamond F$ [Axiom E]

Step 2: $\vdash_{S5} F \to \Diamond F$ [TT 1]

Step 3: $\vdash_{S5} F \to \Box \Diamond F$ [Step 1, Step 2, & PL]

TS5 3: $\vdash_{S5} \Box F \to \Box \Box F$

Step 1: $\vdash_{S5} \Box \Diamond \Box F \to \Box \Box F$ [TS5 1 & TK 1]

Step 2: $\vdash_{S5} \Box F \to \Box \Diamond \Box F$ [TS5 2 & RS]

Step 3: $\vdash_{S5} \Box F \to \Box \Box F$ [Step 1, Step 2, & PL]

8.3.8   Theorems of S5'

TS5' 1: $\vdash_{S5'} \Diamond F \rightarrow \Box \Diamond F$

Step 1: $\vdash_{S5'} \Diamond\Diamond F \rightarrow \Diamond F$		[TS4 2]
Step 2: $\vdash_{S5'} \Box\Diamond\Diamond F \rightarrow \Box\Diamond F$		[TK 1]
Step 3: $\vdash_{S5'} \Diamond F \rightarrow \Box\Diamond\Diamond F$		[B & RS]
Step 4: $\vdash_{S5'} \Diamond F \rightarrow \Box\Diamond F$		[Step 2, Step 3, & PL]

8.4   Modality

A *modality* is a finite sequence of symbols $\Diamond$, $\Box$, and $\neg$. The null sequence represents the *null modality*, denoted as φ. Examples of modalities are $\Diamond$, $\Box$, $\Box\neg\Diamond$ and, $\neg\Box\neg\Box\Diamond\neg$. If a modality contains an even or zero number of occurrences of $\neg$ then it is called *affirmative*; otherwise, the modality is *negative*. Due to the definition of $\Box$ as $\neg\Diamond\neg$, each affirmative modality can be reduced to a modality without the negation symbol. Also, each negative modality can be reduced to a modality of the form $M\neg$, where M does not contain any negation symbol. We define the following four kinds of modalities:

- Type A: Affirmative A-modalities of the form M beginning with $\Box$
- Type B: Affirmative B-modalities of the form M beginning with $\Diamond$
- Type C: Negative A-modalities of the form $M\neg$ beginning with $\Box$
- Type D: Negative B-modalities of the form $M\neg$ beginning with $\Diamond$

The *degree* of a modality is the number of times the symbol $\Diamond$ or $\Box$ occurs. For example, the degree of $\neg\Box\neg\Box\Diamond\neg$ is 3 and that of P or $\neg P$ is 0. A *proper modality* is a modality of degree higher than zero.

<u>Example</u>

Results TS4 1 and TS4 2 assert that in every modality in *S4* two of $\Diamond$ or $\Box$ in immediate succession can be replaced respectively by one $\Diamond$ or $\Box$. Based on these results, there are only 3 Type A modalities in *S4*: $\Box$ (degree 1), $\Box\Diamond$

(degree 2), $\square\lozenge\square$ (degree 3). An affirmative A-modality of the form $\square\lozenge\square\lozenge$ reduces to $\square\lozenge$ by TS4 5. Similarly, there are only three Type B modalities in *S4*, namely $\lozenge$, $\lozenge\square$, and $\lozenge\square\lozenge$, that are duals of three Type A modalities. The set of 12 irreducible proper modalities of *S4* are listed below:

Type A Modality of *S4*: $\square$, $\square\lozenge$, $\square\lozenge\square$

Type B Modality of *S4*: $\lozenge$, $\lozenge\square$, $\lozenge\square\lozenge$

Type C Modality of *S4*: $\square\neg$, $\square\lozenge\neg$, $\square\lozenge\square\neg$

Type D Modality of *S4*: $\lozenge\neg$, $\lozenge\square\neg$, $\lozenge\square\lozenge\neg$

The following implications hold between modalities in *S4*:

$$\square P \rightarrow \square\lozenge\square P \rightarrow \begin{Bmatrix} \lozenge\square P \\ \square\lozenge P \end{Bmatrix} \rightarrow \lozenge\square\lozenge P \rightarrow \lozenge P$$

$$\square\neg P \rightarrow \square\lozenge\square\neg P \rightarrow \begin{Bmatrix} \lozenge\square\neg P \\ \square\lozenge\neg P \end{Bmatrix} \rightarrow \lozenge\square\lozenge\neg P \rightarrow \lozenge\neg P$$

$$\square P \rightarrow P \rightarrow \lozenge P$$

$$\square\neg P \rightarrow \neg P \rightarrow \lozenge\neg P$$

The 14 modalities of *S4*, 12 irreducible ones plus φ and $\neg$ cannot be further reduced.

8.5 Decidability and Matrix Method

A matrix method can be used to decide whether a given formula in a modal system is valid or not. This method is a generalization of the truth-table method in propositional calculus. The method defines a matrix $\mathcal{M}$ as a quintuple $\langle \mathbf{K}, \mathbf{D}, -, *, \supset \rangle$, where $\mathbf{K}$ is a set of possible *values* and $\mathbf{D}$ is a non-empty proper subset of $\mathbf{K}$, called the *designated values*, $-$ and $*$ are unary functions defined over $\mathbf{K}$, and $\supset$ is a binary function defined over $\mathbf{K}$. Intuitively, the symbols $-$, $*$, and $\supset$ correspond to negation, possibility, and implication of a modal system, respectively. The triplet $\langle \mathbf{K}, -, \supset \rangle$ is a Boolean algebra defined as follows:

BL1: $\mathbf{K}$ contains at least two elements,

with two distinguished elements 0 and 1

BL2: If $a, b \in \mathbf{K}$ then $-a, a \supset b, a \times b, a + b \in \mathbf{K}$,

where $a \times b = -(a \supset -b)$, $a + b = -a \supset b$

BL3: $a \times (b + c) = a \times b + a \times c$, for all $a, b, c \in \mathbf{K}$

BL4: $a + 0 = a$ and $a \times 1 = a$, for all $a \in \mathbf{K}$

BL5: $a + {-a} = 1$ and $a \times {-a} = 0$, for all $a \in \mathbf{K}$

Intuitively, the symbols $\times$ and $+$ correspond to conjunction and disjunction of a logical system, respectively. The relation between $\langle \mathbf{K}, -, \supset \rangle$ and the propositional logic can be described as follows.

We first define a valuation or assignment V of a propositional formula F as some member $V(F)$ of $\mathbf{K}$. V assigns each variable P in F some member of $V(P)$ of $\mathbf{K}$ and then evaluates $V(F)$ satisfying the following properties:

V1: If $V(F) = a$ then $V(\neg F) = -a$

V2: If $V(F) = a$ and $V(G) = b$ then $V(F \rightarrow G) = a \supset b$

The algebra $\langle \mathbf{K}, -, \supset \rangle$ validates a formula F if and only if for every valuation V to the variables in F, $V(F) = 1$. It can be proven that a propositional formula is valid if and only if every Boolean algebra validates it. We shall present a similar result for modal systems by developing extended Boolean algebras.

We add to Boolean algebra $\langle \mathbf{K}, -, \supset \rangle$ the unary operator $*$ corresponding to the modal notion of possibility, and then define the $\mathcal{K}$ algebra or matrix $\langle \mathbf{K}, \mathbf{D}, -, *, \supset \rangle$ with the following additional properties:

BL6: If $a \in \mathbf{K}$ then $*a, \#a \in \mathbf{K}$, where $\#a = -*-a$

BL7: If $a \in \mathbf{D}$ then $\#a \in \mathbf{D}$

BL8: If $a, b \in \mathbf{D}$ then if $\#(a \supset b) \in \mathbf{D}$ then $\#a \supset \#b \in \mathbf{D}$

Intuitively, the symbol $\#$ corresponds to the modal notion of necessity. The property BL7 above is due to the rule of necessitation of the modal system $\mathcal{K}$ and the property BL8 corresponding to the axiom K of $\mathcal{K}$. In the case of propositional logic, $\mathbf{K}$ is $\{0, 1\}$ and $\mathbf{D}$ is $\{1\}$.

As for the valuation of the formula in a given modal system, we first choose the sets $\mathbf{K}$ and $\mathbf{D}$, and define operator matrices for each primitive operator. The valuation V is extended for modal formulae by including the following:

V3: If $V(F) = a$ then $V(\Diamond F) = *a$

These operator matrices are then used to compute the value of any formulae of the system. If the value of a formula is not a member of the designated set, then it is considered an invalid formula. The set of matrices for the operators have the following properties:

- Every axiom of the system takes one of the designated values for every assignment of values to the variables. For example, if $\Box(F \to G) \to (\Box F \to \Box G)$ is an axiom of the modal system under consideration, then for every assignment of values from **K** to F and G, the value of the axiom should be a member from **D**.

- Every inference rule of the system can be verified for every assignment of values to the variables. For example, if "If $\vdash F$ and $\vdash F \to G$ then $\vdash G$" is an inference rule of the underlying modal system, then for every assignment of values from **K** to F and G for which the values of F and $F \to G$ are members of the designated set **D**, the value of G should also be a member of **D**.

The above two conditions together ensure that every formula derived from the axioms and inference rules will also take one of the designated values for any assignment of values to the variables.

Validity of a formula can be guaranteed by applying McKinsey's theorem, which states that a formula F containing n sub-sentences is valid in the modal system $\mathcal{K}$ if and only if it is verified by every $\mathcal{K}$ algebra containing at most $2^{2^{n}}$ elements. By a sub-sentence G of a modal formula F we mean a well-formed part of F written in its primitive notations $\neg$, $\to$ and $\Diamond$. For example, the formula $P \to (\neg \Diamond Q \to \Diamond P)$ has the following seven well-formed parts: $P, Q, \Diamond P, \Diamond Q, \neg \Diamond Q, \neg \Diamond Q \to \Diamond P$, and $P \to (\neg \Diamond Q \to \Diamond P)$. To verify the validity of this formula, we need to consider every $\mathcal{K}$ algebra containing not more than $2^{2^{7}}$ elements, which is simply not practical. But the procedure is useful to check non-validity of formulae in the context of a modal system by constructing simple, small matrices as counter examples.

Example

Consider a matrix $\mathcal{M} = \langle \mathbf{K}, \mathbf{D}, -, *, \supset \rangle$, where $\mathbf{K} = \{0, 1, 2, 3\}$ and $\mathbf{D} = \{1\}$. The operators are defined via tables shown in Figure 8-2. It can be easily verified that $\mathcal{M}$ satisfies BL1-BL8. Therefore, $\mathcal{M}$ is a $\mathcal{K}$ algebra.

$$\mathcal{M} = \langle \mathbf{K}, \mathbf{D}, -, *, \supset \rangle$$

$$\mathbf{K} = \{0, 1, 2, 3\}$$

$$\mathbf{D} = \{1\}$$

a	$-a$	$*a$	$\#a$
0	1	0	3
1	0	2	1
2	3	3	3
3	2	2	2

$a \supset b$	0	1	2	3
0	1	1	1	1
1	0	1	2	3
2	3	1	1	3
3	2	1	2	1

$a \times b$	0	1	2	3
0	0	0	0	0
1	0	1	2	3
2	0	2	2	0
3	0	3	0	3

$a + b$	0	1	2	3
0	0	1	2	3
1	1	1	1	1
2	2	1	2	1
3	3	1	1	3

Figure 8-2: Matrices defining $\mathcal{K}$ algebra

Applying McKinsey's result as stated above, it can be verified easily that the formula $\square P \rightarrow \Diamond P$ is not valid in $\mathcal{K}$. Since $\square P \rightarrow \Diamond P$ has 4 well-formed parts, if it is valid in $\mathcal{K}$, then the formula should be verified by every $\mathcal{K}$ algebra containing at most 2^{2^4} elements. In particular, the formula should be verified by the above example algebra with only 4 elements. But considering 1 as an assignment for P, the valuation of the formula is $\#1 \supset *1$, that is, 2, which is not a member of the designated set $\{1\}$.

We shall rely on this algebraic approach to establish relationships among various modal systems. But first we introduce a few more algebras corresponding to various modal systems.

BL9: If $a \in \mathbf{K}$ then if $\#a \in \mathbf{D}$ then $*a \in \mathbf{D}$

BL10: If $a \in \mathbf{K}$ then if $\#a \in \mathbf{D}$ then $a \in \mathbf{D}$

BL11: If $a \in \mathbf{K}$ then if $\#a \in \mathbf{D}$ then $\#\#a \in \mathbf{D}$

BL12: If $a \in \mathbf{K}$ then if $*a \in \mathbf{D}$ then $\#*a \in \mathbf{D}$

BL13: If $a \in \mathbf{K}$ then if $a \in \mathbf{D}$ then $\#*a \in \mathbf{D}$

A $\mathcal{D}$ algebra is an algebra satisfying BL1-BL9, and a $\mathcal{T}$ algebra is the one which also satisfies BL10. An $\mathcal{S4}$ algebra is an algebra satisfying BL1-BL11, an

S5 algebra is an algebra satisfying BL1-BL10 and BL12, and finally, a *B* algebra is an algebra satisfying BL1-BL10 and BL13.

8.6 Relationships among Modal Systems

The system *D* is obtained from *K* by adding the axiom D into *K*. The system *T* is obtained by adding T into *K*. The systems that are obtained from *T* by adding separately three axioms 4, E, and B are *S4*, *S5*, and *B*, respectively. The system *S5'* is obtained from *T* by adding axioms 4 and E. The results TS5 2 and TS5 3 prove that *S5'* is a subsystem of *S5* and the result TS5' 1 proves that *S5* is a subsystem of *S5'*. Therefore, the two systems *S5* and *S5'* are equivalent. We have the following relationship:

$$\mathcal{K} \subseteq \mathcal{D} \subseteq \mathcal{T} \subseteq \left\{ \begin{array}{c} S4 \\ \mathcal{B} \end{array} \right\} \subseteq S5$$

where $\subseteq$ represents the subsystem relationship, that is, $X \subseteq Y$ means every theorem of X is also a theorem of Y. Note that a logical system is defined as the set of all theorems that are deducible using its axioms and inference rules. Next, we prove by constructing examples that the subsystem relationships among the modal systems *K*, *D*, *T*, *S4*, *B*, and *S5* as presented above are "proper" or $\subset$ in the sense that there is a theorem of *D* which is not a theorem of *K*, and there is a theorem of *T* which is not a theorem of *D*, and so on. To prove $\mathcal{K} \subseteq \mathcal{D}$, we will construct, for example, a *K* algebra and show that the algebra does not satisfy axiom D. Then McKinsey's result proves that D is not a theorem of *K*.

Consider a matrix $\mathcal{M} = \langle \mathbf{K}, \mathbf{D}, -, *, \supset \rangle$, where $\mathbf{K} = \{0, 1, 2, 3\}$ and $\mathbf{D} = \{1\}$. A set of * operators are defined via tables shown in Figure 8-3, of which (a) is the *K* algebra shown in Figure 8-2.

- $\mathcal{K} \subset \mathcal{D}$: Consider the * operator as defined in Figure 8-3(a). $\langle \mathbf{K}, \mathbf{D}, -, *, \supset \rangle$ is a *K* algebra. Assigning P to 1, the valuation of the axiom $\Box P \rightarrow \Diamond P$ of *D* is 2, which is not a member of the designated set $\{1\}$. Therefore, $\Box P \rightarrow \Diamond P$ is not a theorem of *K*.

- $\mathcal{D} \subset \mathcal{T}$: Consider the * operator as defined in Figure 8-3(b). For every assignment of P, the valuation of $\Box P \rightarrow \Diamond P$ always takes the value 1. Therefore, $\langle \mathbf{K}, \mathbf{D}, -, *, \supset \rangle$ is a *D* algebra. But, assigning P to 2, the valuation of the axiom $\Box P \rightarrow P$ of *T* is 2, which is not a member of the designated set $\{1\}$. Therefore, $\Box P \rightarrow P$ is not a theorem of *D*.

a	$*a$	$\#a$
0	0	3
1	2	1
2	3	3
3	2	2

a	$*a$	$\#a$
0	0	0
1	1	1
2	1	3
3	2	0

a	$*a$	$\#a$
0	3	0
1	1	2
2	1	0
3	1	0

a	$-a$
0	1
1	0
2	3
3	2

(a) $\mathcal{K}$ Algebra   (b) $\mathcal{D}$ Algebra   (c) $\mathcal{T}$ Algebra

$\mathcal{K} \subset \mathcal{D}$ $\mathcal{D} \subset \mathcal{T}$ $\mathcal{T} \subset S4, \mathcal{B}$

$a \supset b$	0	1	2	3
0	1	1	1	1
1	0	1	2	3
2	3	1	1	3
3	2	1	2	1

a	$*a$	$\#a$
0	0	0
1	1	1
2	2	0
3	1	3

a	$*a$	$\#a$
0	0	0
1	1	1
2	2	0
3	1	3

a	$*a$	$\#a$
0	0	0
1	1	1
2	2	0
3	1	3

(d) $S4$ Algebra   (e) $S4$ Algebra   (f) $\mathcal{B}$ Algebra

$S4 \neq \mathcal{B}$ $S4 \subset S5$ $\mathcal{B} \subset S5$

Figure 8-3: Example algebras for establishing relationships among modal
systems

- $\mathcal{T} \subset S4, \mathcal{B}$: Consider the * operator as defined in Figure 8-3(c). For every assignment of P, the valuation of $\Box P \rightarrow P$ always has the value 1. Therefore, $\langle \mathbf{K}, \mathbf{D}, -, *, \supset \rangle$ is a $\mathcal{T}$ algebra. But, assigning P to 1, the valuation of the axiom $\Box P \rightarrow \Box \Box P$ of $S4$ is 3, which is not a member of the designated set $\{1\}$. Therefore, $\Box P \rightarrow \Box \Box P$ is not a theorem of $\mathcal{T}$. Similarly, assigning P to 1, the valuation of the axiom $P \rightarrow \Box \Diamond P$ of $\mathcal{B}$ is 2, which is not a member of the designated set $\{1\}$. Therefore, $P \rightarrow \Box \Diamond P$ is not a theorem of $\mathcal{T}$.

- $S4 \neq \mathcal{B}$: Left as an exercise

- $S4 \subset S5$: Left as an exercise

- $\mathcal{B} \subset S5$: Left as an exercise

Figure 8-4 shows the relationship among the modal systems $\mathcal{K}$, $\mathcal{D}$, $\mathcal{T}$, $S4$, $\mathcal{B}$, and $S5$ introduced so far.

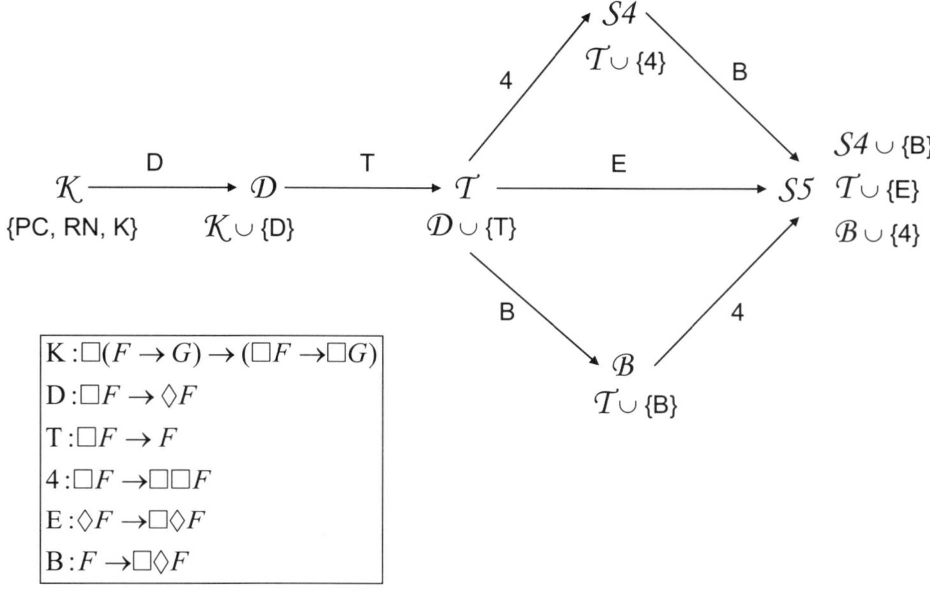

Figure 8-4: Relationships among modal systems

In Figure 8-4, an arrow $\rightarrow$ from X to Y with label A means that the system Y can be obtained from the system X by adding axiom A, and that every theorem of X is also a theorem of X but not vice-versa. See (Chellas, 1980) for more relationships that hold among the subsets of the six axioms K, D, T, 4, E, and B. For example, $\mathcal{KTE} = \mathcal{KTB}4 = \mathcal{KDB}4 = \mathcal{KDBE}$, where $\mathcal{KTE}$ means the modal system containing axioms K, T, and E, and so on.

8.7   Possible World Semantics

For the purpose of reasoning with agent intelligence, an agent's knowledge is represented by a finite set of modal sentences from an underlying epistemological theory. These sentences constitute the proper axioms of the theory. The model theoretic semantics of such modal sentences are given in terms of possible worlds.

By a real world we do *not* mean the planet earth on which we live, but all those items existing in the universe, events that have occurred, and propositions and relations that characterize the items and events. An item is either a physical object such as a Pyramid, tree, Empire State building, moon, or tiger; an abstract object such as the set of prime numbers or classical logic; an individual such as Lincoln, Gandhi, Evita, or Brando. Example events are the assassination of John

Lennon or the Gulf War. Properties such as being supersonic or being tall characterize items (for example, Concord is supersonic and the Eiffel Tower is tall) Relations, such as being a father to, being a colder place, or being located, characterize items (for example, London is colder than Cairo).

Now we begin defining the term "possible world" by including the real or current world as one of the possible worlds based on the fact that if something actually exists then it is possible that it exists. Every possible world, which is not the real world, will possess one or more of the following properties:

- It has more objects, events, and individuals than the real one, such as a cancer cure, a bridge between the earth and the moon, a third world war, a ten feet tall person, a Mars expedition, or Superman.

- It has less objects, events, and individuals than the real one, such as no Pyramids, no Gulf war, and no Shakespeare.

- It differs with respect to properties, for example, the Sydney opera house is red, and I am a multi-millionaire.

- It differs with respect to relations, for example, the Great Wall of China is located in the Middle East.

The set of possible worlds is defined depending on the circumstances or applications at hand. For example, when logical reasoning occurs about what can be done with the current technology, the set of all possible worlds is the set of all technologically possible worlds. In this case, the worlds other than the technologically possible worlds will be considered impossible. Similarly, when logical reasoning occurs in the domain of physics, the set of all possible worlds is the set of all physically possible worlds. The set of all worlds in each of which only valid mathematical theories that exist today are used is the set of all mathematically possible worlds. Therefore, the set of all mathematically possible worlds contains the set of all physically possible worlds. A world where it is assumed that the 3-satisfiability problem is P-complete is mathematically possible. A world in which straightforward inconsistency (P and *not P*) can be detected is considered inconsistent and impossible. Therefore, irrespective of the circumstance or application, the set of all impossible worlds contains the set of all inconsistent worlds. The relations between various types of possible worlds as discussed here are shown in Figure 8-5.

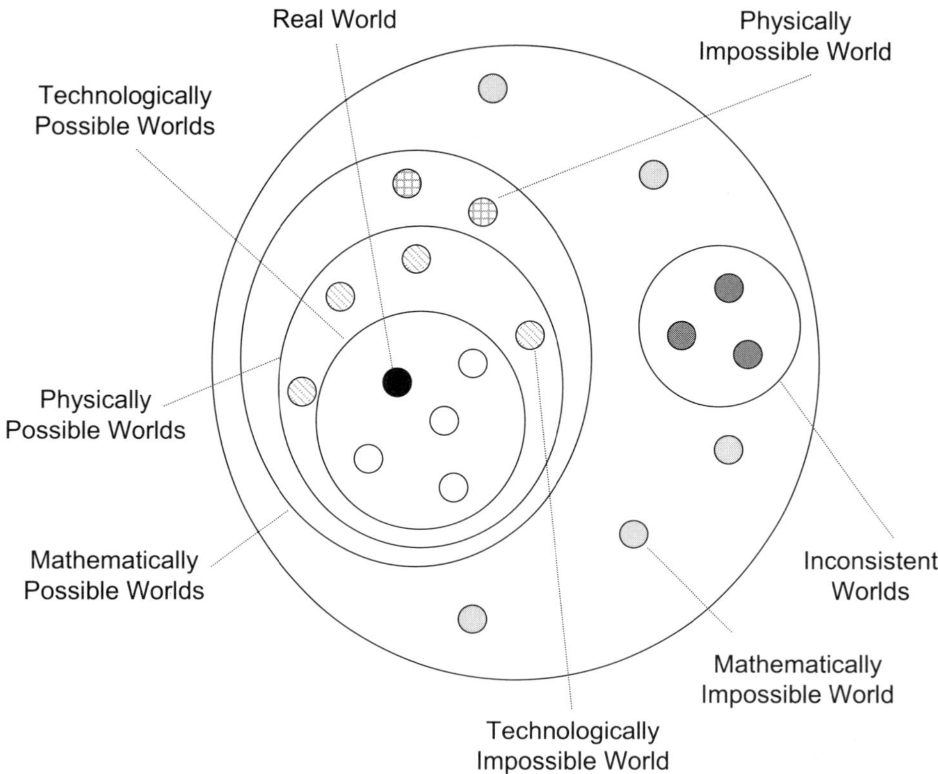

Figure 8-5: Categorization of worlds

A *normal model* $\mathcal{M}$ is a structure of the following form:

$$\mathcal{M} = \langle \mathbf{W}, R, B \rangle$$

where $\mathbf{W}$ is a non-empty set of elements, to be called *worlds*; R is any binary relation on $\mathbf{W}$, to be called the *accessibility relation* for $\mathbf{W}$; and B is a *valuation* function that determines the truth value of a proposition in a world as follows:

$$B : \mathbf{P} \times \mathbf{W} \to \{\top, \bot\}$$

Thus if $B(P,w)$ is $\top$ or $\bot$ then we say that the proposition P is true or false, respectively, at the world w in a model $\mathcal{M}$, and this is written as follows:

$$\vDash_w^{\mathcal{M}} P$$

Equivalently, B can be defined as a function to assign a set of propositions to each world w that are true in w:

$$B : \mathbf{W} \to \mathcal{P}(\mathbf{P})$$

A third equivalent alternative, which we have adopted here, defines B as a function which determines for each proposition P a set $B(P)$ of worlds in each of which P is true:

$$B : \mathbf{P} \to \mathcal{P}(\mathbf{W})$$

For an arbitrary formula F, the *satisfiability* of a closed formula F with respect to a model $\mathcal{M} = \langle \mathbf{W}, R, B \rangle$ and a world $w \in \mathbf{W}$, written as $\models_w^{\mathcal{M}} F$, is recursively defined as follows:

$\models_w^{\mathcal{M}} P$ iff $w \in B(P)$

$\models_w^{\mathcal{M}} \neg F$ iff $\not\models_w^{\mathcal{M}} F$

$\models_w^{\mathcal{M}} F \to G$ iff $\models_w^{\mathcal{M}} F$ implies $\models_w^{\mathcal{M}} G$

$\models_w^{\mathcal{M}} \Box F$ iff for every $w' \in \mathbf{W}$ such that wRw', $\models_{w'}^{\mathcal{M}} F$

Proposition 8-1: $\models_w^{\mathcal{M}} \Diamond F$ if and only if there exists w' such that wRw' and $\models_{w'}^{\mathcal{M}} F$.

Proof: $\models_w^{\mathcal{M}} \Diamond F$

 iff $\models_w^{\mathcal{M}} \neg \Box \neg F$, by the definition of $\Diamond$

 iff $\not\models_w^{\mathcal{M}} \Box \neg F$, by the definition of satisfiability

 iff there is a $w' \in \mathbf{W}$ such that wRw' and $\not\models_w^{\mathcal{M}} \neg F$,

          by the definition of satisfiability

 iff there is a $w' \in \mathbf{W}$ such that wRw' and $\models_w^{\mathcal{M}} \neg \neg F$,

          by the definition of satisfiability

 iff there is a $w' \in \mathbf{W}$ such that wRw' and $\models_w^{\mathcal{M}} F$.

A formula F is *satisfiable* iff $\models_w^{\mathcal{M}} F$, for some model $\mathcal{M} = \langle \mathbf{W}, R, B \rangle$ and $w \in \mathbf{W}$. F is *valid* in $\mathcal{M}$, written as $\models^{\mathcal{M}} F$, iff $\models_w^{\mathcal{M}} F$, for every $w \in \mathbf{W}$. F is *valid* in a class of models Σ, written as $\models^{\Sigma} F$, iff $\models^{\mathcal{M}} F$, for every $\mathcal{M} \in \Sigma$. F is valid, written as $\models F$, iff F is *valid* in every class of models.

We now prepare the development of the soundness and completeness results. The notion of connected models is pertinent in this context. Connected models are defined in terms of ancestral or transitive closure relationships among possible worlds. Suppose ρ_1 and ρ_2 are two binary relations on a set $\mathbf{X}$. The relation ρ_1 is called the *ancestral* of the relation ρ_2 if the following condition is satisfied:

$x\rho_1 y$ iff for all $\mathbf{X}' \subseteq \mathbf{X}$ such that if $x \in \mathbf{X}'$ and $\{z : \exists z'(z' \in \mathbf{X}' \,\&\, z'\rho_2 z)\} \subseteq \mathbf{X}$ then $y \in \mathbf{X}'$. Alternatively, the *transitive closure* or *ancestral* ρ^* of a binary relation ρ on a set $\mathbf{X}$ is recursively defined as follows:

- $x\rho^* y$ if $x\rho y$

- $x\rho^* y$ if $\exists z$ such that $x\rho z$ and $z\rho^* y$

If $x\rho^* y$ then we call y a *descendant* of x. A model $\mathcal{M} = \langle \mathbf{W}, R, B \rangle$ is *connected* if and only if there exists $w \in \mathbf{W}$ such that $\mathbf{W} = \{w\} \cup \{w' : wRw'\}$. Then a model is connected if we can find a world w such that $\mathbf{W}$ is the collection of w together with all of its descendants.

Consider model $\mathcal{M} = \langle \mathbf{W}, R, B \rangle$ and suppose $w \in \mathbf{W}$. Then the *connected model* $\mathcal{M}_w = \langle \mathbf{W}_w, R_w, B_w \rangle$ of $\mathcal{M}$ generated by $\mathcal{M}$ is defined as follows:

$$\mathbf{W}_w = \{w\} \cup \{w' : wR^* w'\}$$

$$R_w = \{\langle w, w' \rangle : wRw' \,\&\, w' \in \mathbf{W}_w\}$$

$$B_w(P) = B(P) \cap \mathbf{W}_w$$

Proposition 8-2: $\mathcal{M}_w = \langle \mathbf{W}_w, R_w, B_w \rangle$ is connected.

Proof: Follows from the definition.

We shall now show that only those worlds that are descendants to w are relevant in determining the truth value of a formula F wrt to a world w.

Theorem 8-1: Suppose $\mathcal{M} = \langle \mathbf{W}, R, B \rangle$ is a model and $w \in \mathbf{W}$. Then $\models_w^{\mathcal{M}} F$ iff $\models_w^{\mathcal{M}_w} F$.

Proof: We prove the theorem by induction on the number of connectives of F.

Case I: $F = P$, that is, F is a proposition without any connective. $\models_w^{\mathcal{M}} P$ iff $w \in B(P)$, and $\models_w^{\mathcal{M}_w} P$ iff $w \in B_w(P)$ iff $w \in B(P) \cap \mathbf{W}_w$. By definition, $w \in \mathbf{W}_w$ and thus $w \in B(P)$ iff $w \in B(P) \cap \mathbf{W}_w$. Therefore, $\models_w^{\mathcal{M}} P$ iff $\models_w^{\mathcal{M}_w} P$.

Case II: $F = \neg G$. $\models_w^{\mathcal{M}} F$ iff $\not\models_w^{\mathcal{M}} G$, and $\models_w^{\mathcal{M}_w} F$ iff $\not\models_w^{\mathcal{M}_w} G$. By induction hypothesis, $\models_w^{\mathcal{M}} G$ iff $\models_w^{\mathcal{M}_w} G$, that is, $\not\models_w^{\mathcal{M}} G$ iff $\not\models_w^{\mathcal{M}_w} G$. Therefore, $\models_w^{\mathcal{M}} F$ iff $\models_w^{\mathcal{M}_w} F$.

Case III: $F = G \rightarrow H$. $\models_w^{\mathcal{M}} F$ iff $\models_w^{\mathcal{M}} G \rightarrow H$ iff $\models_w^{\mathcal{M}} G$ implies $\models_w^{\mathcal{M}} H$, and $\models_w^{\mathcal{M}_w} F$ iff $\models_w^{\mathcal{M}_w} G$ implies $\models_w^{\mathcal{M}_w} H$. By induction hypothesis, $\models_w^{\mathcal{M}} G$ iff $\models_w^{\mathcal{M}_w} G$ and $\models_w^{\mathcal{M}} H$ iff $\models_w^{\mathcal{M}_w} H$. Therefore, $\models_w^{\mathcal{M}} F$ iff $\models_w^{\mathcal{M}_w} F$.

Case IV: $F = \Box G$. $\models_w^{\mathcal{M}} F$ iff $\models_w^{\mathcal{M}} \Box G$ iff for every $w' \in \mathbf{W}$ such that wRw', $\models_{w'}^{\mathcal{M}} G$, and $\models_w^{\mathcal{M}_w} F$ iff $\models_w^{\mathcal{M}_w} \Box G$ iff for every $w' \in \mathbf{W}$ such that $wR_w w'$, $\models_{w'}^{\mathcal{M}_w} G$. We prove $\models_w^{\mathcal{M}} F$ iff $\models_w^{\mathcal{M}_w} F$ by establishing the equivalence between "for every $w' \in \mathbf{W}$ such that wRw', $\models_{w'}^{\mathcal{M}} G$" and "for every $w' \in \mathbf{W}$ such that $wR_w w'$, $\models_{w'}^{\mathcal{M}_w} G$".

Suppose, for every $w' \in \mathbf{W}$ such that wRw', $\models_{w'}^{\mathcal{M}} G$, and $wR_w w'$, for some $w' \in \mathbf{W}$. Since, $wR_w w'$, by definition, wRw' and $w' \in \mathbf{W}_w$. Therefore, $\models_{w'}^{\mathcal{M}} G$. By induction hypothesis, $\models_{w'}^{\mathcal{M}_w} G$.

Conversely, suppose, for every $w' \in \mathbf{W}$ such that $wR_w w'$, $\models_{w'}^{\mathcal{M}_w} G$, and wRw', for some $w' \in \mathbf{W}$. Since $w \in \mathbf{W}$ and wRw', by definition, $w' \in \mathbf{W}_w$, that is, $wR_w w'$. Therefore, $\models_{w'}^{\mathcal{M}_w} G$, and by induction hypothesis, $\models_{w'}^{\mathcal{M}} G$.

Thus, $\models_w^{\mathcal{M}} \Box G$ iff $\models_w^{\mathcal{M}_w} \Box G$, that is, $\models_w^{\mathcal{M}} F$ iff $\models_w^{\mathcal{M}_w} F$.

Corollary 8-1: $\models F$ iff $\models^{\Sigma} F$, for the class Σ of connected models.

Proof: The forward half is straightforward. To prove the converse, suppose, $\not\models F$. Then there exists a model $\mathcal{M} = \langle \mathbf{W}, R, B \rangle$ and $w \in \mathbf{W}$ such that $\not\models_w^{\mathcal{M}} F$. The above theorem states that $\models_w^{\mathcal{M}} F$ iff $\models_w^{\mathcal{M}_w} F$. Therefore, $\not\models_w^{\mathcal{M}} F$ iff $\not\models_w^{\mathcal{M}_w} F$, which yields a contradiction as $\mathcal{M}_w$ is connected. Therefore, we have $\models F$ iff $\models^{\Sigma} F$.

Theorem 8-2: If $\vDash F$ then $\vDash \Box F$

Proof: Consider a model $\mathcal{M} = \langle \mathbf{W}, R, B \rangle$ and suppose $\vDash^{\mathcal{M}} F$. Therefore, $\vDash^{\mathcal{M}}_w F$, for every $w \in \mathbf{W}$. To prove $\vDash^{\mathcal{M}}_w \Box F$, suppose wRw', for some w'. By the definition of $\vDash F$, we have $\vDash^{\mathcal{M}}_{w'} F$. Hence, $\vDash^{\mathcal{M}}_w \Box F$, and therefore, $\vDash \Box F$, since w and $\mathcal{M}$ are chosen arbitrarily.

Theorem 8-3: $\vDash \Box(F \to G) \to (\Box F \to \Box G)$

Proof: Consider a model $\mathcal{M} = \langle \mathbf{W}, R, B \rangle$, and suppose $\vDash^{\mathcal{M}}_w \Box(F \to G)$ and $\vDash^{\mathcal{M}}_w \Box F$, for some $w \in \mathbf{W}$. To prove $\vDash^{\mathcal{M}}_w \Box G$, consider world $w' \in W$ such that wRw'. Then, $\vDash^{\mathcal{M}}_{w'} F \to G$ and $\vDash^{\mathcal{M}}_{w'} F$, yielding $\vDash^{\mathcal{M}}_{w'} G$. Therefore, $\vDash^{\mathcal{M}}_w \Box G$. $\vDash^{\mathcal{M}}_w \Box(F \to G) \to (\Box F \to \Box G)$, and thus, $\vDash \Box(F \to G) \to (\Box F \to \Box G)$ follows from the deduction theorem in $\mathcal{PL}$.

Theorem 8-4: Consider a model $\mathcal{M} = \langle \mathbf{W}, R, B \rangle$ and a world $w \in \mathbf{W}$. Then the following properties hold:

If R is serial, $\vDash^{\mathcal{M}}_w \Box F \to \Diamond F$

If R is reflexive, $\vDash^{\mathcal{M}}_w \Box F \to F$

If R is reflexive and symmetric, $\vDash^{\mathcal{M}}_w F \to \Box \Diamond F$

If R is reflexive and transitive, $\vDash^{\mathcal{M}}_w \Box F \to \Box \Box F$

If R is reflexive and euclidean, $\vDash^{\mathcal{M}}_w \Diamond F \to \Box \Diamond F$

Proof:

$\vDash^{\mathcal{M}}_w \Box F \to \Diamond F$: Suppose $\vDash^{\mathcal{M}}_w \Box F$. Then for every world $w' \in \mathbf{W}$ such that wRw', $\vDash^{\mathcal{M}}_{w'} F$. Since R is serial, there exists a world $w' \in \mathbf{W}$ such that wRw'. Therefore, there is a world $w' \in \mathbf{W}$ such that wRw', $\vDash^{\mathcal{M}}_{w'} F$, that is, $\vDash^{\mathcal{M}}_w \Diamond F$.

$\vDash^{\mathcal{M}}_w \Box F \to F$: Suppose $\vDash^{\mathcal{M}}_w \Box F$. Then for every world $w' \in \mathbf{W}$ such that wRw', $\vDash^{\mathcal{M}}_{w'} F$. Since R is reflexive, wRw, and thus $\vDash^{\mathcal{M}}_w F$.

$\vDash^{\mathcal{M}}_w F \to \Box \Diamond F$: Suppose $\vDash^{\mathcal{M}}_w F$, and, if possible, assume $\nvDash^{\mathcal{M}}_w \Box \Diamond F$. Since $\nvDash^{\mathcal{M}}_w \Box \Diamond F$, there is a world $w' \in \mathbf{W}$ such that wRw' and $\nvDash^{\mathcal{M}}_{w'} \Diamond F$. Since $\nvDash^{\mathcal{M}}_{w'} \Diamond F$, there is no world $w'' \in \mathbf{W}$ such that $w'Rw''$ and $\vDash^{\mathcal{M}}_{w''} F$. But R is symmetric, and

wRw' means $w'Rw$. Moreover, $\models_w^{\mathcal{M}} F$. Thus, there is a world $w''=w$ such that $w'Rw''$ and $\models_{w''}^{\mathcal{M}} F$, yielding a contradiction. Therefore, $\models_w^{\mathcal{M}} \Box\Diamond F$.

$\models_w^{\mathcal{M}} \Box F \rightarrow \Box\Box F$: Suppose $\models_w^{\mathcal{M}} \Box F$, and, if possible, assume $\not\models_w^{\mathcal{M}} \Box\Box F$. Since, $\not\models_w^{\mathcal{M}} \Box\Box F$, there is a world $w' \in \mathbf{W}$ such that wRw' and $\not\models_{w'}^{\mathcal{M}} \Box F$. Since $\not\models_{w'}^{\mathcal{M}} \Box F$, there is a world $w'' \in \mathbf{W}$ such that $w'Rw''$ and $\not\models_{w''}^{\mathcal{M}} F$. But R is transitive, and wRw' and $w'Rw''$ mean wRw''. Therefore, $\models_w^{\mathcal{M}} \Box F$ means $\models_{w''}^{\mathcal{M}} F$, yielding a contradiction. Therefore, $\models_w^{\mathcal{M}} \Box\Box F$.

$\models_w^{\mathcal{M}} \Diamond F \rightarrow \Box\Diamond F$: Suppose $\models_w^{\mathcal{M}} \Diamond F$, and, if possible, assume $\not\models_w^{\mathcal{M}} \Box\Diamond F$. Since $\models_w^{\mathcal{M}} \Diamond F$, there is a world $w' \in \mathbf{W}$ such that wRw' and $\models_{w'}^{\mathcal{M}} F$. Since $\not\models_w^{\mathcal{M}} \Box\Diamond F$, there is a world $w'' \in \mathbf{W}$ such that wRw'' and $\not\models_{w''}^{\mathcal{M}} \Diamond F$. Since, $\not\models_{w''}^{\mathcal{M}} \Diamond F$, there is no world in relation to w'' where F is true. But R is euclidean, and wRw'' and wRw' mean $w''Rw'$. Therefore, there is a world w' in $\mathbf{W}$ such that $w''Rw'$ and $\models_{w'}^{\mathcal{M}} F$, yielding a contradiction. Therefore, $\models_w^{\mathcal{M}} \Box\Diamond F$.

8.8   Soundness and Completeness Results

Theorem 8-5 and Theorem 8-6 below provide the soundness results for the systems $\mathcal{K}$, $\mathcal{D}$, $\mathcal{T}$, $\mathcal{B}$, $\mathcal{S}4$, and $\mathcal{S}5$.

Theorem 8-5: If $\vdash_{\mathcal{K}} F$ then $\models F$

Proof: The MP rule preserves the truth, that is, if $\models F$ and $\models F \rightarrow G$ then $\models G$. Taking this and Theorem 8-2 and Theorem 8-3 into consideration, the rest of the proof can be established easily by applying induction on the number of connectives in a formula.

Theorem 8-6: Consider model $\mathcal{M} = \langle \mathbf{W}, R, B \rangle$. Then the following properties hold:

If $\vdash_{\mathcal{D}} F$ then $\models^{\mathcal{M}} F$, if R is serial

If $\vdash_{\mathcal{T}} F$ then $\models^{\mathcal{M}} F$, if R is reflexive

If $\vdash_{\mathcal{B}} F$ then $\models^{\mathcal{M}} F$, if R is reflexive and symmetric

If $\vdash_{S4} F$ then $\models^{\mathcal{M}} F$, if R is reflexive and transitive

If $\vdash_{S5} F$ then $\models^{\mathcal{M}} F$, if R is reflexive and euclidean

Proof: Follows from Theorem 8-4.

To prove the completeness for S, where S is one of $\mathcal{K}$, $\mathcal{D}$, $\mathcal{T}$, $\mathcal{B}$, $S4$, and $S5$, we consider the following *canonical model* $\mathcal{M}_S = \langle \mathbf{W}_S, R_S, B_S \rangle$ of S:

- $\mathbf{W}_S$ is the set of all maximal consistent extensions of S

- $wR_S w'$ iff $\{F : \Box F \in w\} \subseteq w'$

- $B_S(P) = \{w : P \in w\}$

Proposition 8-3: $wR_S w'$ **iff** $\{\Diamond F : F \in w'\} \subseteq w$

Proof: Suppose, $wR_S w'$ and $F \in w'$, and, if possible, let $\Diamond F \notin w$. Since w is maximal consistent, and therefore complete, $\neg \Diamond F \in w$, that is, $\Box \neg F \in w$. Then, by the definition of $wR_S w'$, $\neg F \in w'$. Therefore, both F and $\neg F$ are in w', making w' inconsistent, which is a contradiction. Therefore, $\Diamond F \in w$.

Conversely, suppose $\{\Diamond F : F \in w'\} \subseteq w$, and, if possible, let $\{F : \Box F \in w\} \subsetneq w'$. This means that there is F for which $\Box F \in w$ but $F \notin w'$. Since $F \notin w'$ and w' is maximal consistent, $\neg F \in w'$. Thus, $\Diamond \neg F \in w$, that is, $\neg \Box F \in w$. This contradicts with $\Box F \in w$ as w is consistent. Therefore, $\{F : \Box F \in w\} \subseteq w'$, that is, $wR_S w'$.

Theorem 8-7: In a canonical model $\mathcal{M}_S = \langle \mathbf{W}_S, R_S, B_S \rangle$ of a system S, $\vdash_S F$ iff $\models^{\mathcal{M}_S} F$.

Proof: $\vdash_S F$ iff, for all $w \in \mathbf{W}_S$, $F \in w$. Thus, to prove the theorem we only need to prove that $\models_w^{\mathcal{M}_S} F$ iff $F \in w$, for every $w \in \mathbf{W}_S$. We apply the induction principle on the number of connectives in F.

Suppose, $\models_w^{\mathcal{M}_S} F$.

Case I: F is P. Then $\models_w^{\mathcal{M}_S} P$ implies that $w \in B_S(P)$, that is, $P \in w$.

Case II: F is $\neg G$. Then $\models_w^{\mathcal{M}_S} \neg G$ implies that $\not\models_w^{\mathcal{M}_S} G$, that is, $G \notin w$, by induction hypothesis. Since w is maximal consistent, $\neg G \in w$.

Case III: F is $G \rightarrow H$. Then $\models_w^{\mathcal{M}_S} G \rightarrow H$ implies that if $\models_w^{\mathcal{M}_S} G$ then $\models_w^{\mathcal{M}_S} H$. Therefore, by induction hypothesis, if $G \in w$ then $H \in w$. Since w is maximal consistent, $G \rightarrow H \in w$.

Case IV: F is $\Box G$. Then $\models_w^{\mathcal{M}_S} \Box G$ implies that for all $w' \in \mathbf{W}_S$ such that $wR_S w'$, $\models_{w'}^{\mathcal{M}_S} G$. By induction hypothesis, $\models_w^{\mathcal{M}_S} F$ implies that for all $w' \in \mathbf{W}_S$ such that $wR_S w'$, $G \in w'$. Now, $wR_S w'$ iff $\{H : \Box H \in w\} \subseteq w'$. Thus, G belongs to every maximal consistent extension of $\{H : \Box H \in w\}$ within the system S. Therefore, $\{H : \Box H \in w\} \vdash_S G$, by applying the maximal consistency property. Thus, there exist $H_1, ..., H_n$ in $\{H : \Box H \in w\}$ such that $\Box H_1, ..., \Box H_n \in w$ and $\{H_1, ..., H_n\} \vdash_S G$. Therefore, $\vdash_S H_1 \wedge ... \wedge H_n \rightarrow G$, that is, $\vdash_S \Box H_1 \wedge ... \wedge \Box H_n \rightarrow \Box G$, by a generalization of TK 4. Since $\Box H_1, ..., \Box H_n \in w$ and w is maximal consistent, $\Box G \in w$.

Conversely, suppose, $F \in w$.

Case I: F is P. Then $P \in w$ implies that $w \in B_S(P)$, that is, $\models_w^{\mathcal{M}_S} P$

Case II: F is $\neg G$. Then $\neg G \in w$ implies that $G \notin w$, by the maximal consistency of w. By induction hypothesis, $\not\models_w^{\mathcal{M}_S} G$, that is, $\models_w^{\mathcal{M}_S} \neg G$, that is, $\models_w^{\mathcal{M}_S} F$.

Case III: F is $G \rightarrow H$. Then $G \rightarrow H \in w$ implies that if $G \in w$ then $H \in w$, by the maximal consistency of w. By induction hypothesis, if $\models_w^{\mathcal{M}_S} G$ then $\models_w^{\mathcal{M}_S} H$, that is, $\models_w^{\mathcal{M}_S} G \rightarrow H$, that is, $\models_w^{\mathcal{M}_S} F$.

Case IV: F is $\Box G$. Then $\Box G \in w$. Suppose, $wR_S w'$, for some $w' \in \mathbf{W}_S$. By definition of canonical model, $\{H : \Box H \in w\} \subseteq w'$. Therefore, $G \in w'$, that is, $\models_{w'}^{\mathcal{M}_S} G$. Thus, $\models_w^{\mathcal{M}_S} \Box G$, that is, $\models_w^{\mathcal{M}_S} F$.

Theorem 8-8: In a canonical model $\mathcal{M}_S = \langle \mathbf{W}_S, R_S, B_S \rangle$ of a system S, where S ranges over $\mathcal{K}$, $\mathcal{D}$, $\mathcal{T}$, $\mathcal{B}$, $S4$, or $S5$, the following holds:

$R_{\mathcal{D}}$ is serial

$R_{\mathcal{T}}$ is reflexive

$R_{\mathcal{B}}$ is reflexive and symmetric

R_{S4} is reflexive and transitive

R_{S5} is reflexive and euclidean

Proof:

$\mathcal{M}_{\mathcal{D}} = \langle \mathbf{W}_{\mathcal{D}}, R_{\mathcal{D}}, B_{\mathcal{D}} \rangle$: We have to prove the existence of $w' \in \mathbf{W}_{\mathcal{D}}$ such that $wR_{\mathcal{D}}w'$. Now, $wR_{\mathcal{D}}w'$ iff $\{F : \Box F \in w\} \subseteq w'$. Consider the set $\Gamma = \{F : \Box F \in w\}$. If Γ is not $\mathcal{D}$-consistent, there exists $F_1, ..., F_n \in \Gamma$ such that each $\Box F_1, ..., \Box F_n \in w$ and $\vdash_{\mathcal{D}} \neg F$, where $F = F_1 \wedge ... \wedge F_n$. By RN, $\vdash_{\mathcal{D}} \Box \neg F$, that is, $\vdash_{\mathcal{D}} \neg \Diamond F$. Since w is a maximal consistent extension of $\mathcal{D}$, $\neg \Diamond F \in w$. By a generalization of TK 1, $\Box F = \Box F_1 \wedge ... \wedge \Box F_n$. Since each $\Box F_i \in w$ and w is maximally consistent, $\Box F \in w$. Since w is a maximal consistent set containing $\mathcal{D}$, $\Box F \to \Diamond F \in w$. Therefore, $\Diamond F \in w$, which violates the consistency of w. Therefore, Γ is consistent and there will be a maximal consistent extension w' of Γ such that $\Gamma \subseteq w'$, that is, $\{F : \Box F \in w\} \subseteq w'$, that is, $wR_{\mathcal{D}}w'$.

$\mathcal{M}_{\mathcal{T}} = \langle \mathbf{W}_{\mathcal{T}}, R_{\mathcal{T}}, B_{\mathcal{T}} \rangle$: We have to prove that for every $w \in \mathbf{W}_{\mathcal{T}}$, $wR_{\mathcal{T}}w$, that is, $\{F : \Box F \in w\} \subseteq w$. Since $\mathcal{T}$ (and therefore w as well) contains $\Box F \to F$, $F \in w$ for each $\Box F$ in w. Thus, $\{F : \Box F \in w\} \subseteq w$ and therefore, $wR_{\mathcal{T}}w$.

$\mathcal{M}_{\mathcal{B}} = \langle \mathbf{W}_{\mathcal{B}}, R_{\mathcal{B}}, B_{\mathcal{B}} \rangle$: We have to prove that for every $w, w' \in \mathbf{W}_{\mathcal{B}}$, if $wR_{\mathcal{B}}w'$ then $w'R_{\mathcal{B}}w$. Suppose, $wR_{\mathcal{B}}w'$, that is, $\{F : \Box F \in w\} \subseteq w'$. To prove $w'R_{\mathcal{B}}w$, we have to show that $\{F : \Box F \in w'\} \subseteq w$. Consider the set $\Gamma = \{F : \Box F \in w'\}$, and, if possible, let $\Box F \in w'$ and $F \in \Gamma$ but $F \notin w$. Since w is maximal consistent, $\neg F \in w$. Also, w contains $\mathcal{B}$ means $\Diamond \Box F \to F \in w$ by TB 1, and thus $\neg \Diamond \Box F \in w$, that is, $\Box \neg \Box F \in w$. Since $\{F : \Box F \in w\} \subseteq w'$, $\neg \Box F \in w'$, which yields a contradiction. Therefore, $\{F : \Box F \in w'\} \subseteq w$, that is, $w'R_{\mathcal{B}}w$.

$\mathcal{M}_{S4} = \langle \mathbf{W}_{S4}, R_{S4}, B_{S4} \rangle$: Since $S4$ contains $\mathcal{T}$, R_{S4} is reflexive. We only have to prove that R_{S4} is transitive, that is, for every $w, w', w'' \in \mathbf{W}_{S4}$, if $wR_{S4}w'$ and $w'R_{S4}w''$ then $wR_{S4}w''$. Since $wR_{S4}w'$ and $w'R_{S4}w''$, $\{F : \Box F \in w\} \subseteq w'$ and

$\{F : \Box F \in w'\} \subseteq w''$. Consider $\Gamma = \{F : \Box F \in w\}$. To prove, $wR_{S4}w''$, that is, $\Gamma \subseteq w''$, suppose $F \in \Gamma$. By definition of Γ, $\Box F \in w$. Also, w contains $S4$ means $\Box F \rightarrow \Box\Box F \in w$, and thus $\Box\Box F \in w$, since w is maximal consistent. Since $\{F : \Box F \in w\} \subseteq w'$, $\Box F \in w'$. Since $\{F : \Box F \in w'\} \subseteq w''$, $F \in w''$. Therefore, $\Gamma \subseteq w''$, that is, $\{F : \Box F \in w\} \subseteq w''$, that is, $wR_{S4}w''$.

$\mathcal{M}_{S5} = \langle \mathbf{W}_{S5}, R_{S5}, B_{S5} \rangle$: Since $S5$ contains T, R_{S5} is reflexive. We only have to prove that R_{S5} is euclidean, that is, for every $w, w', w'' \in \mathbf{W}_{\mathcal{B}}$, if $wR_{S5}w'$ and $wR_{S5}w''$ then $w'R_{S5}w''$. Since $wR_{S5}w'$ and $wR_{S5}w''$, $\{F : \Box F \in w\} \subseteq w'$ and $\{F : \Box F \in w\} \subseteq w''$. Consider $\Gamma = \{F : \Box F \in w'\}$. To prove, $w'R_{S5}w''$, that is, $\Gamma \subseteq w''$, suppose $F \in \Gamma$. By definition of Γ, $\Box F \in w'$, that is, $\neg\Box\neg F \in w'$, that is, $\Box\neg F \notin w'$, since w' is maximal consistent containing Axiom D. If $F \notin w''$ then $F \notin \{F : \Box F \in w\}$, since $\{F : \Box F \in w\} \subseteq w''$. Therefore, $\neg\Box F \in w$, since w is maximal consistent. By TS5 1 and maximal consistency of w containing $S5$, $\Diamond\Box F \rightarrow \Box F \in w$. Thus, $\neg\Diamond\Box F \in w$, that is, $\Box\Diamond\neg F \in w$, that is, $\Diamond\neg F \in w'$, since $\{F : \Box F \in w\} \subseteq w'$. Therefore, $\neg\Box F \in w'$, which contradicts the consistency of w'. Therefore, $F \in w''$, that is, $\{F : \Box F \in w'\} \subseteq w''$, that is, $w'R_{S5}w''$.

Theorem 8-9: The following properties hold in modal systems:

- $\vdash_{\mathcal{D}} F$ iff $\vDash^{\mathcal{M}} F$, for every model $\mathcal{M} = \langle \mathbf{W}, R, B \rangle$ in which if R is serial

- $\vdash_{T} F$ iff $\vDash^{\mathcal{M}} F$, for every model $\mathcal{M} = \langle \mathbf{W}, R, B \rangle$ in which if R is reflexive

- $\vdash_{\mathcal{B}} F$ iff $\vDash^{\mathcal{M}} F$, for every model $\mathcal{M} = \langle \mathbf{W}, R, B \rangle$ in which if R is reflexive and symmetric

- $\vdash_{S4} F$ iff $\vDash^{\mathcal{M}} F$, for every model $\mathcal{M} = \langle \mathbf{W}, R, B \rangle$ in which if R is reflexive and transitive

- $\vdash_{S5} F$ iff $\vDash^{\mathcal{M}} F$, for every model $\mathcal{M} = \langle \mathbf{W}, R, B \rangle$ in which if R is reflexive and euclidean

Proof: Theorem 8-6 provides the "if" part of each of the above five properties. To prove the "only if" part of the first of the above five properties, suppose $\vDash^{\mathcal{M}} F$, for every model $\mathcal{M} = \langle \mathbf{W}, R, B \rangle$ in which if R is serial. In particular,

$\models^{\mathcal{M}_{\mathcal{D}}} F$, for each canonical model $\mathcal{M}_{\mathcal{D}} = \langle \mathbf{W}_{\mathcal{D}}, R_{\mathcal{D}}, B_{\mathcal{D}} \rangle$, where $R_{\mathcal{D}}$ is serial. From Theorem 8-7, $\vdash_{\mathcal{D}} F$. The rest of the proof follows in a similar manner.

Generalized versions of Theorem 8-4 and Theorem 8-8 have been proved by Lemmon (1977), who considered a general axiom scheme of the following form:

$$\Diamond^m \Box^n F \rightarrow \Box^p \Diamond^q F$$

where m, n, p, and q are non-negative integers. Note that the each of the axioms D, T, B, 4, and E can be obtained from the general representation by varying the values of m, n, p, and q. For example, $m = 1$, $n = 0$, $p = 1$, and $q = 1$ give axiom E, which is $\Diamond F \rightarrow \Box \Diamond F$. In a similar manner, the following general relation scheme was considered:

> For all w, w_1, w_2, if $wR^m w_1$ and $wR^p w_2$ then there exists w' such
> that $w_1 R^n w'$ and $w_2 R^q w'$, where $w_i R^0 w_j \equiv w_i = w_j$

Various instantiations of R can be obtained by choosing different values for m, n, p, and q. For example, choosing $m = 0$, $n = 1$, $p = 0$, and $q = 0$ yields the serial relation R as "For all w_1 there exists w' such that $w_1 R w'$." A result analogous to Theorem 8-6 is that $\models^{\mathcal{M}} F$ for the relation of $\mathcal{M}$ satisfying this general relation.

8.9 Complexity and Decidability of Modal Systems

We say that a modal system is *decidable* if there is a decision procedure that determines whether or not a formula is a theorem or not. In this section, we shall prove the decidability results only for the systems $\mathcal{K}$, $\mathcal{D}$, $\mathcal{T}$, $\mathcal{S}4$, $\mathcal{B}$, and $\mathcal{S}5$ that we have been focusing on in this chapter.

To prove whether an arbitrary formula F is a theorem of a modal system $\mathcal{S}$ or not, our approach is to construct an abstract finite model $\mathcal{M}$ based on F and the canonical model $\mathcal{M}_{\mathcal{S}}$. We can then show that if F is not a theorem of $\mathcal{S}$, then the negation of F is satisfiable in $\mathcal{M}$. Since the size of $\mathcal{M}$ is finite, a decision procedure exists to determine the satisfiability of F with respect to $\mathcal{M}$. Our procedure is not practical, but it establishes the decidability results. Practical procedures based on the resolution principle for determining unsatisfiability will be developed later in this chapter.

Ladner (1977) proved that the complexity of the satisfiability problem in the modal logics $\mathcal{K}$, $\mathcal{D}$, $\mathcal{T}$, $\mathcal{B}$, and $\mathcal{S}4$ is PSPACE-complete, and in $\mathcal{S}5$ is NP-complete.

Halpern (1995b) also studied the effect of bounding modal depth on the complexity of modal logics.

Suppose we have a formula F in a modal system S whose canonical model is $\mathcal{M}_S = \langle \mathbf{W}_S, R_S, B_S \rangle$. Define an equivalence relation ρ on $\mathbf{W}_S$ as follows:

$$w \rho w' \text{ iff for every subformula } G \text{ of } F, \; G \in w \text{ iff } G \in w'$$

It can be easily verified that ρ is an equivalence relation on $\mathbf{W}_S$. Moreover, if n is the number of subformulae of F then there are 2^n possible combinations of the subformulae that both w and w' can contain. Therefore, the relation ρ partitions $\mathbf{W}_S$ into at most 2^n equivalence classes (from a result presented earlier in the mathematical prelimary section). Each class containing a world w is denoted by $[w/\rho]$. We construct a new model $\overline{\mathcal{M}}_S = \langle \overline{\mathbf{W}}_S, \overline{R}_S, \overline{B}_S \rangle$ based on $\mathcal{M}_S = \langle \mathbf{W}_S, R_S, B_S \rangle$ as follows:

$$\overline{\mathbf{W}}_S = \{ [w/\rho] : w \in \mathbf{W}_S \}$$

$$[w/\rho] \overline{R}_S [w'/\rho] \text{ iff for every subformula } \Box G \text{ of } F, \text{ if } \Box G \in w \text{ then } G \in w'$$

$$\overline{B}_S = \{ [w/\rho] : G \text{ is subformula of } F \text{ and } G \in w \}$$

The above definition is independent of the choice of representatives w and w' from classes $[w/\rho]$ and $[w'/\rho]$, respectively. To prove this, let $[w/\rho] \overline{R}_S [w'/\rho]$ and $w \rho w_1$ and $w' \rho w_2$. We need to show that $[w_1/\rho] \overline{R}_S [w_2/\rho]$. Since $[w/\rho] \overline{R}_S [w'/\rho]$, for every subformula $\Box G$ of F, if $\Box G \in w$ then $G \in w'$. Suppose, $\Box G \in w_1$ for a subformula $\Box G$ of F. Since $w \rho w_1$, $\Box G \in w_1$ implies $\Box G \in w$ and thus $G \in w'$. Since $w' \rho w_2$, $G \in w'$ implies that $G \in w_2$. Thus, for every subformula $\Box G$ of F, if $\Box G \in w_1$ then $G \in w_2$. Therefore, $[w_1/\rho] \overline{R}_S [w_2/\rho]$.

Proposition 8-4: If $w R_S w'$ then $[w/\rho] \overline{R}_S [w'/\rho]$

Proof: If $w R_S w'$ then $\{ G : \Box G \in w \} \subseteq w'$. If $\Box G$ is a subformula of F and $\Box G \in w$ then $G \in w'$. Therefore, by definition of $\overline{R}_S$, $[w/\rho] \overline{R}_S [w'/\rho]$.

Theorem 8-10: Suppose S is the modal system $\mathcal{K}$. Then, for any formula F in S and for every subformula G of F, $\models^{\overline{\mathcal{M}_S}}_{[w/\rho]} G$ iff $\models^{\mathcal{M}_S}_w G$.

Proof: We prove by applying the induction principle on the length of the subformula G.

Case I: G is P. Then $\models^{\overline{\mathcal{M}_S}}_{[w/\rho]} P$ iff $[w/\rho] \in \overline{B}_S(P)$. Since P is a subformula of itself, $[w/\rho] \in \overline{B}_S(P)$ iff $P \in w$. By definition, $P \in w$ iff $\models^{\mathcal{M}_S}_w P$. Therefore, $\models^{\overline{\mathcal{M}_S}}_{[w/\rho]} P$ iff $\models^{\mathcal{M}_S}_w P$.

Case II: G is $\neg H$. Since H is a subformula of F of length less than the length of H, by induction hypothesis, $\models^{\overline{\mathcal{M}_S}}_{[w/\rho]} H$ iff $\models^{\mathcal{M}_S}_w H$, that is, $\not\models^{\overline{\mathcal{M}_S}}_{[w/\rho]} H$ iff $\not\models^{\mathcal{M}_S}_w H$, that is, $\models^{\overline{\mathcal{M}_S}}_{[w/\rho]} \neg H$ iff $\models^{\mathcal{M}_S}_w \neg H$, that is, $\models^{\overline{\mathcal{M}_S}}_{[w/\rho]} G$ iff $\models^{\mathcal{M}_S}_w G$.

Case III: G is $H_1 \to H_2$. By induction hypothesis, $\models^{\overline{\mathcal{M}_S}}_{[w/\rho]} H_1$ iff $\models^{\mathcal{M}_S}_w H_1$ and $\models^{\overline{\mathcal{M}_S}}_{[w/\rho]} H_2$ iff $\models^{\mathcal{M}_S}_w H_2$. Therefore, $\models^{\overline{\mathcal{M}_S}}_{[w/\rho]} H_1 \to H_2$ iff $\models^{\mathcal{M}_S}_w H_1 \to H_2$, that is, $\models^{\overline{\mathcal{M}_S}}_{[w/\rho]} G$ iff $\models^{\mathcal{M}_S}_w G$.

Case IV: G is $\Box H$. Suppose $\models^{\overline{\mathcal{M}_S}}_{[w/\rho]} \Box H$. Then, for all $[w'/\rho]$ such that $[w/\rho]\overline{R}_S[w'/\rho]$, $\models^{\overline{\mathcal{M}_S}}_{[w'/\rho]} H$. To show that $\models^{\mathcal{M}_S}_w \Box H$, suppose $wR_S w'$. Then $[w/\rho]\overline{R}_S[w'/\rho]$, by Proposition 8-4, and thus $\models^{\overline{\mathcal{M}_S}}_{[w'/\rho]} H$. By induction hypothesis, $\models^{\overline{\mathcal{M}_S}}_{[w'/\rho]} H$ iff $\models^{\mathcal{M}_S}_{w'} H$. Therefore, $\models^{\mathcal{M}_S}_w \Box H$.

Conversely, suppose $\models^{\mathcal{M}_S}_w \Box H$. Then, for all w' such that $wR_S w'$, $\models^{\mathcal{M}_S}_{w'} H$. To show that $\models^{\overline{\mathcal{M}_S}}_{[w/\rho]} \Box H$, suppose $[w/\rho]\overline{R}_S[w'/\rho]$. Since $\models^{\mathcal{M}_S}_w \Box H$ and w is maximal consistent, $\Box H \in w$. Since $[w/\rho]\overline{R}_S[w'/\rho]$ and $\Box H \in w$, by the definition of $\overline{R}_S$, $H \in w'$, that is, $\models^{\mathcal{M}_S}_{w'} H$, since w' is maximal consistent. By induction hypothesis, $\models^{\overline{\mathcal{M}_S}}_{[w'/\rho]} H$. Therefore, $\models^{\overline{\mathcal{M}_S}}_{[w/\rho]} \Box H$.

Theorem 8-11: Systems $\mathcal{K}$, $\mathcal{D}$, $\mathcal{T}$, $S4$, $\mathcal{B}$, and $S5$ are decidable.

Proof: Suppose F represents a formula, with n subformulae, of the modal system S under consideration, $\mathcal{M}_S = \langle \mathbf{W}_S, R_S, B_S \rangle$ is the canonical model of S, and w is a world in $\mathbf{W}_S$.

$\mathcal{K}$ is decidable: Since F is a subformula of itself, from Theorem 8-7 and Theorem 8-10, $\vdash_{\mathcal{K}} F$ iff $\models^{\mathcal{M}_{\mathcal{K}}} F$ iff $\models^{\overline{\mathcal{M}_{\mathcal{K}}}} F$. Since $\mathcal{M}_{\mathcal{K}}$ is finite (at most 2^n), we can determine whether $\models^{\overline{\mathcal{M}_{\mathcal{K}}}} F$ or not. Therefore, $\mathcal{K}$ is decidable.

$\mathcal{D}$ is decidable: Since $R_{\mathcal{D}}$ is serial, $wR_{\mathcal{D}}w'$, for some w'. Since $wR_{\mathcal{D}}w'$, $\{F : \Box F \in w\} \subseteq w'$. In particular, for any subformula $\Box G$ of F, if $\Box G \in w$ then $G \in w'$. Thus, there exists a world w' such that $[w/\rho]\overline{R}_{\mathcal{D}}[w'/\rho]$. Thus, $\overline{R}_{\mathcal{D}}$ is serial. From Theorem 8-9, $\vdash_{\mathcal{D}} F$ iff $\models^{\overline{\mathcal{M}_{\mathcal{D}}}} F$. Since $\mathcal{M}_{\mathcal{D}}$ is finite, the truth value of a formula with respect to $\overline{\mathcal{M}_{\mathcal{D}}}$ can be determined in a finite time. Therefore, the decidability problem in $\mathcal{D}$ can be equivalently transformed to a satisfiability problem which can be solved in a finite time.

$\mathcal{T}$ is decidable: Since $R_{\mathcal{T}}$ is reflexive, $wR_{\mathcal{T}}w$, for every w. Since $wR_{\mathcal{T}}w$, $\{F : \Box F \in w\} \subseteq w$. In particular, for any subformula $\Box G$ of F, if $\Box G \in w$ then $G \in w$. Thus, for every world w, $[w/\rho]\overline{R}_{\mathcal{T}}[w/\rho]$. Thus, $\overline{R}_{\mathcal{T}}$ is serial. The decidability of $\mathcal{T}$ follows as per the argument above.

The rest of the cases are left as exercises.

8.10   Modal First-Order Logics

This section briefly introduces modal first-order logics. The syntax of modal first-order logics can be obtained from the classical first-order logic by the introduction of modalities into the language. Then we can produce modal first-order systems analogous to the modal propositional systems $\mathcal{K}$, $\mathcal{D}$, $\mathcal{T}$, $\mathcal{S4}$, $\mathcal{B}$, and $\mathcal{S5}$ presented in the previous section. In principle, semantics of these first-order modal systems could be obtained by extending the possible world semantics for modal propositional systems. However, complications arise when due consideration is given to issues related to the underlying domain structure and term rigidity. Before addressing these issues, we briefly capture the syntax, axioms, and semantics of first-order modal systems.

A formal axiomatic theory for the first-order version of $\mathcal{K}$, denoted as $\mathcal{FOL}{+}\mathcal{K}$, can be obtained from the classical first-order logic (presented in chapter 3) as follows:

- The symbols of $\mathcal{K}$ are the first-order alphabet and the symbol '$\Box$' (necessary).

- *Terms* are expressions which are defined inductively as follows:
    - A variable or an individual constant is a term.
    - If f is an n-ary function symbol and $t_1, t_2, ..., t_n$ are terms, then $f(t_1, t_2, ..., t_n)$ is a term.
    - An expression is a term if it can be shown to be so only on the basis of the above two conditions.

- *Well-formed formulae* (*wffs*) or formulae of modal first-order logics are recursively defined as follows:
    - If P is an n-ary predicate symbol and $t_1, t_2, ..., t_n$ are terms, then $P(t_1, t_2, ..., t_n)$ is an *atomic* formula.
    - If F is a formula, then $\neg F$ and $\Box F$ are formulae.
    - If F is a formula and x is a variable, then $\forall x(F)$ is a formula.
    - If F and G are formulae, then $F \to G$ is a formula.
    - An expression is a formula only if it can be generated by the above four conditions.

- If F, G, H are formulae, then the following are axioms of the theory:
    - Logical axioms from first-order logic
    - Modal Axioms:

      K: $\Box(F \to G) \to (\Box F \to \Box G)$

- The following are the rules of inference of $\mathcal{K}$:
    - Modus Ponens (MP): If $\vdash F$ and $\vdash F \to G$ then $\vdash G$
    - Generalization: If $\vdash F$ then $\vdash \forall x(F)$
    - Rule of Necessitation (RN): If $\vdash F$ then $\vdash \Box F$

- The definitions in $\mathcal{K}$ are the usual definitions of the connectives $\wedge, \vee, \leftrightarrow$, and the following equivalencies of the existential and the modal operator for possibility:

$$\exists x(F) \equiv \neg \forall x(\neg F)$$
$$\Diamond F \equiv \neg \Box \neg F$$

The first-order versions of the systems $\mathcal{D}$, $\mathcal{T}$, $\mathcal{S4}$, $\mathcal{B}$, and $\mathcal{S5}$ can be obtained from $\mathcal{FOL}+\mathcal{K}$ as follows:

$$\mathcal{FOL}+\mathcal{D} = \mathcal{FOL}+\mathcal{K}+ \text{D} : \Box F \to \Diamond F$$

$$\mathcal{FOL}+\mathcal{T} = \mathcal{FOL}+\mathcal{K}+\mathrm{T}:\Box F \to F$$

$$\mathcal{FOL}+\mathcal{S4} = \mathcal{FOL}+\mathcal{K}+\mathrm{T}:\Box F \to F + 4:\Box F \to \Box\Box F$$

$$\mathcal{FOL}+\mathcal{B} = \mathcal{FOL}+\mathcal{K}+\mathrm{T}:\Box F \to F + \mathrm{B}:F \to \Box\Diamond F$$

$$\mathcal{FOL}+\mathcal{S5} = \mathcal{FOL}+\mathcal{K}+\mathrm{T}:\Box F \to F + \mathrm{E}:\Diamond F \to \Box\Diamond F$$

A *normal model* $\mathcal{M}$ is a structure of the following form:

$$\mathcal{M} = \langle \mathbf{W}, \mathbf{D}, R, B, \delta, \phi \rangle$$

where $\mathbf{W}$ is a non-empty set of elements, to be called *worlds*; $\mathbf{D}$ is a non-empty set, called the *domain of interpretation*; R is any binary relation on $\mathbf{W}$, to be called the *accessibility relation* for $\mathbf{W}$; and B is a *valuation* function determines the truth value of a ground atom in a world as shown below:

$$B : \mathbf{P} \times \mathbf{W} \to \{\top, \bot\}$$

where $\mathbf{P}$ is the set of all ground atoms defined as follows:

$$\mathbf{P} = \{p(a_1,...,a_n) \mid p \text{ is an n-ary predicate symbol } \& \ (a_1,...,a_n) \in \mathbf{D}^n\}$$

Thus if $B(p(a_1,...,a_n),w)$ is $\top$ or $\bot$, then we say that the atom $p(a_1,...,a_n)$ is *true* or *false*, respectively, at the world w in a model $\mathcal{M}$, and this is written as follows:

$$\vDash_w^{\mathcal{M}} p(a_1,...,a_n)$$

Equivalently, B can be defined as a function to assign a set of ground atoms to each world w that is true in w:

$$B : \mathbf{W} \to \mathcal{P}(\mathbf{P})$$

δ is a function that assigns to each world w a non-empty subset of $\mathbf{D}$, called the domain of $\mathbf{W}$. Thus,

$$\delta(w) \subseteq \mathbf{D}$$

Finally, φ is a function that represents the interpretation of constants and function symbols. It maps each pair of world w and n-ary function symbol f to a function from $\mathbf{D}^n \to \mathbf{D}$. Thus,

$$\phi(w,f) \in \mathbf{D}^n \to \mathbf{D}$$

Note that a constant a is a 0-ary function symbol and therefore $\phi(w,a)$ is a fixed member of $\mathbf{D}$. The assignment of a term t with respect to the model is recursively defined as follows:

- If t is the individual constant a, then its assignment is $\phi(w,a)$.

- If t is of the form $f(t_1,...,t_n)$, then its assignment $\phi(w,t)$ is $\phi(w,f)(\phi(w,t_1),...,\phi(w,t_n))$.

For an arbitrary formula F, the *satisfiability* of a formula F with respect to a model $\mathcal{M} = \langle \mathbf{W}, \mathbf{D}, R, B, \delta, \phi \rangle$ and a world $w \in \mathbf{W}$, written as $\vDash_w^{\mathcal{M}} F$, is recursively defined as follows:

$\vDash_w^{\mathcal{M}} P(t_1,...,t_n)$ iff $P(\phi(w,t_1),...,\phi(w,t_n)) \in B(w)$

$\vDash_w^{\mathcal{M}} \neg F$ iff $\nvDash_w^{\mathcal{M}} F$

$\vDash_w^{\mathcal{M}} F \rightarrow G$ iff $\vDash_w^{\mathcal{M}} F$ implies $\vDash_w^{\mathcal{M}} G$

$\vDash_w^{\mathcal{M}} \forall x F$ iff for all $d \in \delta(w)$ $\vDash_w^{\mathcal{M}} F[x/d]$

$\vDash_w^{\mathcal{M}} \exists x F$ iff there exists $d \in \delta(w)$ such that $\vDash_w^{\mathcal{M}} F[x/d]$

$\vDash_w^{\mathcal{M}} \Box F$ iff for every $w' \in \mathbf{W}$ such that wRw', $\vDash_{w'}^{\mathcal{M}} F$

$\vDash_w^{\mathcal{M}} \Diamond F$ iff for there is a $w' \in \mathbf{W}$ such that wRw' and $\vDash_{w'}^{\mathcal{M}} F$

A formula F is *satisfiable* iff $\vDash_w^{\mathcal{M}} F$, for some model $\mathcal{M} = \langle \mathbf{W}, \mathbf{D}, R, B, \delta, \phi \rangle$ and $w \in \mathbf{W}$. F is *valid* in $\mathcal{M}$, written as $\vDash^{\mathcal{M}} F$, iff $\vDash_w^{\mathcal{M}} F$, for every $w \in \mathbf{W}$. F is *valid* in a class of models Σ, written as $\vDash^{\Sigma} F$, iff $\vDash^{\mathcal{M}} F$, for every $\mathcal{M} \in \Sigma$. F is valid, written as $\vDash^{\mathcal{M}} F$, iff F is valid in every class of models.

There are at least two aspects of the model that should be discussed regarding the two functions δ and ϕ that were added to the normal model for propositional logic to handle object domains and term designation. For any two worlds w and w', if wRw', we have the following three possibilities for the relationship between $\delta(w)$ and $\delta(w')$:

- $\delta(w) = \delta(w')$: This is the *fixed domain* case where every world has the fixed set of objects.

- $\delta(w) \subseteq \delta(w')$: This is the *cumulative domain* case where the objects in the current world continue to exist in every accessible world, but there may be new objects in the accessible worlds. In biology, for example, new species are being discovered all the time.

- $\delta(w) \supseteq \delta(w')$: This is the case where the objects in the current world are not guaranteed to exist in accessible worlds.

- No restrictions on the relation between $\delta(w)$ and $\delta(w')$

From the practical application standpoint, both the fixed domain and cumulative domain cases are important, though the fixed-domain interpretation has the advantages of simplicity and familiarity. As for the function, for any two worlds w and w', and for every function symbol f, we have the following two cases:

- $\phi(w, f) = \phi(w', f)$: This is the *rigid designation* case where the denotation of terms is the same in every world.

- No restrictions on the relation between $\phi(w, f)$ and $\phi(w', f)$

In the case of rigid designation, $\Box p(c)$ follows from the axiom $(\forall x F) \rightarrow G$ and the two formulae $p(c)$ and $\forall x(p(x) \rightarrow \Box p(x))$. But this is not the case when the denotation of c is allowed to vary from world to world.

The soundness and completeness results for each of the first-order versions of the systems $\mathcal{D}$, $\mathcal{T}$, $\mathcal{S4}$, $\mathcal{B}$, and $\mathcal{S5}$ can be proven with respect to the model $\mathcal{M} = \langle \mathbf{W}, \mathbf{D}, R, B, \delta, \phi \rangle$, where the usual restriction on R applies, the domain is cumulative, and the term designation is rigid. Adding the Barcan Formula (BF) (Barcan 1946) can usually axiomatize first-order modal logics that are adequate for applications with a fixed domain:

$$\text{BF: } \forall x \Box F[x] \rightarrow \Box \forall x F[x]$$

where $F[x]$ means F has zero or more free occurrences of the variable x. The above axiom means that if everything that exists necessarily possesses a certain property F then it is necessarily the case that everything possesses F. But in the cumulative domain case there might be some additional objects in accessible worlds that may not possess the property F. On the other hand, the converse $\Box \forall x F[x] \rightarrow \forall x \Box F[x]$ of the Barcan formula, however, can be validated in first-order modal logics with cumulative domain and rigid designation:

TFML 1: $\vdash_{\mathcal{K}} \Box \forall x F[x] \rightarrow \forall x \Box F[x]$

    Step 1: $\vdash_{\mathcal{K}} \forall x F[x] \rightarrow F[x/y]$

        [Axiom $(\forall x F) \rightarrow G$, where y is free for x in F]

    Step 2: $\vdash_{\mathcal{K}} \Box(\forall x F[x] \rightarrow F[x/y])$ [Rule of Necessitation]

    Step 3: $\vdash_{\mathcal{K}} \Box \forall x F[x] \rightarrow \Box F[x/y])$

        [Axiom $\Box(F \rightarrow G) \rightarrow (\Box F \rightarrow \Box G)$ and MP]

Step 4: $\vdash_{\mathcal{K}} \forall y(\Box \forall x F[x] \to \Box F[x/y]))$ [Generalization Rule]

Step 5: $\vdash_{\mathcal{K}} \Box \forall x F[x] \to \forall y \Box F[x/y])$

     [Axiom $\forall x(F \to G) \to (F \to \forall x G)$]

Step 6: $\vdash_{\mathcal{K}} \Box \forall x F[x] \to \forall x \Box F[x])$ [Variable renaming]

The Barcan formula BF is not provable in any of the systems $\mathcal{FOL}+\mathcal{D}$, $\mathcal{FOL}+\mathcal{T}$, and $\mathcal{FOL}+\mathcal{S4}$ (Lemmon, 1960), but provable in each of the systems $\mathcal{FOL}+\mathcal{B}$ and $\mathcal{FOL}+\mathcal{S5}$.

TFML 2: $\vdash_{S5} \forall x \Box F[x] \to \Box \forall x F[x]$

Step 1: $\vdash_{S5} \forall x \Box F[x] \to \Box F[x/y]$

     [Axiom $(\forall x F) \to G$, where y is free for x in F]

Step 2: $\vdash_{S5} \Diamond \forall x \Box F[x] \to \Diamond \Box F[x/y]$

     [TK 2: If $\vdash F \to G$ then $\vdash \Diamond F \to \Diamond G$]

Step 3: $\vdash_{S5} \Diamond \forall x \Box F[x] \to F[x/y]$ [TB 1: $\vdash_{\mathcal{B}} \Diamond \Box F \to F$]

Step 4: $\vdash_{S5} \forall y(\Diamond \forall x \Box F[x] \to F[x/y])$ [Generalization Rule]

Step 5: $\vdash_{S5} \Diamond \forall x \Box F[x] \to \forall y F[x/y]$

     [Axiom $\forall x(F \to G) \to (F \to \forall x G)$]

Step 6: $\vdash_{S5} \Diamond \forall x \Box F[x] \to \forall x F[x]$ [Variable renaming]

Step 7: $\vdash_{S5} \Box \Diamond \forall x \Box F[x] \to \Box \forall x F[x]$

     [Axiom $\Box(F \to G) \to (\Box F \to \Box G)$ and MP]

Step 8: $\vdash_{S5} \forall x \Box F[x] \to \Box \forall x F[x]$ [Axiom B $: F \to \Box \Diamond F$]

Every model in the context of the systems $\mathcal{FOL}+\mathcal{B}$, and $\mathcal{FOL}+\mathcal{S5}$ is fixed as they derive the Bracan formula BF. The soundness and completeness results for each of the systems, namely $\mathcal{FOL}+\mathcal{D}$, $\mathcal{FOL}+\mathcal{T}$, and $\mathcal{FOL}+\mathcal{S4}$, with BF can be established with respect to the model $\mathcal{M} = \langle \mathbf{W}, \mathbf{D}, R, B, \delta, \phi \rangle$, where the usual restriction on R applies and the domain is fixed.

8.11   Resolution in Modal First-Order Logics

Each formula in first-order logic can be transformed to an equivalent set of clauses in the form $L_1 \vee \ldots \vee L_n$, where each L_i is an atomic formula of the form P or $\neg P$. The resolution principle that we presented in the mathematical logic chapter is applied on a transformed set of clauses. If a set of clauses is unsatisfiable, then the resolution process derives contradiction. To determine whether a formula F is a theorem of a set of clauses $\mathbf{S}$ or not, we first transformed $\neg F$ into a set of clauses and then combined that set with $\mathbf{S}$. If no contradiction can be derived from the combined set using the resolution procedure, then F is a theorem of $\mathbf{S}$.

The main reason why an equivalent procedure cannot be developed in a modal system context is that, in general, an arbitrary modal formula cannot be transformed to an equivalent clause where each component is one of the atomic modal formulae P, $\neg P$, $\Diamond P$, or $\Box P$. For example, there is no equivalent clausal representation of modal formulae $\Box(P \vee Q)$ or $\Diamond(P \wedge Q)$ in the systems that we have considered. But, using the equivalence $\Box P \equiv \neg \Diamond \neg P$ and other equivalences between the logical connectives, an arbitrary modal formula can be transformed to an equivalent *modal clause* representation where only negation, disjunction, and conjunction symbols appear, and a negation appears only as a prefix of a proposition. An example of this transformation of the formula $\Diamond P \rightarrow \Box \Diamond P$ is as follows:

$$\Diamond \Box P \rightarrow \Box P \equiv \neg \Diamond \Box P \vee \Box P$$
$$\equiv \neg \Diamond \neg \Diamond \neg P \vee \Box P$$
$$\equiv \Box \Diamond \neg P \vee \Box P$$

The modal resolution procedure presented here first transforms modal formulae into these kinds of modal clauses. The procedure then eliminates modalities by introducing appropriate world symbols and quantifiers, thus transforming a modal clause to a formula in classical logic (not necessarily first-order), which can then be transformed to a set of clauses for applying an enhanced resolution procedure. We first explain the transformation procedure below in the context of a few examples:

- $\Diamond \forall x P(x)$ to $\forall x P([0a], x)$: From the initial world 0 there is a world a accessible from 0 such that for all x $P(x)$ holds, where P is interpreted in the world a.

- $\forall x \Diamond P(x)$ to $\forall x P([0a(x)], x)$: From the initial world 0 there is a world a accessible from 0 but depending on x such that for all x $P(x)$ holds, where P is interpreted in the world a.

- $\Box \forall x P(x)$ to $\forall w \forall x P([0w], x)$: From the initial world 0 and for every world w that is accessible from 0, for all x $P(x)$ holds, where P is interpreted in the world w.

- $\forall x \Box P(x)$ to $\forall x \forall w P([0w], x)$: From the initial world 0 and for every world w that is accessible from 0 and therefore independent of x, for all x $P(x)$ holds, where P is interpreted in the world w.

- $\Diamond \Box \forall x P(x)$ to $\forall w \forall x P([0aw], x)$: From the initial world 0 there is a world a accessible from 0, and for all worlds w which are accessible from a, for all x $P(x)$ holds, where P is interpreted in the world w.

- $\Box \Diamond \forall x P(x)$ to $\forall w \forall x P([0wa], x)$: From the initial world 0 and for every world w accessible from 0 there is a world a accessible from w such that for every x $P(x)$ holds, where P is interpreted in the world a.

- $\Box \Diamond \forall x \exists y P(x, y)$ to $\forall w \forall x P([0wa], x, f([0wa], x))$: Same as above except f is a skolem function that denotes a y that must exist in the world a in the form dependent on x.

Our objective is to transform a modal formula into an equivalent formula in world-path syntax, which can then be further transformed into a set of clauses in the same way as in first-order logic. As in the above examples, a skolem function needs to have a world-path as its first argument because every term unification has to be carried out in a world that is indicated by the world-path in the context. But the standard skolemization for first-order logic cannot introduce such world-paths. Skolemization from the process of transforming a world-path formula into a set of clauses can be avoided by considering modal formulae that are in *negation normal form* without the implication and equivalence sign, and with all negation signs moved in front of atoms. For example, $\neg \forall x (\Box P(x) \vee Q(x))$ is not in negation normal form, but its equivalent formula $\exists x (\Diamond \neg P(x) \wedge \neg Q(x))$ is in negation normal form. Any modal formula can be brought into this form using the appropriate tranformation rules of first-order logic and the two additional equivalences $\neg\Box F \equiv \Diamond\neg F$ and $\neg\Diamond F \equiv \Box\neg F$.

8.11.1 Transformation Algorithm

Notation

- WP: world-path of the form $[w_1...w_m]$, where each w_i is a world symbol

- VL: variable list of the form $(x_1...x_n)$, where each x_i is a world symbol

- $|VL|$: length of the list VL, that is, n if VL is $(x_1...x_n)$

- $WP + w$: world-path $[w_1...w_m w]$ obtained by appending the word w to $WP = [w_1...w_m]$

- $VL + x$: variable list $(x_1...x_n x)$ obtained by appending the variable x to $VL = (x_1...x_n)$

- $f(VL)$: the term $f(x_1,...,x_n)$, where $VL = (x_1...x_n)$

Input

- Modal formula F in negation normal form

Output

- The formula in the world-path syntax obtained by applying repeatedly the following set of transformation rules on the top-level call $\mathcal{T}(F,[0],())$, where 0 denotes the initial world and () is the empty variable list

Transformation Rules

Rule 1: $\mathcal{T}(\neg P,WP,VL) := \neg\mathcal{T}(P,WP,VL)$

Rule 2: $\mathcal{T}(F \wedge G,WP,VL) := \mathcal{T}(F,WP,VL) \wedge \mathcal{T}(G,WP,VL)$

Rule 3: $\mathcal{T}(F \vee G,WP,VL) := \mathcal{T}(F,WP,VL) \vee \mathcal{T}(G,WP,VL)$

Rule 4: $\mathcal{T}(\forall x F,WP,VL) := \forall x \mathcal{T}(F,WP,VL + x)$

Rule 5: $\mathcal{T}(\Box F,WP,VL) := \forall w \mathcal{T}(F,WP + w,VL)$, where w is an added new variable which does not occur in F

Rule 6: $\mathcal{T}(\exists xF, WP, VL) := \mathcal{T}(F\{x/f(VL)\}, WP, VL)$, where f is an added new $|VL|$-place function symbol and $F\{x/f(VL)\}$ means substitute x by $f(VL)$

Rule 7: $\mathcal{T}(\lozenge F, WP, VL) := \mathcal{T}(F, WP + a(VL), VL)$, where a is an added new $|VL|$-place function symbol

Rule 8: $\mathcal{T}(P(t_1,...,t_n), WP, VL) := P(WP, \mathcal{T}(t_1, WP, VL),...,\mathcal{T}(t_n, WP, VL))$, where P is an n-place predicate symbol

Rule 9: $\mathcal{T}(f(t_1,...,t_n), WP, VL) := f(WP, \mathcal{T}(t_1, WP, VL),...,\mathcal{T}(t_n, WP, VL))$, where f is an n-place function symbol

Rule 10: $\mathcal{T}(x, WP, VL) := x$

Since a formula in world-path syntax contains no existential quantifier and no modal operator, it can be transformed to an equivalent set of clauses in the same way as in first-order logic but without the need for skolemiztion.

Example

$\mathcal{T}(\forall x \square (P(x) \vee \lozenge Q(x)), [0], ())$	Given
$\forall x \mathcal{T}(\square (P(x) \vee \lozenge Q(x)), [0], (x))$	Rule 4
$\forall x \mathcal{T}(P(x) \vee \lozenge Q(x), [0w], (x))$	Rule 5
$\forall x(\mathcal{T}(P(x), [0w], (x)) \vee \mathcal{T}(\lozenge Q(x), [0w], (x)))$	Rule 3
$\forall x(\mathcal{T}(P(x), [0w], (x)) \vee \mathcal{T}(Q(x), [0wa(x)], (x)))$	Rule 7
$\forall x(P([0w], x) \vee Q([0wa(x)], x))$	Rules 8, 10

Example

$\mathcal{T}(\square \forall x \forall y \lozenge (\square \neg P(x,y) \wedge \exists z Q(x,y,z)), [0], ())$	Given
$\forall w \mathcal{T}(\forall x \forall y \lozenge (\square \neg P(x,y) \wedge \exists z Q(x,y,z)), [0w], ())$	Rule 5
$\forall w \forall x \forall y \mathcal{T}(\lozenge (\square \neg P(x,y) \wedge \exists z Q(x,y,z)), [0w], (xy))$	Rule 4
$\forall w \forall x \forall y \mathcal{T}(\square \neg P(x,y) \wedge \exists z Q(x,y,z), [0wa(x,y)], (xy))$	Rule 7
$\forall w \forall x \forall y (\mathcal{T}(\square \neg P(x,y), [0wa(x,y)], (xy)) \wedge \mathcal{T}(\exists z Q(x,y,z), [0wa(x,y)], (xy)))$	Rule 2

$$\forall w \forall x \forall y (\forall u \mathcal{T}(\neg P(x,y),[0wa(x,y)u],(xy)) \wedge$$
$$\mathcal{T}(Q(x,y,f(x,y)),[0wa(x,y)],(xy)))$$

Rules 5, 6

$$\forall w \forall x \forall y (\forall u \neg \mathcal{T}(P(x,y),[0wa(x,y)u],(xy)) \wedge$$
$$\mathcal{T}(Q(x,y,f(x,y)),[0wa(x,y)],(xy)))$$

Rule 1

$$\forall w \forall x \forall y (\forall u \neg P([0wa(x,y)u],x,y) \wedge$$
$$Q([0wa(x,y)],x,y,f([0wa(x,y)],x,y)))$$

Rules 8-10

8.11.2 *Unification*

The very first argument of each predicate or function symbol in a transformed modal formula is a world-path. The unification of these world-paths is the only difference from the unification of first-order logic terms. The world-path unification algorithm is dependent on the properties of the accessibility relation by which the worlds are connected. In the following list, we informally describe the algorithm in each of the cases of the accessibility relation.

- **Serial**: In this case, two world-paths are unifiable when they are of equal length and the world-paths can be treated like ordinary terms. Therefore, world-paths $[0ua]$ and $[0bv]$ are unifiable with a unifier $\{u/b, v/a\}$, whereas the world-paths $[0ua]$ and $[0bvw]$ are not. There is at most one *mgu* in this case.

- **Non-Serial**: When the accessibility relation is not serial, $\Box(P \wedge \neg P)$ is not necessarily inconsistent because no worlds may be accessible from the current world. But the transformation algorithm on $\Box(P \wedge \neg P)$ produces the formula $\forall w(P([0w]) \wedge \neg P([0w]))$, which yields inconsistency at the world w by a simple world-path matching. Therefore, we must assume that the accessibility relation is at least serial.

- **Reflexive**: The meaning of the formula $P([aw],x)$ is that $P(x)$ is true in every world w accessible from the current world a. Now, if the accessibility relation is reflexive, then a itself can substitute for w and in this case $P(x)$ is true in a. Thus, for the purpose of unifying $[aw]$ and another world-path, say $[a]$, the world w can simply be removed from the world-path $[aw]$ (or be substituted with the empty path $[]$) and the formula written as $P([a],x)$. Therefore, if the accessibility relation is reflexive, then the unification algorithm must consider all possibilities to remove a world variable w by the substitution $\{w/[]\}$ as well as all possibilities to unify two world-paths pair-wise as is the case when the accessibility relation is serial. For example, the two world-paths $[0ua]$ and $[0bvw]$ are unifiable with the two unifiers $\{u/b, v/a, w/[]\}$ and $\{u/b, v/[], w/a\}$.

- **Symmetric**: The meaning of the formula $P([abw], x)$ is that $P(x)$ is true in every world w accessible from the world b which is accessible from the current world a. Now, if the accessibility relation is symmetric, then a is accessible from b, and thus a can substitute for w and in this case $P(x)$ is true in a. Thus, for the purpose of unifying $[abw]$ and another world-path, say $[a]$, bw can simply be removed from the world-path $[abw]$ (or we can substitute b^{-1} for w) and the formula can be written as $P([a], x)$. Therefore, if the accessibility relation is symmetric, then the unification algorithm must consider all possibilities to remove a world variable w and its predecessor u from the world-path by the substitution $\{w/u^{-1}\}$ and must consider all possibilities to unify two world-paths pair-wise as is the case when the accessibility relation is serial. For example, the two world-paths $[0u]$ and $[0bvw]$ are unifiable with the unifier $\{u/b, v/w^{-1}\}$.

- **Transitive**: The meaning of the formula $P([abc], x)$ is that $P(x)$ is true in the world c accessible from the world b which is accessible from the current world a. Now, if the accessibility relation is transitive then c is accessible from a. Thus, for the purpose of unifying $[abc]$ and another world-path, say $[aw]$, bc can simply be collapsed to c (or substitute $[bc]$ for w) to be unifiable with w. For example, the two world-paths $[0wab]$ and $[0cdub]$ are unifiable with the unifier $\{w/[cd], u/a\}$, but the substitution $\{w/[cdu_1], u/[u_1u_2], u_2/a\}$, that is, $\{w/[cdu_1], u/[u_1a]\}$ is also a unifier by splitting u via the introduction of u_1. Therefore, if the accessibility relation is transitive, then the unification algorithm must consider both the collapsing of world-paths and the splitting of world variables.

- **Euclidean**: The meaning of the formula $P([abw], x)$ is that $P(x)$ is true in every world w accessible from the world b which is accessible from the current world a. The meaning of the formula $P([ac], x)$ is that $P(x)$ is true in the world c accessible from the world a. Therefore, the unification of the two world-paths $[abw]$ and $[ac]$ is possible provided the accessibility relation is Euclidean, guaranteeing c is accessible from b. Therefore, if the unification $\{[bw]/c\}$ is right, then c must be accessible from b. The unification algorithm must keep track of this kind of information.

- **Reflexive and Symmetric**: The ideas of removing world variables for reflexivity and for symmetry can be combined. Therefore, if the accessibility relation is reflexive and symmetric then the unification algorithm must consider all possibilities to remove a world variable w by the substitution $\{w/[]\}$, to remove a world variable w and its predecessor u in the world-path by the substitution $\{w/u^{-1}\}$, and to unify two world-paths pair-wise. For example, the two world-paths $[0u]$ and $[0bcvw]$ are unifiable with the unifier $\{u/b, v/c^{-1}, w/[]\}$.

- **Reflexive and Transitive**: The ideas of removing world variables for reflexivity and collapsing world-paths and splitting world variables for transitivity can be combined. Therefore, if the accessibility relation is reflexive and transitive, then the unification algorithm must consider all possibilities to remove a world variable w by the substitution $\{w/[]\}$, collapse a world-path, split a world variable, and unify two world-paths pair-wise. For example, the two world-paths $[0ude]$ and $[0bcvw]$ are unifiable with the two unifiers $\{u/b, v/c^{-1}, w/[de]\}$ and $\{u/[bc], v/d, w/e\}$.

- **Equivalence**: The ideas of removing world variables for reflexivity and for symmetry, and collapsing world-paths and splitting world variables for transitivity can be combined. Moreover, when the accessibility relation is an equivalence relation, the modal system is $S5$, which has six irreducible modalities φ, $\neg$, $\Diamond$, $\Box$, $\Diamond\neg$, and $\Box\neg$. Therefore, world-paths are expected to be shorter in length.

The following example illustrates the transformation and unification algorithms. Specifically, we prove via refutation that in the modal system $S4$, in which the accessibility relation is reflexive and transitive, the formula $\Box P \rightarrow \Box \Diamond \Box P$ is a theorem.

__Example__

$\Box P \to \Box \Diamond \Box P$	Given
$\neg(\Box P \to \Box \Diamond \Box P)$	Negation
$\Box P \wedge \Diamond \Box \Diamond \neg P$	Negation normal form
$\forall u P([0u]) \wedge \forall v \neg P([0avb])$	World-path syntax
$\forall u P([0u]) \wedge \forall v \neg P([0avb])$	Conjunctive normal form

$$\begin{cases} C1: & P([0u]) \\ C2: & \neg P([0avb]) \end{cases}$$
Clausal form

$C1 \,\&\, C2 : empty$

$unifier = \{v/[\,], u/[ab]\}$
Resolution refutation

8.12 Modal Epistemic Logics

Epistemic logic is concerned with the ideas of an agent's or decision-maker's "knowledge" and "belief" within a context. Modal epistemic logic is a logic for knowledge and belief and is viewed as an instance of modal logic by interpreting necessity and possibility in an epistemic manner. Therefore, the soundness and completeness results and the resolution principle that we have presented before can be adopted for modal epistemic logics.

Traditionally, logic of knowledge is based on the modal system $S5$. The logic, which will be denoted as $\mathcal{L}_K$, can be obtained from $S5$ by interpreting the notion of necessity as knowledge. Thus $\langle know \rangle F$ is written instead of $\Box F$ to represent "the agent knows that F." The axioms and inference rules of $\mathcal{L}_K$ are analogous to those in system $S5$.

Axioms of $\mathcal{L}_K$:

LK 1: All propositional axioms

LK 2: $\langle know \rangle (F \to G) \to (\langle know \rangle F \to \langle know \rangle G)$

LK 3: $\langle know \rangle F \to F$

LK 4: $\langle know \rangle F \to \langle know \rangle \langle know \rangle F$

LK 5: $\neg \langle know \rangle F \to \langle know \rangle \neg \langle know \rangle F$

Inference Rules of $\mathcal{L}_K$:

> Modus Ponens (MP): If $\vdash F$ and $\vdash F \rightarrow G$ then $\vdash G$
>
> Rule of Necessitation (RN): If $\vdash F$ then $\vdash \langle know \rangle F$

Axiom LK 3 (axiom T in $\mathcal{S}5$) states that the agent's known facts are true. Axioms LK 4 (axiom 4 in $\mathcal{S}5$) and LK 5 (a version of axiom E in $\mathcal{S}5$) are positive and negative introspection, respectively. Axiom LK 4 states that an agent knows that it knows something whereas axiom LK 5 states that an agent knows that it does not know something. Epistemic formulae in $\mathcal{L}_K$ are interpreted on Kripke's possible world model structures in a manner exactly like the formulae in system $\mathcal{S}5$.

The idea of an agent's belief differs from knowledge only in that an agent may believe in something that is not true (as opposed to Axiom LK 3) but will not believe in anything that is demonstrably false. Thus $\langle bel \rangle F$ is written specifically to represent "the agent believes that F" and the relation between the modal operators $\langle know \rangle$ and $\langle bel \rangle$ is the following:

$$\langle know \rangle F \equiv F \wedge \langle bel \rangle F$$

Therefore $\langle know \rangle F \rightarrow \langle bel \rangle F$ holds. This gives rise to the logic of belief, denoted as $\mathcal{L}_B$, based on the modal system $\mathcal{KD}4\mathcal{E}$. The axioms and inference rules of $\mathcal{L}_B$ are analogous to those in system $\mathcal{KD}4\mathcal{E}$.

Axioms of $\mathcal{L}_B$:

> LB 1: All propositional axioms
>
> LB 2: $\langle bel \rangle (F \rightarrow G) \rightarrow (\langle bel \rangle F \rightarrow \langle bel \rangle G)$
>
> LB 3: $\neg \langle bel \rangle \perp$
>
> LB 4: $\langle bel \rangle F \rightarrow \langle bel \rangle \langle bel \rangle F$
>
> LB 5: $\neg \langle bel \rangle F \rightarrow \langle bel \rangle \neg \langle bel \rangle F$

Inference Rules of $\mathcal{L}_B$:

> Modus Ponens (MP): If $\vdash F$ and $\vdash F \rightarrow G$ then $\vdash G$
>
> Rule of Necessitation (RN): If $\vdash F$ then $\vdash \langle bel \rangle F$

Axiom LB 2 states that an agent believes all the logical consequences of its beliefs, that is, an agent's beliefs are closed under logical deduction. For example, belief in *Cancelled* generates belief in *Cancelled* $\vee$ *Delayed* due to the knowledge *Cancelled* $\rightarrow$ *Cancelled* $\vee$ *Delayed*. Axiom LB 3, which is equivalent to axiom D in $\mathcal{KD4E}$, states that an agent does not believe in inconsistency. The derivation of the symbol $\perp$ from the database implies inconsistency. The two facts that an agent believes that it believes in something and an agent believes that it does not believe in something are expressed by Axioms LB 4 and LB 5, respectively.

Proposition 8-5: The following are theorems of $\mathcal{L}_\mathbf{B}$:

$$\langle bel \rangle (F \wedge G) \leftrightarrow \langle bel \rangle F \wedge \langle bel \rangle G$$

$$\langle bel \rangle F \vee \langle bel \rangle G \rightarrow \langle bel \rangle (F \vee G)$$

$$\langle bel \rangle F \rightarrow \neg \langle bel \rangle \neg F$$

We can now construct a modal epistemic logic of knowledge and belief, denoted $\mathcal{L}_\mathbf{KB}$, that has two modal operators $\langle know \rangle$ and $\langle bel \rangle$ with the relationship $\langle know \rangle F \equiv F \wedge \langle bel \rangle F$, and whose axioms are the combined set of axioms of $\mathcal{L}_\mathbf{K}$ and $\mathcal{L}_\mathbf{B}$. Axiom LK 4 can be excluded from this combined set as TS5 3 suggests that it is derivable. Although $\langle know \rangle F \rightarrow \langle bel \rangle F$ is derivable from the relationship $\langle know \rangle F \equiv F \wedge \langle bel \rangle F$, it can be added as an axiom for the sake of convenience. Consequently, one can discard the rule of necessitation of $\mathcal{L}_\mathbf{B}$ as it is superfluous in presence of the rule of necessitation of $\mathcal{L}_\mathbf{K}$.

8.13 Logic of Agents Beliefs ($\mathcal{LAB}$)

As mentioned in the introduction of this chapter, modal logics are not suitable to represent an agent's degrees of support about an assertion. In this section, we extend the syntax of traditional modal epistemic logic to include an indexed modal operator $\langle sup_d \rangle$ to represent these degrees. The proposed Logic of Agents Beliefs ($\mathcal{LAB}$) can model support to an assertion (to some extent d), but does not necessarily warrant an agent committing to believe in that assertion. The extended modal logic is given a uniform possible world semantics by introducing the concept of an accessibility *hyperelation*. Within our proposed framework, levels of support may be taken from one of a variety of different qualitative, semi-qualitative, and quantitative "dictionaries" which may be supplied with aggregation operations for combining levels of support.

There are two approaches to handling degrees of supports about assertions: considering degrees of supports as terms of the underlying logic (Fagin et al.,

1990), and handling degrees of supports as modalities (Fattorosi-Barnaba and Amati, 1987; Halpern and Rabin, 1987; Alechina, 1994; Chatalic and Froidevaux, 1992). The former approach requires axioms to handle the dictionary representing degrees of supports, and an associated algebra (for example, axioms of arithmetic and probability if the probabilistic dictionary is considered). On the other hand, the modality-based approach, which $\mathcal{LAB}$ has adopted, requires modal axioms for the properties of supports, and the degrees of these supports are manipulated via a meta-level reasoning. The main advantage of this approach is that it keeps the logic generic, capable of multiple interpretations for various qualitative and quantitative dictionaries. The modal operator for support in $\mathcal{LAB}$ is interpreted via hyperelation, and is a generalization of the accessibility relation concept of possible world semantics. The extended hyperelation-based possible world semantics coincides with the probabilistic Kripke model (Voorbraak; 1996) when the probabilistic dictionary is considered to introduce the support modal operator. The uniqueness of $\mathcal{LAB}$ is that its approach is not grounded to any fixed dictionary, and it has the ability to deal with inconsistent supports. Both of these features are highly desirable when reasoning with uncertain agent knowledge from multiple sources.

8.13.1 Syntax of $\mathcal{LAB}$

Suppose **P** is the set of all propositions that includes the special symbol $\top$ (true) and **D** is an arbitrary *dictionary* of symbols that will be used to label supports in propositions and assertions. In general, a dictionary will be a semi-lattice with a partial order relation $\leq$ and top element Δ. For the dictionary of probability, the lattice is a chain and Δ is 1. $\mathcal{LAB}$ is essentially a propositional logic extended with certain modal operators. The modal operator $\langle bel \rangle$ of $\mathcal{LAB}$ corresponds to "belief." In addition, for each dictionary symbol $d \in$ **D**, we have the modal "support" operator $\langle sup_d \rangle$ in $\mathcal{LAB}$. The *formulae* (or *assertions*) of $\mathcal{LAB}$ extend the domain of propositional formulae to the domain of *formulae* as follows:

- *propositions* are formulae.

- $\langle bel \rangle F$ is a formula, where F is a formula.

- $\langle sup_d \rangle F$ is a formula, where F is a formula and d is in the dictionary D.

- $\neg F$ and $F \wedge G$ are formulae, where F and G are formulae.

We take $\bot$ (false) to be an abbreviation of $\neg \top$. Other logical connectives and the existential quantifier are defined using $\neg$ and $\wedge$ in the usual manner.

Example

Consider a situation when the agent (or decision-maker) observes heavy rain while preparing to go to town for the game. The newly discovered weather status then becomes the agent's knowledge. The agent then infers from its common sense knowledge that there are three mutually exclusive and exhaustive possibilities or "candidate" states of the game, *On*, *Cancelled*, and *Delayed*, and so constructs arguments for and against these alternatives. These arguments use other beliefs, based on observations, such as transport disruption due to rain, the time of the game, and terrorist attack threats. Here is how a part of this problem is modeled as various types of sentences in the syntax of $\mathcal{LAB}$:

Arguments:

$\langle bel \rangle Transport\ Disruption \rightarrow \langle sup_{0.6} \rangle Delayed$

$\langle bel \rangle Last\ Game \rightarrow \langle sup_{0.9} \rangle (On \vee Cancelled)$

$\langle bel \rangle Terrorist\ Threat \rightarrow \langle sup_{0.7} \rangle Delayed$

Rules:

$\langle bel \rangle (Heavy\ Rain \rightarrow Transport\_Disruption)$

Integrity Constraints:

$\langle know \rangle (On \vee Delayed \vee Cancelled)$

$\langle know \rangle (\neg (On \wedge Delayed) \wedge \neg (On \wedge Cancelled) \wedge \neg (Delayed \wedge Cancelled))$

Beliefs:

$\langle know \rangle Heavy\ Rain$

$\langle bel \rangle Last\ Game$

In this model, informally, 1) arguments are for and against decision options at hand; 2) rules help to deduce implicit knowledge; 3) integrity constraints (Das, 1992) are properties that the knowledge base must satisfy; and 4) beliefs reflect the agent's understanding of the current situation. In the above representation, arguments encode the agent's own supports for various candidates, the only rule infers the believability of transport disruption from heavy rain, the integrity constraints encode three mutually exclusive and exhaustive candidates, and beliefs are the current beliefs and knowledge of the agent. Note that the presence of the knowledge *Heavy Rain* means it is currently raining. On the other hand,

$\langle bel \rangle Last\,Game$ means the agent believes that the game is the last game of the day.

8.13.2   Axioms of $\mathcal{LAB}$

The axioms of $\mathcal{LAB}$ include all the axioms and inference rules of the logic of belief $\mathcal{L_B}$ relating the modal operator $\langle bel \rangle$. We now present the axioms and inference rules relating to the modal operator $\langle sup_d \rangle$. The following inference rule states that the support operator is closed under implication. In other words, if F has support d and $F \rightarrow G$ is valid in $\mathcal{LAB}$ then G too has support d.

    LAB IR 1: *if* $\vdash F \rightarrow G$ *then* $\vdash \langle sup_d \rangle F \rightarrow \langle sup_d \rangle G,\ for\ every\ d \in D$

An assertion that is believed by an agent always has the highest support (at least from the agent itself as a subjective judgment), and the following axiom reflects this behavior:

    LAB 1: $\langle bel \rangle F \rightarrow \langle sup_\Delta \rangle F$

Proposition 8-6: *if* $\vdash F$ *then* $\vdash \langle sup_\Delta \rangle F$

Support operators can be combined to obtain a single support operator by using the following axiom:

    LAB 2: $\langle sup_{d_1} \rangle F \wedge \langle sup_{d_2} \rangle G \rightarrow \langle sup_{d_1 \otimes d_2} \rangle (F \wedge G)$

where $\otimes : \mathbf{D} \times \mathbf{D} \rightarrow \mathbf{D}$ is the function for combining supports for two assertions. The axiom states that if d_1 and d_2 are supports for F and G respectively then $\otimes(d_1, d_2)$ (or $d_1 \otimes d_2$ in infix notation) is a derived support for $F \wedge G$. Note that $d \otimes \Delta = d$, for every d in $\mathbf{D}$. The function takes into account the degrees of dependency between the two supports it is combining via their common source of evidence. In the Dempster-Shafer interpretation of $\mathcal{LAB}$ to be presented later in this subsection, the function becomes Dempster's combination rule for combining two masses on focal element.

Proposition 8-7: The following are theorems of $\mathcal{LAB}$:

$$\langle sup_\Delta \rangle \top$$

$$\langle sup_{d_1} \rangle F \wedge \langle sup_{d_2} \rangle (F \to G) \to \langle sup_{d_1 \otimes d_2} \rangle (F \wedge G)$$

$$\langle sup_d \rangle (F \to G) \to (\langle sup_d \rangle F \to \langle sup_d \rangle G)$$

<u>Example</u>

Here is a simple example derivation of support for the game to be delayed:

$\vdash \langle know \rangle \textit{Heavy Rain}$ Given

$\vdash \langle bel \rangle \textit{Heavy Rain}$ Axiom $\langle know \rangle F \to \langle bel \rangle F$

$\vdash \langle bel \rangle (\textit{Heavy Rain} \to \textit{Transport Disruption})$ Given

$\vdash \langle bel \rangle \textit{Transport Disruption}$ Axiom LB 2 and MP

$\vdash \langle sup_{0.6} \rangle \textit{Delayed}$ Given first argument and MP

8.13.3 Possible World Semantics of $\mathcal{LAB}$

This section presents a coherent possible world semantics that clarifies the concept of support within the concepts of knowledge and belief. A *model* of $\mathcal{LAB}$ is a 4-tuple

$$\langle \mathbf{W}, V, R, R_s \rangle$$

in which $\mathbf{W}$ is a set of possible worlds. A world consists of a set of qualified assertions outlining what is true in the world. V is a valuation that associates each world with a subset of the set of propositions. In other words,

$$V : \mathbf{W} \to \Pi(\mathbf{P})$$

where $\mathbf{P}$ is the set of propositions and $\Pi(\mathbf{P})$ is the power set of $\mathbf{P}$. The image of the world w under the mapping V, written as $V(w)$, is the set of all propositions that are true in the world w. This means that p holds in w for each p in $V(w)$.

The relation R is an accessibility relation that relates a world w to a set of worlds considered possible (satisfying all the integrity constraints) by the decision-maker from w. In a decision-making context, if there are n mutually exclusive and exhaustive candidates for a decision that are active in a world w, then there are n possible worlds.

We introduce semantics for our support operator by introducing the concept of a *hyperelation* that relates sets of worlds to the current world as opposed to relating only one set of worlds to the current world by the accessibility relation for the belief operator. The hyperelation R_s is a subset of the set

$$\mathbf{W} \times \mathbf{D} \times \Pi(\mathbf{W})$$

Semantically, if $\langle w, d, \mathbf{W}' \rangle \in R_s$, then there is an amount of support d for moving to one of the possible worlds in $\mathbf{W}'$ from the world w. If $\mathbf{W}'$ is empty, then the support is for an inconsistent world.

An assertion is a *belief* of a decision-maker at a world w if and only if it is true in all possible worlds that are accessible from the world w. Note that the members of R_s have been considered to be of the form $\langle w, d, \mathbf{W}' \rangle$ rather than $\langle w, d, w' \rangle$. The main reason is that the derivability of $\langle sup_d \rangle F$ means F is true only in a subset of all possible worlds accessible from w. If F is true in all possible worlds accessible from w then we would have had $\langle bel \rangle F$, which implies the highest form of support for F that is greater than or equal to d.

Due to the axioms related to the modal operator $\langle bel \rangle$, the standard set of model properties that will be possessed by the accessibility relation R is the following:

LAB MP 1: R is serial, transitive, and Euclidean.

The hyperelation R_s satisfies the following properties due to the axioms related to the modal operator $\langle sup_d \rangle$:

LAB MP 2: For every w, w_1, w_2 in W and d_1, d_2 in D, the relation R_s satisfies the following:

- $\langle w, \Delta, \mathbf{W} \rangle \in R_s$.

- If $\langle w, d_1, \mathbf{W}_1 \rangle$ and $\langle w, d_2, \mathbf{W}_2 \rangle$ are in R_s, then $\langle w, d_1 \otimes d_2, \mathbf{W}_1 \cap \mathbf{W}_2 \rangle \in R_s$.

Given the current world w, a decision-maker either stays in the current world if nothing happens (non-occurrence of events) or makes a transition to one of the possible worlds accessible from w by R_s. In either case, the new current world always belongs to $\mathbf{W}$, and thus $\langle w, \Delta, \mathbf{W} \rangle \in R_s$. The other property simply composes supports for worlds using the function $\otimes$ defined earlier, but checks dependencies between $\mathbf{W}_1$ and $\mathbf{W}_2$.

The semantics of supports and beliefs are as follows. Given a model $\mathcal{M} = \langle \mathbf{W}, V, R, R_s \rangle$, the truth values of formulae with respect to a world w are determined by the rules given below:

- $\models_{\langle \mathcal{M}, w \rangle} \top$

- $\models_{\langle \mathcal{M}, w \rangle} p$ if and only if $p \in V(w)$

- $\models_{\langle \mathcal{M}, w \rangle} \langle sup_d \rangle F$ if and only if there exists $\langle w, d, \mathbf{W}' \rangle$ in R_s such that $\models_{\langle \mathcal{M}, w' \rangle} F$, for every $w' \in \mathbf{W}'$

- $\models_{\langle \mathcal{M}, w \rangle} \langle bel \rangle F$ if and only if for every $w' \in \mathbf{W}$ such that wRw', $\models_{\langle \mathcal{M}, w' \rangle} F$

- $\models_{\langle \mathcal{M}, w \rangle} \neg F$ if and only if $\not\models_{\langle \mathcal{M}, w \rangle} F$

- $\models_{\langle \mathcal{M}, w \rangle} F \wedge G$ if and only if $\models_{\langle \mathcal{M}, w \rangle} F$ and $\models_{\langle \mathcal{M}, w \rangle} G$

A formula F is said to be *true* in model $\mathcal{M}$ (written as $\models_{\mathcal{M}} F$ or simply $\models F$ when $\mathcal{M}$ is clear from the context) if and only if $\models_{\langle \mathcal{M}, w \rangle} F$, for every w in $\mathbf{W}$. A formula F is said to be *valid* if F is true in every model.

Example

Continuing with our earlier example, the set $\mathbf{P}$ of propositions is the following set of 7 elements:

{On, Delayed, Cancelled, Heavy Rain, Transport Disruption, Last Game,
Terrorist Threat}

Therefore, the set of all possible worlds can potentially contain 2^7 elements, each corresponding to an element of the power set of $\mathbf{P}$. But the set of integrity constraints, along with the beliefs and knowledge, will make some of worlds impossible. For example, a world containing the elements *On* and *Cancelled* or a world that does not contain *Heavy Rain*. Consequently, the set $\mathbf{W}$ contains 48 elements, instead of 128, two of which are shown below as examples:

{Heavy Rain, Last Game, Delayed}

{Heavy Rain, Transport Disruption, Last Game, Cancelled}

Given the current state of world (say, w_0) as shown in the example, the relation R will not relate the first of these examples as a believable world as it does not contain a proposition that is currently believed (for example, *Transport Disruption*). The following set has to be a part of every possible world from w_0:

$$w_b = \{Heavy\ Rain, Last\ Game, Transport\ Disruption\}$$

The relation R relates the following set of worlds to the current world:

$$w_1 = w_b \cup \{On\}$$
$$w_2 = w_b \cup \{Delayed\}$$
$$w_3 = w_b \cup \{Cancelled\}$$

The relations R and R_s in the model definition from w_0 are defined as follows (see Figure 8-6):

$$R(w_0) = \{\langle w_0, w_1 \rangle, \langle w_0, w_2 \rangle, \langle w_0, w_3 \rangle\}$$
$$R_s(w_0) = \{\langle w_0, 0.6, \{w_2\} \rangle, \langle w_0, 0.9, \{w_1, w_3\} \rangle\}$$

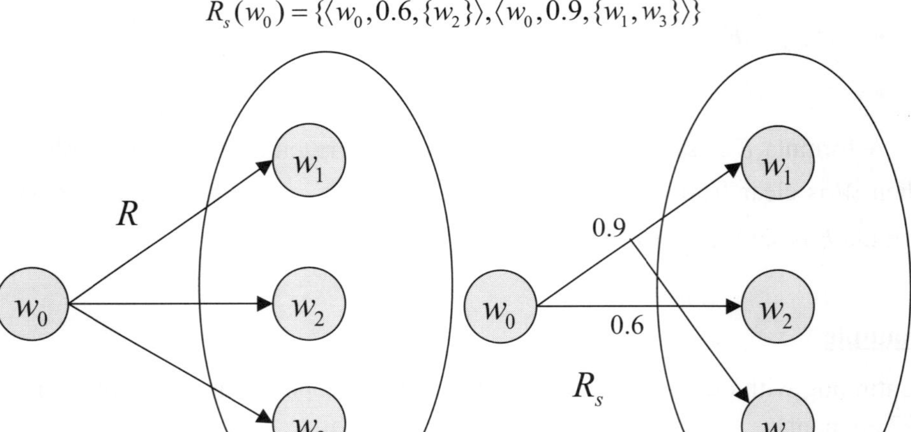

Figure 8-6: Illustration of accessibility relation and hyperelation

8.13.4 Soundness and Completeness of $\mathcal{LAB}$

First of all, the following two propositions prove that the validity in a class of models is preserved by the use of the rules of inference and the axioms of $\mathcal{LAB}$.

Proposition 8-8: For all formulae F and G and model $\mathcal{M}$:

- If $\models F$ then $\models \langle bel \rangle F$

- If $\models F \rightarrow G$ then $\models \langle sup_d \rangle F \rightarrow \langle sup_d \rangle G$

- If $\vDash F$ then $\vDash \langle sup_\Delta \rangle F$

Proof: Let $\mathcal{M} = \langle \mathbf{W}, V, R, R_s \rangle$ be a model and $w \in \mathbf{W}$.

- Suppose $\vDash F$. Then for every $w' \in \mathbf{W}$, $\vDash_{\langle \mathcal{M},w' \rangle} F$. In particular, for every $w' \in \mathbf{W}$ such that wRw', $\vDash_{\langle \mathcal{M},w' \rangle} F$. Therefore, $\vDash_{\langle \mathcal{M},w \rangle} \langle bel \rangle F$.

- Suppose $\vDash_{\langle \mathcal{M},w \rangle} \langle sup_d \rangle F$. Then there is $\langle w, d, \mathbf{W}' \rangle \in R_s$ such that $\vDash_{\langle \mathcal{M},w' \rangle} F$, for every $w' \in \mathbf{W}'$. Since $\vDash F \to G$, for every $w' \in \mathbf{W}'$, $\vDash_{\langle \mathcal{M},w' \rangle} F \to G$. Therefore, for every $w' \in \mathbf{W}'$, $\vDash_{\langle \mathcal{M},w' \rangle} G$. Therefore, $\vDash_{\langle \mathcal{M},w \rangle} \langle sup_d \rangle G$.

- Suppose $\vDash_{\mathcal{M}} F$. From the property of R_s, $\langle w, \Delta, W \rangle \in R_s$. Since $\vDash_{\mathcal{M}} F$, for every $w' \in \mathbf{W}$, $\vDash_{\langle \mathcal{M},w' \rangle} F$. Thus, by definition, $\vDash_{\mathcal{M}} \langle sup_\Delta \rangle F$.

Proposition 8-9: For all formulae F and G, axioms of $\mathcal{L}_B$ and LAB 1 and LAB 2 are valid in $\mathcal{LAB}$.

Proof: Proofs of validity of the axioms related to the modal operator $\langle bel \rangle$ can be found above in the section on modal logic.

To prove $\langle bel \rangle F \to \langle sup_\Delta \rangle F$, suppose $\vDash_{\langle \mathcal{M},w \rangle} \langle bel \rangle F$. Therefore, for every $w' \in \mathbf{W}$ such that wRw', $\vDash_{\langle \mathcal{M},w' \rangle} F$. But $\langle w, \Delta, \mathbf{W} \rangle \in R_s$ and $\vDash_{\langle \mathcal{M},w' \rangle} F$, for every $w' \in \mathbf{W}$. Therefore, by definition, $\vDash_{\langle \mathcal{M},w \rangle} \langle sup_\Delta \rangle F$.

To prove $\langle sup_{d_1} \rangle F \wedge \langle sup_{d_2} \rangle G \to \langle sup_{d_1 \otimes d_2} \rangle (F \wedge G)$, suppose $\vDash_{\langle \mathcal{M},w \rangle} \langle sup_{d_1} \rangle F$ and $\vDash_{\langle \mathcal{M},w \rangle} \langle sup_{d_2} \rangle G$. Therefore, there exists $\mathbf{W}_1 \in \Pi(\mathbf{W})$ such that $\langle w, d_1, \mathbf{W}_1 \rangle \in R_s$ and for every $w' \in \mathbf{W}_1$, $\vDash_{\langle \mathcal{M},w' \rangle} F$. Similarly, there exists $\mathbf{W}_2 \in \Pi(\mathbf{W})$ such that $\langle w, d_2, \mathbf{W}_2 \rangle \in R_s$ and for every $w' \in \mathbf{W}_2$, $\vDash_{\langle \mathcal{M},w' \rangle} G$. Thus, for every $w' \in \mathbf{W}_1 \cap \mathbf{W}_2$, $\vDash_{\langle \mathcal{M},w' \rangle} F \wedge G$. By LAB MP 2, $\langle w, d_1 \otimes d_2, \mathbf{W}_1 \cap \mathbf{W}_2 \rangle \in R_s$. Thus, there exists $\mathbf{W}_1 \cap \mathbf{W}_2 \in \Pi(\mathbf{W})$ such that $\langle w, d_1 \otimes d_2, \mathbf{W}_1 \cap \mathbf{W}_2 \rangle \in R_s$ and for every $w' \in \mathbf{W}_1 \cap \mathbf{W}_2$, $\vDash_{\langle M,w' \rangle} F \wedge G$. Therefore, by definition, $\vDash_{\langle \mathcal{M},w \rangle} \langle sup_{d_1 \otimes d_2} \rangle (F \wedge G)$.

Proposition 8-8 and Proposition 8-9 establish the basis of the soundness result. Next, we outline the completeness proof along the line of (Chellas, 1980). For the completeness result, the following class of models is relevant.

A model $\mathcal{M} = \langle \mathbf{W}, V, R, R_s \rangle$ of $\mathcal{LAB}$ is called a *canonical model*, written as $\mathcal{M}_c$, if and only if:

- $\mathbf{W} = \{w: w$ is a maximal consistent set in logic $\mathcal{LAB}\}$.
- For every w, $\langle bel \rangle F \in w$ if and only if for every $w' \in \mathbf{W}$ such that wRw', $F \in w'$.
- For every w and d, $\langle sup_d \rangle F \in w$ if and only if there exists $\langle w, d, \mathbf{W}' \rangle \in R_s$ such that $F \in w'$, for every $w' \in \mathbf{W}$.
- For each proposition P, $\vDash_{\langle \mathcal{M}, w \rangle} P$ if and only if $P \in w$.

Proposition 8-10: Let $\mathcal{M}$ be a canonical model of $\mathcal{LAB}$. Then for every w in W, $\vDash_{\langle \mathcal{M}, w \rangle} F$ if and only if $F \in w$.

Therefore, the worlds in a canonical model $\mathcal{M}$ for $\mathcal{LAB}$ will always verify just those sentences they contain. In other words, the sentences that are true in such a model are precisely the theorems of $\mathcal{LAB}$. In other words, $\vdash F$ if and only if $\vDash_{\mathcal{M}} F$.

Existence of a canonical model for $\mathcal{LAB}$ is shown by the existence of a proper canonical model defined as follows.

A model $\mathcal{M} = \langle \mathbf{W}, V, R, R_s \rangle$ of $\mathcal{LAB}$ is called a *proper canonical model*, if and only if:

- $\mathbf{W} = \{w: w$ is a maximally consistent set in $\mathcal{LAB}\}$.
- For every w and w', $wR_b w'$ if and only if $\{F : \langle bel \rangle F \in w\} \subseteq w'$.
- For every w, d and $\mathbf{W}'$, $\langle w, d, \mathbf{W}' \rangle \in R_s$ if and only if $\{\langle sup_d \rangle F : F \in \cap \mathbf{W}'\} \subseteq w$.
- For each proposition P, $\vDash_{\mathcal{M}} P$ if and only if $P \in w$.

By definition, a proper canonical model exists and a proper canonical model is a canonical model.

Proposition 8-11: If $\mathcal{M}$ is a proper canonical model of $\mathcal{LAB}$ then the model satisfies LAB MP 1 and LAB MP 2.

Suppose Γ is the class of all models satisfying LAB MP 1 and LAB MP 2. Then the following soundness and completeness theorem establishes the fact that $\mathcal{LAB}$ is determined by Γ.

Theorem 8-12: For every formula F in $\mathcal{LAB}$, $\vdash F$ if and only if $\vDash F$.

8.13.5 *Rational Extension of $\mathcal{LAB}$*

A *rational system*, denoted as $\mathcal{LAB}(R)$, extends $\mathcal{LAB}$ by some axioms reflecting the behavior of a *rational decision-making agent*. First of all, a rational agent always believes in anything that has support with the top element of the dictionary. Thus, the following axiom should also be considered for $\mathcal{LAB}(R)$:

LAB 3: $\langle sup_\Delta \rangle F \to \langle bel \rangle F$

This axiom, of course, derives that an assertion and its negation are not simultaneously derivable with the top element as support. That is, we have an integrity constraint of the following form:

Proposition 8-12: $\langle sup_\Delta \rangle F \wedge \langle sup_\Delta \rangle \neg F \to \bot$

It is difficult to maintain consistency of a database in the presence of the above axiom, particularly when the database is constructed from different sources; mutual inconsistency and mistakes sometimes need to be tolerated. In these circumstances, it might be left to the decision-maker to arbitrate over which assertions to believe and which not to believe.

Consider the following restriction on the hyperelation R_s, which says that the believable worlds are a subset of the worlds with the highest support:

LAB MP 3: For every w in W, if $\langle w, \Delta, \mathbf{W}' \rangle \in R_s$ then $\mathbf{W}' = R(w)$, where $R(w)$ is $\{w' \in \mathbf{W} : wRw'\}$.

If an agent is rational, then the concepts of believability and support with the top element coincide. In other words,

LAB IR 1: $\langle bel \rangle F \leftrightarrow \langle sup_\Delta \rangle F$

The soundness and completeness theorems for the rational extension can easily be established when the corresponding model property is taken into account.

8.13.6   Goals in $\mathcal{LAB}$

This subsection extends $\mathcal{LAB}$ with axioms dealing with the goals of an agent. We adopt the following two standard axioms for goals (Cohen and Levesque, 1990; Wainer, 1994):

> LAB G 1: $\neg\langle goal\rangle \perp$

The above axiom states that something that is impossible to achieve cannot be a goal of a decision-maker.

> LAB G 2: $\langle goal\rangle F \wedge \langle goal\rangle(F \to G) \to \langle goal\rangle G$

The above axiom states that all the logical consequences of an agent's goals are also goals. For technical reasons, all worlds compatible with an agent's goals must be included in those compatible with the agent's beliefs (Cohen and Levesque, 1990). This is summarized in the following axiom:

> LAB G 3: $\langle bel\rangle F \to \langle goal\rangle F$

Consequently, many redundant goals will be generated due to the presence of the above axiom. A goal will be considered *achieved* (resp. *active*) in a state if it is *believed* (resp. *not believed*) in the state. Thus, a goal F generated through the above axiom will be considered achieved and therefore not pursued actively. But the presence of a goal of this kind is a gentle reminder to the agent for establishing the truth of F, not just believing in F.

The possible world semantic of $\mathcal{LAB}$ and the soundness and completeness results can be extended incorporating the above modal concepts of goals. See (Fox and Das, 2000) for details.

8.13.7   *Dempster-Shafer Interpretation of* $\mathcal{LAB}$

We introduced the Dempster-Shafer theory of belief function earlier in the chapter on logical rules. In the Dempster-Shafer interpretation of $\mathcal{LAB}$, we consider the dictionary **D** as the probabilistic dictionary $[0,1]$ and the modal operator $\langle sup_d\rangle$ is used to represent the basic probability assignment (BPA) m. But an agent may obtain evidence, often inconsistent, from various sources, each of which will represent a BPA.

Continuing with our rainy game example in which $\Omega = \{On,\ Delayed,\ Cancelled\}$, handling each incoming piece of evidence as a belief provides a BPA. Suppose m_1, m_2, and m_3 are BPAs for evidence of transport disruption, last game of the day, and terrorist threats, respectively. These BPAs have to be combined in the order they arrive via a meta-level reasoning outside the scope of the logic. So, Dempster's combination rule for m_1 and m_2 yields m_{12}, which is then combined with m_3, giving m. LAB 2 formalizes this "static" combination process at the logic level to yield a combined BPA without updating the knowledge base. As an example of this combination, $m_1(\{Delayed\})$ and $m_2(\{On,Cancelled\})$ yield a certain amount of belief committed to inconsistency, producing $m_{12}(\{\bot\}) = 0.54$. Moreover, multiple evidences on a single assertion are obtained that need to be combined. For example, $m_1(\{Delayed\})$ and $m_1(\{Delayed\})$. The belief function that is obtained from the resulting BPA m is then modeled via another modal operator $\langle bel_d \rangle$.

We add a new modal operator $\langle bel_d \rangle$ into $\mathcal{LAB}$ for every $d \in \mathbf{D}$ to represent the notion of the belief function *Bel*. Our argument for not having this operator in the core $\mathcal{LAB}$ is based on the assumption that an agent only finds evidence that "supports" (captured via modal operator $\langle sup_d \rangle$) a focal element locally without thinking about aggregated evidence for the focal element. The agent's thought process for aggregation is a meta-level reasoning, results of which are "degrees of belief" (captured via $\langle bel_d \rangle$). A *model* of $\mathcal{LAB}$ is extended as a quintuple $\langle W, V, R, R_s, R_b \rangle$, where W, V, R, R_s are defined as before. The hyperelation R_b is a special case of R_s, and the following is the rule for determining truth values of formulae involving the modal operator $\langle bel_d \rangle$:

$$\models_{\langle \mathcal{M}, w \rangle} \langle bel_d \rangle F \quad \text{iff there exists only one } \langle w, d, \mathbf{W}' \rangle \text{ in } R_s \text{ such that } \mathbf{W}'$$
$$\text{uniquely characterizes } F$$

We also define the modal operator for plausibility in terms of the modal operator $\langle bel_d \rangle$ as follows:

$$\langle pl_d \rangle F \equiv \langle bel_{1-d} \rangle \neg F$$

In a pure Bayesian system, the concept of belief amounts to that of probability, and, therefore, $\langle bel_d \rangle F \equiv \langle bel_{1-d} \rangle \neg F$. Thus, the notions of believability and plausibility coincide. Since believability implies plausibility, the following property is expected to hold in $\mathcal{LAB}$.

Proposition 8-13: $\langle bel_d \rangle F \rightarrow \langle pl_{d'} \rangle F$, for some $d' \geq d$

Proof: Suppose $\vDash_{\langle \mathcal{M}, w \rangle} \langle bel_d \rangle F$. Then there exists $\langle w, d, \mathbf{W}' \rangle$ in R_s such that $\vDash_{\langle \mathcal{M}, w' \rangle} F$ for every $w' \in \mathbf{W}$. This means for every $w' \in \mathbf{W} - \mathbf{W}'$, $\vDash_{\langle \mathcal{M}, w' \rangle} \neg F$. But $\mathbf{W} - \mathbf{W}'$ uniquely characterizes $\neg F$. Therefore, for some d' , $\vDash_{\langle \mathcal{M}, w \rangle} \langle bel_{d'} \rangle \neg F$, that is, $\vDash_{\langle \mathcal{M}, w \rangle} \langle bel_{1-(1-d')} \rangle \neg F$, that is, $\vDash_{\langle \mathcal{M}, w \rangle} \langle pl_{1-d'} \rangle F$. But the sum of the degrees of beliefs for F and $\neg F$ cannot exceed 1, which yields $d + d' \leq 1$, that is, $d \leq 1 - d'$. Therefore, $\vDash_{\langle \mathcal{M}, w \rangle} \langle bel_d \rangle \rightarrow \langle pl_{1-d'} \rangle F$.

As per Dempster's combination rule, the operator $\otimes$ in LAB 2 yields the total support for F as $d_1 + d_2 - d_1 \times d_2$ when F and G are equivalent; otherwise, the total support for $F \wedge G$, if $F \wedge G$ is consistent, is just $d_1 \times d_2$. In case of inconsistency, Dempster's combination rule excludes the amount of belief committed to inconsistency during the normalization process for computing degrees of beliefs. Therefore, along the line of axiom LB 3 for $\langle bel \rangle$, we consider the following axiom for $\langle bel_d \rangle$ which states that there can be no degrees of belief for an inconsistency:

LAB 4: $\neg \langle bel_d \rangle \perp$, *for every $d \in \mathbf{D}$*

Now, if a focal element is believed, then the highest element of the dictionary is its total degree of belief. This property yields the following axiom for $\langle bel_d \rangle$:

LAB 5: $\langle bel \rangle F \rightarrow \langle bel_\Delta \rangle F$

Though the degree of belief for F is computed by summing up all the relevant supports for F, $\mathcal{LAB}$ does not provide any axiom to formalize this summation so that it can remain sufficiently generic. Rather, we view the computation of degrees of belief as a meta-level reasoning on the underlying logic. The following are the set of properties that are possessed by the accessibility relation R_s in the new model:

LAB MP 4: For every w, w_1, w_2 in $\mathbf{W}$ and d_1, d_2 in $\mathbf{D}$, the relation R_s satisfies the following conditions:

- $\langle w, \Delta, \mathbf{W} \rangle \in R_s$.

- If $\langle w, d, \mathbf{W} \rangle$ is in R_s then $\mathbf{W} \neq \Phi$.

- If $\langle w, d_1, \mathbf{W}_1 \rangle$ and $\langle w, d_2, \mathbf{W}_2 \rangle$ are in R_s such that $\mathbf{W}_1 \subseteq \mathbf{W}_2$, then $d_1 \leq d_2$.

The first two conditions are due to LAB 4 and LAB 5 respectively. To illustrate the third condition in the context of our rainy game example, suppose $\mathbf{W}_1 = \{w_1\}$ and $\mathbf{W}_2 = \{w_1, w_3\}$. Then $\mathbf{W}_1 \subseteq \mathbf{W}_2$, and $\mathbf{W}_1$ and $\mathbf{W}_2$ uniquely characterize $w_b \wedge On$ and $w_b \wedge (On \vee Cancelled)$, respectively (here w_b stands for the conjunction of its constituent elements). If $\langle w, d_1, \mathbf{W}_1 \rangle, \langle w, d_2, \mathbf{W}_2 \rangle \in R_s$ then

$$\models_{\langle \mathcal{M}, w \rangle} \langle bel_{d_1} \rangle (w_b \wedge On)$$

$$\models_{\langle \mathcal{M}, w \rangle} \langle bel_{d_2} \rangle (w_b \wedge (On \vee Cancelled))$$

Since $w_b \wedge On \rightarrow w_b \wedge (On \vee Cancelled)$, the degrees of belief d_1 in $w_b \wedge On$ is expected to be not more than the degrees of belief d_2 in $w_b \wedge (On \vee Cancelled)$. Unfortunately, it is not possible to axiomatize the third condition in LAB MP 4 without introducing the dictionary symbols and associated algebra into the logic. The soundness and completeness results can be easily extended with respect to the models satisfying the rest of the conditions.

8.14 Further Readings

The two most popular text books on modal logics are (Chellas, 1980) and (Hughes and Cresswell, 1986). A good historical review of modal logics can be found in (Goldblatt, 2001). (Blackburn et al., 2006) is a great source book for modal logics in general. Lewis and Langofrd's book (1932) provides a detailed foundation of modern day modal logics. Another concise book on various modal systems and their interrelationships is by Lemmon (1977). The soundness and completeness results presented in this chapter are largely based on his book. Kripke's paper (1963) is the seminal work on possible world semantics. For an elaborate discussion on modal first order logics, see (Hughes and Cresswell, 1986) and (Fitting and Mendelsohn, 1998). The modal resolution schemes presented here are based on Ohlbach's work (1988). See the authors' work in (Fox and Das, 2000) for a detailed account of a logic along the line of $\mathcal{LAB}$. Two books that are worth mentioning from the point of view of reasoning about agents' knowledge and belief are (Fagin et al., 1995) and (Meyer et al., 2001).

Chapter 9

Symbolic Argumentation for Decision-Making

Chapters 3, 6, and 8 provided foundations in classical logics, probabilistic networks, and modal logics, respectively. All of these foundations were built upon to create reasoning for decision-making by intelligent agents. Modal logics are extensions to classical logics, and probabilistic or belief networks are an efficient computational paradigm for the probabilistic foundation. As shown in chapters 5 and 7, sound and practical decision-making systems can be built via logic-based rules that incorporates probabilities or mass values as strengths, and via influence diagrams based on belief networks. The question then arises on how to combine the strengths of each approach, such as the logic-based declarative representation of agent knowledge, intuitive probability semantics used in everyday language, automatic aggregation of support for decision options via belief networks and Dempster's rule of combination, and the possible world concept that parallels decision-making options. This chapter presents one approach that combines aspects of classical logics, modal logics, and belief networks into an argumentation framework, yielding the $\mathcal{P}3$ (Propositional, Probability, and Possibility) model for agent-based decision-making. The underlying theoretical foundation of $\mathcal{P}3$ is the logic $\mathcal{LAB}$ presented in the last section.

Within an argumentation framework for decision-making, such as the one developed in (Fox and Das, 2000) and adopted here, the agent combines arguments for and against the decision options at hand and then commits to the one with the highest aggregated evidence. In this chapter, we use the syntax of a modal propositional logic to represent arguments, and probabilities to represent the strength of each argument. Axiomatic inferencing schemes within this type of logics are insufficient for aggregating arguments, as the typical aggregation process is a meta-level reasoning involving sets of arguments. We propose two schemes for aggregating arguments: one that uses the Dempster-Shafer theory of belief functions, and the other that uses Bayesian belief networks. The mass distribution concept in the theory of belief functions naturally lends itself to the consideration of possible worlds. Aggregation via belief networks is performed on networks that are "automatically" constructed out of available arguments without requiring any additional knowledge.

Essentially, this aggregation process provides the agent with an overall judgment on the suitability of each decision option. Whether the agent accepts this judgment from a mechanical, meta-level reasoning process depends on the nature of the agent. We shall describe a *rational* agent as one who accepts the judgment and acts and revises its own belief accordingly.

An argumentation-based decision-making framework like the one described here is functionally similar to the classical rule-based framework presented in Chapters 4 and 5, with the following exceptions:

- It deals with more expressive knowledge in the form of arguments "for" and "against" (as opposed to simple rules which deal only with arguments "for"), and can use a variety of dictionaries.

- It incorporates an inference mechanism which is capable of aggregating arguments for and against decision options, and is therefore more general than simple forward chaining.

As for an overall comparison with belief network based decision-making framework presented in Chapter 6, it is easier to acquire a set of arguments from domain experts than construct a belief network. The latter involves a much more methodical approach to knowledge elicitation, and is usually much more time consuming to establish variables and CPTs. In this respect, a major advantage of an argumentation-based framework is that an overall support can be generated to help make a decision even with very few arguments, making the framework highly flexible. But a typical propagation algorithm for a belief network fails to work even if a single entry within a CPT of the network is missing.

The rest of the chapter is organized as follows. We first provide readers with a short background of the argumentation concept. Then we present the domino decision-making model underlying $\mathcal{P}3$, along with the model's knowledge representation language for expressing agents' beliefs and knowledge for making decisions. We then describe the belief function and belief network-based approaches to aggregation of arguments.

9.1 Toulmin's Model of Argumentation

In his book, Toulmin (1958) discussed how difficult it is to cast everyday, practical arguments into classical deductive form. He claimed that arguments needed to be analyzed using a richer format than the simple if-then form of classical logic. He characterizes practical argumentation using the scheme shown in Figure 9-1.

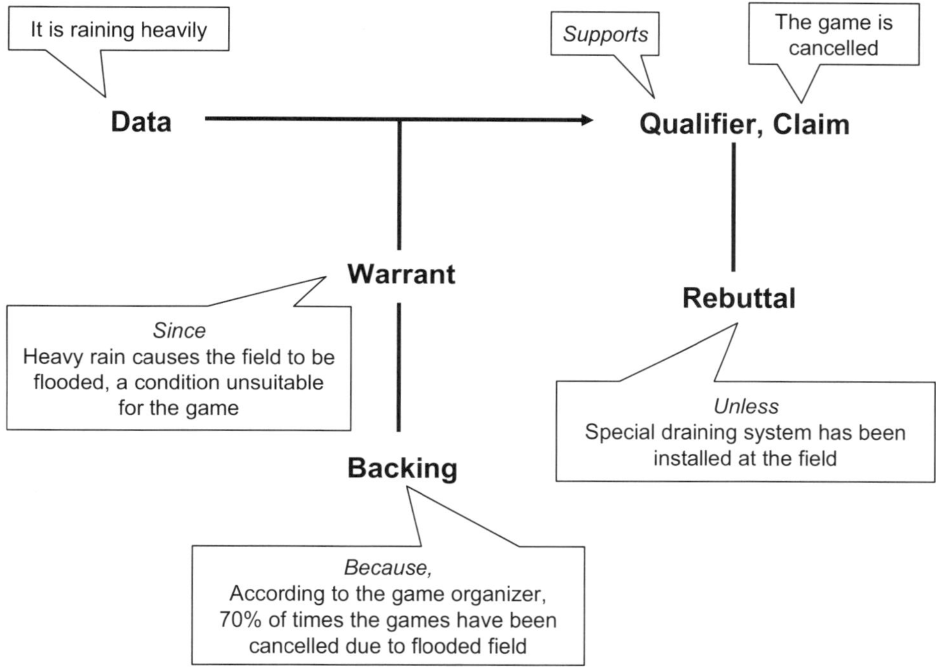

Figure 9-1: Toulmin's model of argumentation

As shown in Figure 9-1, Toulmin's model decomposes an argument into a number of constituent elements:

- **Claim**: The point an agent (or decision-maker) is trying to make
- **Data**: The facts about a situation relied on to clarify a claim
- **Warrant**: Statements indicating general ways of arguing
- **Backing**: Generalizations providing explicit support for an argument
- **Qualifier**: Phrases showing the confidence an argument confers on a claim
- **Rebuttal**: Acknowledges exceptions or limitations to the argument

To illustrate, consider an argument (as annotated in Figure 9-1) claiming that the game, which was to be held today, has been cancelled. The fact or belief (that is, data) on which this claim is made is that there is heavy rain. General principles or rules, such as "heavy rain causes the field to be flooded, a condition unsuitable for the game," warrant the argument, based on statistical research published by the game organizer, which is the backing. Since the argument is not conclusive, we insert the qualifier "supports" in front of the claim, and note the possibility that the conclusion may be rebutted on other grounds, such as installation of a new drainage system at the game venue, or hearing radio commentary (not shown in the picture) that clarifies the game status.

Our approach is to transform Toulmin's work to a more formal setting. We deal with the concepts of warrant and rebuttal, but as very simple propositional arguments for and against. We do not deal with first-order sentences that are more suitable for representing backings in Toulmin's model; rather, we introduce the use of a single qualifier called *support*.

9.2   Domino Decision-Making Model for $\mathcal{P}3$

This section develops a generic architecture for $\mathcal{P}3$'s argumentation based decision-making process as presented above via Toulmin's model. Continuing with our rainy game example, Figure 9-2 shows the underlying process which starts when the decision maker observes heavy rain while preparing to go to town for the game.

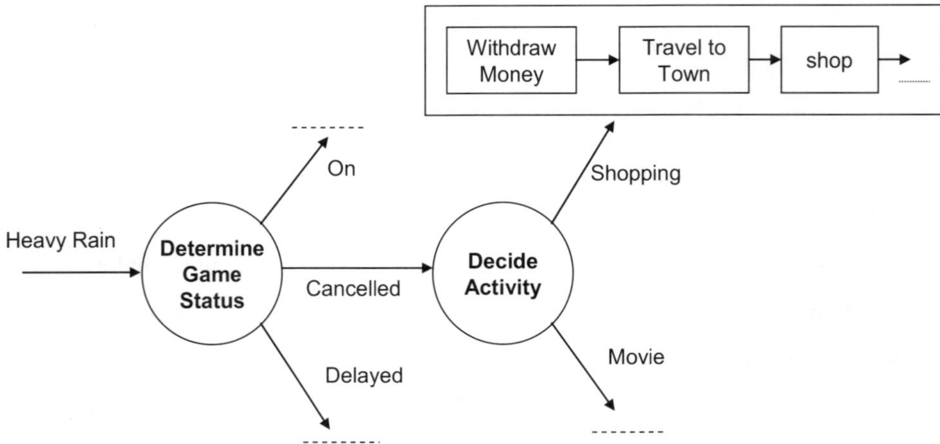

Figure 9-2: Decision-making flow

The newly discovered weather status then becomes the decision maker's belief. Given that the decision maker "believes" that it is raining heavily, it raises a "goal" of finding the status of the game. It then infers from its common sense knowledge that there are three possible or "candidate" states of the game, *On*, *Cancelled*, and *Delayed*, and constructs arguments for and against these alternatives. These arguments use other beliefs of the agent, based on observations such as transport availability and radio commentary. In this case, the balance of "argument" is in favor of the game being cancelled, and this conclusion is added into the decision maker's database of beliefs.

Given this new belief regarding the cancelled status of the game, a new goal is raised, that is, to "plan" for alternative activities. As in determining the status of the game, the agent first enumerates the options. In this case there are two options for alternative activities, shopping and going to a movie, and the decision-maker once again constructs arguments for the alternatives, taking into account the crowded holiday season, transport, and costs, and recommends going shopping as the most preferred alternate activity on the basis of the arguments. The adoption of the shopping "plan" leads to an appropriate scheduling of "actions" involved in shopping, such as withdrawing money, traveling to town, and going to stores. The effects of these actions are recorded in the decision-maker's database, which may lead to further goals, and so on.

Figure 9-3, the Domino model, captures graphically the decision-making framework, where the chain of arrows in the figure represents the previous example decision-making and planning process. Within our proposed framework, a decision schema has several component parts: an evoking situation, a goal, one or more candidates and corresponding arguments for and against, one or more commitment rules, and alternative plans and actions. Please note that temporal planning is out of the scope of this book. See (Fox and Das, 2000) for information on how to deal with a scheduled plan that is committed.

A *situation* describes, as a Boolean expression on the database of beliefs, the situation or event that initiates decision-making. For example, a belief that an abnormality (for example, heavy rain) is present may lead to a choice between alternative possible causes or effects of the abnormality.

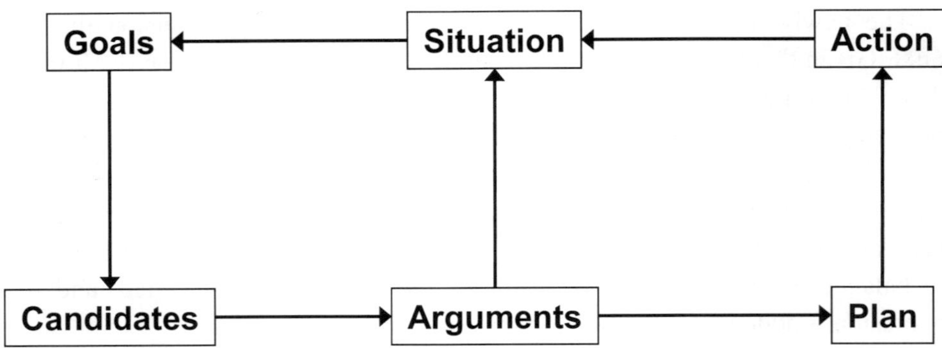

Figure 9-3: Domino process view of the example

A *goal* is raised as soon as the evoking situation occurs. In particular, the belief that an abnormality is present may raise the goal of determining its cause or effects. For example, if it is raining heavily, one of the possible effects is the cancellation of the game, and therefore the goal is to determine game status. Or, if there is no radio commentary, we would again want to determine the status of the game, as its cancellation causes no radio commentary. Typically, a goal is represented by a state that the decision-maker tries to bring about.

Candidates are a set of alternative decision options, such as *On, Cancelled,* and *Delayed.* In principle, the set of candidates may be defined extensionally (as a set of propositions) or intentionally (by rules), but we only consider the former case here.

Arguments are modal-propositional rules that define the arguments appropriate for choosing between candidates for the decision. Argument schemas are typically concerned with evidence when the decision involves competing hypotheses (beliefs), and with preferences and values when the decision is concerned with actions or plans.

Commitment rules (not illustrated in Figure 9-3) define the conditions under which the decision may be recommended, or taken autonomously, by the decision-maker. It may include logical and/or numerical conditions on the argument and belief databases.

9.3   Knowledge Representation Syntax of $\mathcal{P}3$

The concept of the Domino model-based decision scheme and its components is captured in a high-level declarative syntax. Figure 9-4 gives the decision construct (actual syntax of input to the parser) representing the decision circle *Determine Game Status* in Figure 9-2 or the left box of the Domino model in

Figure 9-3. All decisions have an evoking situation, which, if the decision-maker believes it to be true, raises the corresponding goal. The three possible paths from the decision circle go to the following three alternative pathways: *On, Cancelled,* and *Delayed.* These candidates are represented explicitly in the decision construct. The arguments and commitments within a decision construct are also represented explicitly.

```
decision:: game_status
  situation
    heavy_rain
  goal
    determine_game_status
  candidates
    on;
    cancelled;
    delayed
  arguments
    heavy_rain => support(not on, 0.7);
    terrorist_attack_threat => support(not on, 1.0);
    players_injury => support(cancelled, 0.8);
    players_union_strike => support(cancelled, 1.0);
    club_financial_crisis => support(not cancelled, 0.6);
    transport_disruption => support(delayed, 0.7);
  commits
    netsupport(X, U) & netsupport(Y, V) &
    netsupport(Z, W) & U > V & U > W => add(X).
```

Figure 9-4: Example decision construct

The decimal number in an argument represents the probabilistic measure of support given by the argument to the decision candidate. The basic idea is that an argument is a reason to believe something, or a reason to act in some way, and an argument schema is a rule for applying such reasons during decision-making. The more arguments there are for a candidate belief or action, then the more a decision-maker is justified in committing to it. The aggregation function can be a simple "weighing of pros and cons" (such as the `netsupport` function in the example of Figure 9-4), but it represents a family of more or less sophisticated functions by which we may assess the merit of alternative candidates based on the arguments about them.

In general, an argument schema is like an ordinary inference rule with

```
support(⟨candidate⟩, ⟨sign⟩)
```

as its consequent, where $\langle\texttt{sign}\rangle$ is drawn from a set called a dictionary. $\langle\texttt{sign}\rangle$ represents, loosely, the confidence that the inference confers on the candidate. The dictionary may be strictly quantitative (for example, the numbers in the [0,1] interval) or qualitative, such as the symbols {+, -} or {*high, medium, low*}. Here we are dealing with probabilistic arguments, so $\langle\texttt{sign}\rangle$ is drawn from the probability dictionary [0,1]. An example argument from the decision construct in Figure 9-4 is

```
transport_disruption  ⇒  support(delayed, 0.7)
```

where $\langle\texttt{candidate}\rangle$ is $\texttt{delayed}$. Informally, the argument states that if there is transport disruption then there is a 70% chance that the game will be delayed. The rest of the arguments of the decision construct provide support for and against the decision options based on the evidence of radio commentary, lack of radio commentary, terrorist attack threat, player injury, a players' union strike, and the hosting club's financial condition. A knowledge base for the decision-maker consists of a set of definitions of this and various other decision tasks. For the Dempster-Shafer theory to be applicable, one acceptable interpretation of 0.7 is that it is the confidence on the source from which the information about the transport disruption is obtained. Such a source could be a human observer or a sensor placed on a highway. More on this interpretation is discussed in the following section.

A decision-maker considers the decision $\texttt{game\_status}$ in Figure 9-4 for activation when the belief $\texttt{heavy\_rain}$ is added to the database. When the decision-maker detects this, it checks whether any candidate has already been committed. If not, the decision will be activated and the goal $\texttt{determine\_game\_status}$ is raised; otherwise no action is taken. While the goal is raised, further information about the situation (such as the weather) can be examined to determine whether the premises of any argument schemas are instantiated.

A commitment rule is like an ordinary rule with one of

```
add(⟨property⟩)
```

```
schedule(⟨plan⟩)
```

as its consequent. The former adds a new belief to the knowledge base, and the latter causes a plan to be scheduled (see Figure 9-3). The scheduled plan is hierarchically broken down into actions via plan constructs (Please note that temporal planning is out of the scope of this book.). The generated actions from a plan are executed in the environment, thus changing the state of the environment and the decision-maker's beliefs.

When a decision is in progress, as additional arguments become valid, the decision's commitment rules are evaluated to determine whether it is justified to select a candidate. A commitment rule will often make use of an aggregation function, such as `netsupport`, but this is not mandatory. The function evaluates collections of arguments for and against any candidate to yield an overall assessment of confidence and establish an ordering over the set of candidates; this ordering may be based on qualitative criteria or on quantitative assessment of the strength of the arguments. This function has the form:

```
netsupport(⟨candidate⟩, ⟨support⟩)
```

Later in this section, we implement the `netsupport` function based on two approaches: Dempste's rule of combination, and an evidence propagation algorithm in belief networks.

```
decision:: alternative_activity
    situation
        cancelled
    goal
        decide_alternative_ activity
    candidates
        shopping;
        movie
    arguments
        holiday_season =>
                support(no shopping, 0.8);
        ...
    commits
        ... => schedule(shopping).
```

Figure 9-5: Example decision construct

A decision-maker considers the decision `alternative_activity` in Figure 9-5 for activation when the belief about the game cancellation is added to the database. Figure 9-6 illustrates mapping these decision and plan constructs into the Domino model.

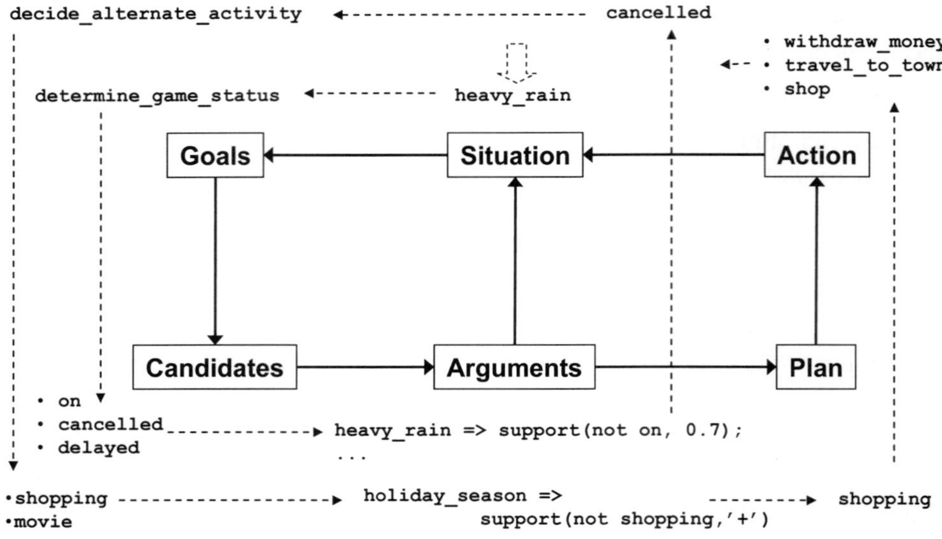

Figure 9-6: Decision and plan constructs mapped into Domino model

9.4   Formalization of $\mathcal{P}3$ via $\mathcal{LAB}$

In this section, we provide some example sentences in $\mathcal{LAB}$ that are translations of the decision construct shown in Figure 9-4. The situation and goal portion in the decision `game_status` in Figure 9-4 is translated to the following modal rule:

$$\langle bel \rangle Transport\ Disruption \rightarrow \langle goal \rangle Determine\ Game\ Status$$

The above $\mathcal{LAB}$ sentence states that if *Transport Disruption* is believed, then a goal is *Determine Game Status*. A goal is considered achieved as soon as it becomes true. In the context of the decision `game_status`, this is reflected in the following formulae:

$$\langle bel \rangle (On \wedge \neg Cancelled \wedge \neg Delayed) \rightarrow \langle bel \rangle Determine\ Game\ Status$$
$$\langle bel \rangle (Cancelled \wedge \neg On \wedge \neg Delayed) \rightarrow \langle bel \rangle Determine\ Game\ Status$$
$$\langle bel \rangle (Postponed \wedge \neg On \wedge \neg Delayed) \rightarrow \langle bel \rangle Determine\ Game\ Status$$

The first of these three sentences states that if it is believed that the game is on, and neither cancelled nor delayed, then *Determine Game Status* is believed. In other words, the earlier goal *Determine Game Status* is considered achieved upon believing that the game is on. The $\mathcal{LAB}$ representations for the arguments in the diagnosis decision are:

$$\langle bel \rangle Heavy\ Rain \rightarrow \langle sup_{0.7} \rangle \neg On$$

$$\langle bel \rangle (Terrorist\ Attack\ Threat \rightarrow \neg On)$$

$$\langle bel \rangle Players\ Injury \rightarrow \langle sup_{0.8} \rangle \neg Cancelled$$

$$\langle bel \rangle (Players\ Union\ Strike \rightarrow Cancelled)$$

$$\langle bel \rangle Club\ Financial\ Crisis \rightarrow \langle sup_{0.6} \rangle \neg Cancelled$$

$$\langle bel \rangle Transport\ Disruption \rightarrow \langle sup_{0.7} \rangle Delayed$$

This straightforward translation mechanism from the knowledge representation syntax of $\mathcal{P}3$ to the syntax of the $\mathcal{LAB}$ logic of belief provides a strong theoretical foundation for $\mathcal{P}3$. Moreover, necessary derivations in $\mathcal{P}3$ for answering queries can be carried out via inferencing in $\mathcal{LAB}$. This is the path we took to develop a prototype implementation of the Domino model.

9.5 Aggregation via Dempster-Shafer Theory

In order to illustrate the Dempster-Shafer theory in the context of our example, we consider only the following subset of three arguments from the set of arguments presented in the last section:

$$\langle bel \rangle Heavy\ Rain \rightarrow \langle sup_{0.7} \rangle \neg On$$

$$\langle bel \rangle Club\ Financial\ Crisis \rightarrow \langle sup_{0.6} \rangle \neg Cancelled$$

$$\langle bel \rangle Players\ Injury \rightarrow \langle sup_{0.8} \rangle \neg Cancelled$$

Note that Dempster-Shafer theory requires that evidences to be combined are independent. In this set of arguments, the potential usable evidences (rainy condition, financial situation, and player injury) are independent. We excluded the argument relating transport disruption because it is causally related to the heavy rain condition. We have also excluded the definitive arguments (with support 1.0) from our consideration to make our evidence aggregation process illustration more interesting. The frame-of-discernment Ω in this example is $\{On,\ Delayed,\ Cancelled\}$.

The first evidence corresponds to the decision-maker's observation of heavy rain while preparing to go to town for the game. The decision maker has 0.7 subjective probability for the game not being on (that is, cancelled or delayed) given this rainy weather. Therefore, evidence of heavy rain alone justifies a 0.7 degree of belief that the game is not on, but only a zero degree of belief (not 0.3) that the game is on. This zero belief does not mean that the decision-maker is sure that the game is not on (as a zero probability would), but states that evidence of heavy rain gives the decision-maker no reason to believe that the game is on.

The values 0.7 and 0 together constitute a belief function; they can be thought of as lower bounds on probabilities $p(Cancelled \, or \, Delayed \, | \, Heavy \, Rain)$ and $p(\neg(Cancelled \, or \, Delayed) \, | \, Heavy \, Rain)$, respectively.

The evidence suggest the focal elements are $\{Cancelled, Delayed\}$ and Ω, and $m_1(\{Cancelled, Delayed\}) = 0.7$. We know nothing about the remaining probability so it is allocated to the whole frame of discernment as $m_1(\Omega) = 0.3$. The decision-maker also believes that the current financial situation of the club is bad, resulting in 0.6 subjective probability that the game will not be cancelled in this situation. The new evidence provides the focal elements $\{On, Delayed\}$ and Ω, with $m_2(\{On, Delayed\}) = 0.6$. The remaining probability, as before, is allocated to the whole frame of discernment as $m_2(\Omega) = 0.4$. Because the current financial situation and transport disruption are independent of each other, Dempster's rule can be used to combine the masses as in Table 9-1.

Table 9-1

$Can - Cancelled$ $Del - Delayed$	$m_2(\{On, Del\}) = 0.6$	$m_2(\Omega) = 0.4$
$m_1(\{Can, Del\}) = 0.7$	$m_{1,2}(\{Del\}) = 0.42$	$m_{1,2}(\{Can, Del\}) = 0.28$
$m_1(\Omega) = 0.3$	$m_{1,2}(\{On, Del\}) = 0.18$	$m_{1,2}(\Omega) = 0.12$

The basic probability assignments m_1 and m_2 are different but consistent, and therefore the degrees of belief in both $\{Cancelled, Delayed\}$ and $\{On, Delayed\}$ being true (that is, the game is delayed) is the product of $m_1(\{Cancelled, Delayed\})$ and $m_2(\{On, Delayed\})$, that is, 0.42. The revised focal elements and their beliefs and plausibilities are shown in Table 9-2.

Table 9-2

Focal Element (A)	$Bel(A)$	$Pl(A)$
$\{Delayed\}$	0.42	1.0
$\{On, Delayed\}$	0.60	1.0
$\{Cancelled, Delayed\}$	0.70	1.0
Ω	1.0	1.0

Finally, the player injury situation suggests the focal elements are $\{Cancelled\}$ and Ω, so that $m_3(\{Cancelled\}) = 0.8$, $m_3(\Omega) = 0.2$. Dempster's rule of combination applies as shown in Table 9-3, but with one modification. When the evidence is inconsistent, their products of masses are assigned to a single measure of inconsistency, say k.

Table 9-3

Can – Cancelled *Del – Delayed*	$m_3(\{Can\}) = 0.8$	$m_3(\Omega) = 0.2$
$m_{1,2}(\{Del\}) = 0.42$	$k = 0.336$	$m(\{Del\}) = 0.084$
$m_{1,2}(\{On, Del\}) = 0.18$	$k = 0.144$	$m(\{On, Del\}) = 0.036$
$m_{1,2}(\{Can, Del\}) = 0.28$	$m(\{Can\}) = 0.224$	$m(\{Can, Del\}) = 0.056$
$m_{1,2}(\Omega) = 0.12$	$m(\{Can\}) = 0.096$	$m(\Omega) = 0.024$

The total mass of evidence assigned to inconsistency k is $0.336 + 0.144 = 0.48$. The normalizing factor is $1 - k = 0.52$. The resulting masses of evidence are as follows:

$$m(\{Cancelled\}) = (0.224 + 0.096)/0.52 = 0.62$$
$$m(\{Delayed\}) = 0.084/0.52 = 0.16$$
$$m(\{On, Delayed\}) = 0.036/0.52 = 0.07$$
$$m(\{Cancelled, Delayed\}) = 0.056/0.52 = 0.11$$
$$m(\Omega) = 0.024/0.52 = 0.04$$

The revised focal elements and their beliefs and plausibilities are shown in Table 9-4.

Table 9-4

Focal Element (A)	$Bel(A)$	$Pl(A)$
$\{Cancelled\}$	0.62	0.77
$\{Delayed\}$	0.16	0.38
$\{On, Delayed\}$	0.23	0.38
$\{Cancelled, Delayed\}$	0.89	1.0
Ω	1.0	1.0

We consider two examples to illustrate two special cases for evidence aggregation. Hypothetically, consider the case when the set of focal elements of the basic probability distribution m_2 is exactly the same as m_1. The evidence combination table is shown in Table 9-5.

Table 9-5

$Can - Cancelled$ $Del - Delayed$	$m_2(\{Can, Del\}) = 0.6$	$m_2(\Omega) = 0.4$
$m_1(\{Can, Del\}) = 0.7$	$m_{1,2}(\{Can, Del\}) = 0.42$	$m_{1,2}(\{Can, Del\}) = 0.28$
$m_1(\Omega) = 0.3$	$m_{1,2}(\{Can, Del\}) = 0.18$	$m_{1,2}(\Omega) = 0.12$

Now,

$$Bel(\{Cancelled, Delayed\}) = 0.42 + 0.18 + 0.28 = 0.88 = 0.6 + 0.7 - 0.6 \times 0.7.$$

In general, when two mass distributions m_1 and m_2 agree on focal elements, the combined degree of belief on a common focal element is $p_1 + p_2 - p_1 \times p_2$, where p_1 and p_2 are evidences on the focal element by the two distributions. This formula coincides with the noisy-or technique in Bayesian belief networks for combining probabilities of variables that have certain properties.

As opposed to agreeing on focal elements, if m_2 is contradictory to m_1 then an example evidence combination is shown in Table 9-6.

Table 9-6

$Can - Cancelled$ $Del - Delayed$	$m_2(\{On\}) = 0.6$	$m_2(\Omega) = 0.4$
$m_1(\{Can, Del\}) = 0.7$	$k = 0.42$	$m_{1,2}(\{Can, Del\}) = 0.28$
$m_1(\Omega) = 0.3$	$m_{1,2}(On) = 0.18$	$m_{1,2}(\Omega) = 0.12$

Now,

$$Bel(\{Cancelled, Delayed\}) = 0.28/(1 - 0.42) = 0.7(1 - 0.6)/(1 - 0.42).$$

In general, when two mass distributions m_1 and m_2 are contradictory, then the combined degree of belief on the focal element for m_1 is $p_1(1 - p_2)/(1 - p_1 \times p_2)$ and the combined degree of belief on the focal element for m_2 is

$p_2(1-p_1)/(1-p_1 \times p_2)$, where p_1 and p_2 are evidences on the focal element by the two distributions.

9.6   Aggregation via Bayesian Belief Networks

In order to illustrate the aggregation process in the context of our example, we consider the following five arguments to ensure that the potential evidences that generate support for individual decision options are independent:

$$\langle bel \rangle Players\ Injury \rightarrow \langle sup_{0.8} \rangle Cancelled$$

$$\langle bel \rangle Heavy\ Rain \rightarrow \langle sup_{0.6} \rangle Cancelled$$

$$\langle bel \rangle Heavy\ Rain \rightarrow \langle sup_{0.75} \rangle Delayed$$

$$\langle bel \rangle Transport\ Disruption \rightarrow \langle sup_{0.9} \rangle Delayed$$

$$\langle bel \rangle Terrorist\ Attack\ Threat \rightarrow \langle sup_{0.7} \rangle Delayed$$

The three decision options are *On*, *Cancelled*, and *Delayed*. Due to differences in the type of knowledge represented and in the formalism used to represent uncertainty, much of the knowledge needed to build an equivalent belief network for aggregation could not be extracted from the above set of arguments. In our approach, we will be able to automatically extract the network structure fully without requiring additional knowledge under the assumption that the conditions (such as *Player Injury* or *Heavy Rain*) based on which arguments for the decision options are generated are independent.

We first construct fragments of networks using the arguments relevant to the decision-making task at hand. Note that, given a network fragment with a variable, and its parents and CPT, the fragment can be equivalently viewed as a set of arguments. For example, consider the network fragment in Figure 9-7, which states that player injury and rain together can determine the status of the game.

Each column of the CPT yields an argument for and an argument against a state of the variable *Game*. For example, if there is player injury and it rains, then there is an argument for a game with support 0.05.

$$\langle bel \rangle (Injury \wedge Rain) \rightarrow \langle sup_{0.05} \rangle Game$$

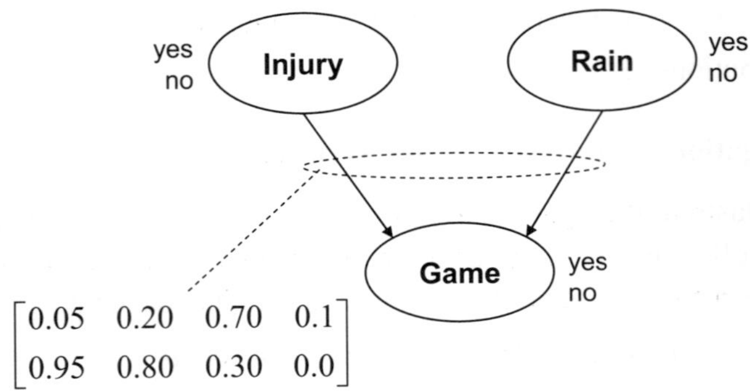

$$\begin{bmatrix} 0.05 & 0.20 & 0.70 & 0.1 \\ 0.95 & 0.80 & 0.30 & 0.0 \end{bmatrix}$$

Figure 9-7: Example belief network fragment

Since the arguments are probabilistic, there will be another argument (corresponding to the argument above) which states that if there is player injury and it rains, then there is an argument against the game with support $1 - 0.05$, that is, 0.95, yielding the following:

$$\langle bel \rangle (Injury \wedge Rain) \rightarrow \langle sup_{0.95} \rangle \neg Game$$

The rest of the entries of the CPT can be translated to arguments in a similar manner.

Continuing with our illustration of the network construction process from the considered set of five arguments, each set of arguments for a decision option is translated to a network fragment containing a random variable corresponding to the option as the child node with as many parent nodes as the number of arguments. The parent nodes are formed using the antecedents in the arguments. Therefore, the set of five arguments in our example for the two decision options *Cancelled* and *Delayed* are translated to the two fragments shown in Figure 9-8.

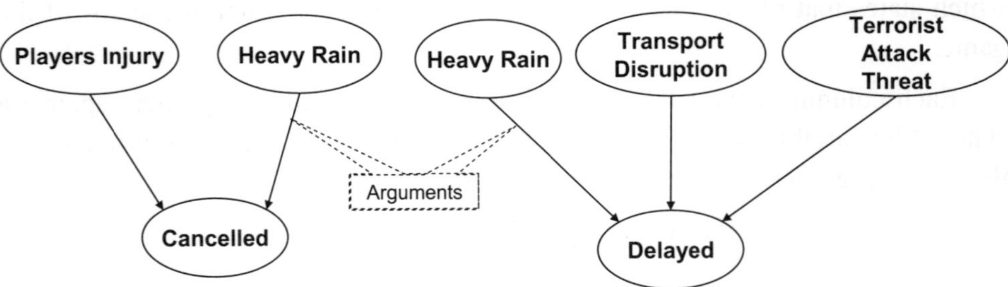

Figure 9-8: Belief network fragments created by converting arguments

Each of the nodes is binary with states *yes* and *no*. Since a particular decision option may occur in consequents of many arguments, their corresponding nodes (for example, *Heavy Rain*) may be shared among the network fragments. The common nodes are collapsed under such circumstances. Figure 9-9 shows the transformed belief network along with the CPTs.

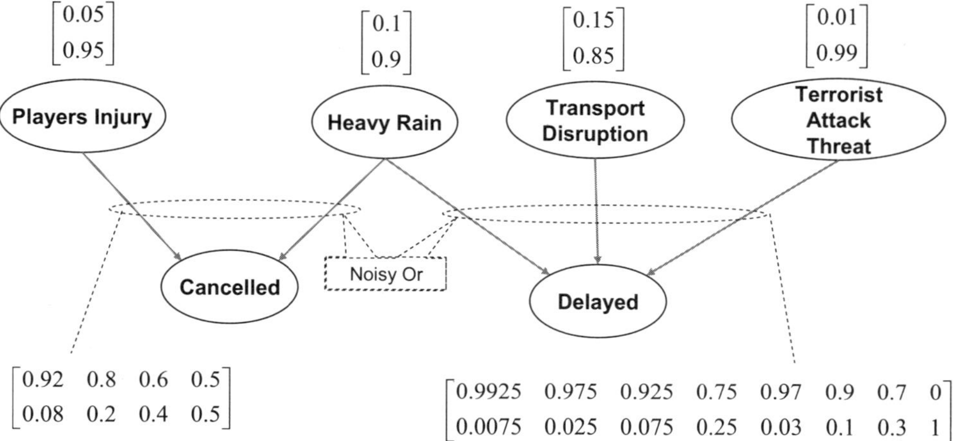

Figure 9-9: Belief network fragments along with CPTs

The entries for the CPTs are computed by applying the noisy-or technique (see the chapter on belief networks). For example, the following entry in the CPT comes directly from the argument:

$$p(Delayed = yes \mid Heavy\,Rain = yes, Transport\,Disruption = yes,$$
$$Terrorist\,Attack\,Threat = no) = 0.75 + 0.9 - 0.75 \times 0.9 = 0.975$$

where the arguments for the decision option *Delayed* provide the following probabilities:

$$p(Delayed = yes \mid Heavy\,Rain = yes) = 0.75$$
$$p(Delayed = yes \mid Transport\,Disruption = yes) = 0.9$$

The prior probabilities for the parent nodes are chosen arbitrarily. These prior probabilities can be assumed by default to be equally distributed.

Now that we have network fragments for arguments for and against individual decision options, we need to combine these arguments to rank the decision options. Since evidence for and against each decision option was combined individually, they may not necessarily sum to one, which is necessary as we assume that decision options are mutually exclusive. Therefore, we need to perform a normalization step. For this, we create a random variable with the states corresponding to the decision options for the task at hand. In the context of our example, we create a random variable called *Game* with three states: *on*, *cancelled*, and *delayed*. The variable has three parents corresponding to the three decision options. The decision options are ranked based on the aggregation of arguments for and against the decision options; the values of the CPT are determined accordingly. For example, if we have aggregated evidence for each of the three decision options *On*, *Cancelled*, and *Delayed*, then the probability distribution of the variable *Game* is evenly distributed as follows:

$$p(Game = on \,|\, On, Cancelled, Delayed) = 0.33$$

$$p(Game = cancelled \,|\, On, Cancelled, Delayed) = 0.33$$

$$p(Game = delayed \,|\, On, Cancelled, Delayed) = 0.33$$

Note that we have the same probability distribution as when we aggregated evidence against each of the three decision options (that is, *not On*, *not Cancelled*, and *not Delayed*). On the other hand, for example, if we have aggregated evidence for each of the two decision options *On* and *Cancelled*, and aggregated evidence against the decision option *Delayed*, then the probability distribution on the states of the variable *Game* is as follows:

$$p(Game = on \,|\, On, Cancelled, not\ Delayed) = 0.5$$

$$p(Game = cancelled \,|\, On, Cancelled, not\ Delayed) = 0.5$$

$$p(Game = delayed \,|\, On, Cancelled, not\ Delayed) = 0.0$$

Figure 9-10 illustrates the CPT for *Game*.

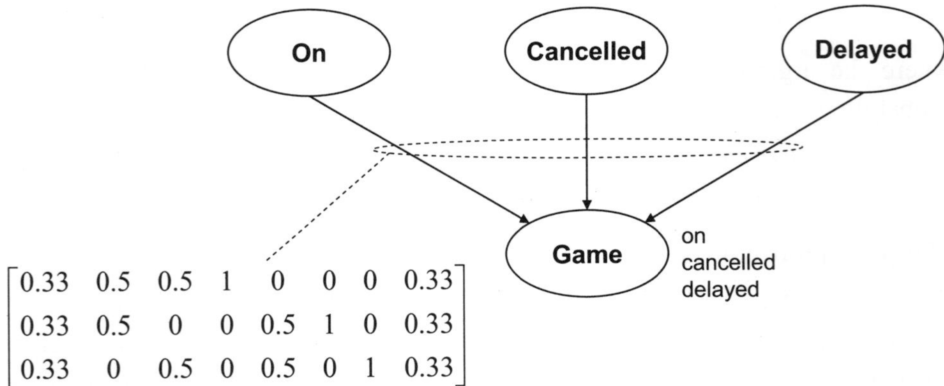

$$\begin{bmatrix} 0.33 & 0.5 & 0.5 & 1 & 0 & 0 & 0 & 0.33 \\ 0.33 & 0.5 & 0 & 0 & 0.5 & 1 & 0 & 0.33 \\ 0.33 & 0 & 0.5 & 0 & 0.5 & 0 & 1 & 0.33 \end{bmatrix}$$

Figure 9-10: Belief network fragment for aggregating arguments for and against decision options

Consider the case when two variables (not corresponding to decision options) are causally connected by two rules in the knowledge base as follows:

$$\langle bel \rangle Power\,Outage \rightarrow \langle sup_{0.95} \rangle Transport\,Disruption$$

$$\langle bel \rangle \neg Power\,Outage \rightarrow \langle sup_{0.9} \rangle \neg Transport\,Disruption$$

These two yield the following two conditional probabilities:

$$P(Transport\,Disruption \mid Power\,Outage) \;=\; 0.95$$

$$P(not\,Transport\,Disruption \mid not\,Power\,Outage) \;=\; 0.90$$

Thus, the rules together can be transformed to a network fragment along with a complete CPT as shown on the left hand side in Figure 9-11.

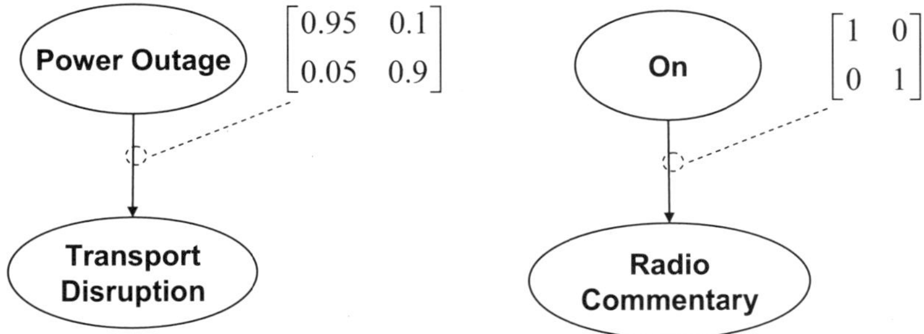

Figure 9-11: Belief network fragments corresponding to domain knowledge

Consider the case when a decision option causally influences another variable as follows:

$$\langle bel \rangle (On \rightarrow Radio\,Commentary)$$

The rule is an absolute one in the sense that there will always be radio commentary if the game is on. This rule is translated to the belief network fragment shown on the right hand side in Figure 9-11. In this fragment, evidence of radio commentary will increase the belief in that the game is on. The belief network formalism produces this additional diagnostic information though it is not explicitly specified in the above argument from which the fragment is constructed. We can assume some form of completed definition (see predicate completion in the logic programming chapter) of the variable *RadioCommentary* to correspond to this diagnostic reasoning.

Figure 9-12 shows the combined network for aggregating the arguments. Such a network has three blocks: the Argument Block, the Aggregation Block, and the Rule Block. The Argument Block (shown in Figure 9-9) is constructed out of the network fragments obtained by translating the arguments in the decision construct. The Aggregation Block (shown in Figure 9-10) implements the normalization step. The rest of the network, the Rule Block, is based on the rules from the underlying knowledge base.

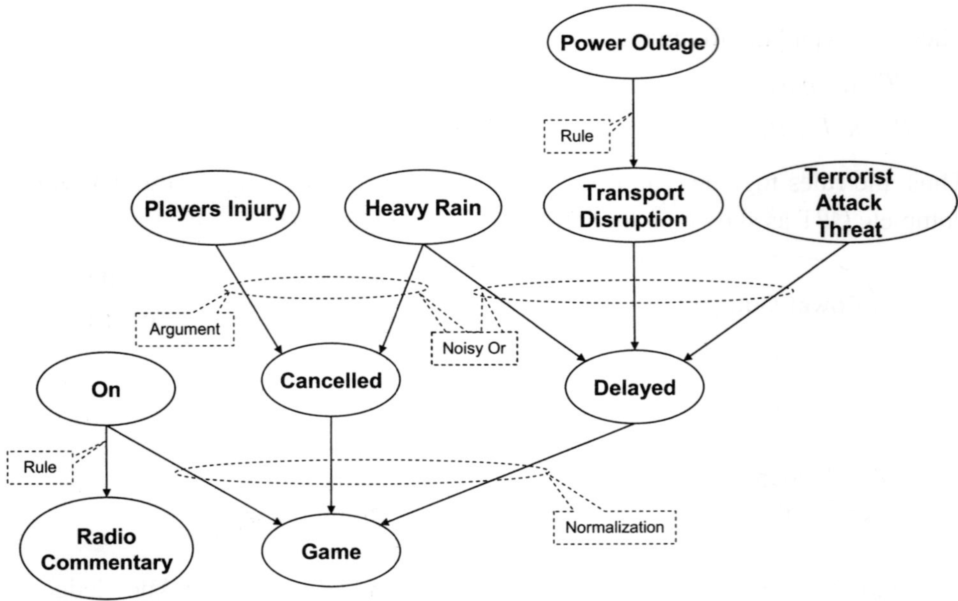

Figure 9-12: Combined belief network for argument aggregation

In the absence of any evidence, no arguments are generated and the *a priori* probabilities of the decision options are as follows:

$$P(Game \ = \ on) \ = \ 0.81$$

$$P(Game \ = \ cancelled) \ = \ 0.07$$

$$P(Game \ = \ delayed) \ = \ 0.12$$

No evidence in the network has been posted at this stage, not even for any prior beliefs on the variables. Now, given that there is transport disruption and heavy rain, the network ranks the decision options based on the following posterior probabilities:

$$P(Game \ = \ on \ | \ Transport \ Disruption, \ Heavy \ Rain) = \ 0.36$$

$$P(Game \ = \ cancelled \ | \ Transport \ Disruption, \ Heavy \ Rain) \ = \ 0.22$$

$$P(Game \ = \ delayed \ | \ Transport \ Disruption, \ Heavy \ Rain) \ = \ 0.42$$

The chances for game to be on have gone down dramatically. If we now receive information the rain has almost subsided (evidence (0.1, 0.9) for the variable *Heavy Rain*) then the network ranks the decision options as follows:

$$P(Game = on \mid Disruption, Heavy\,Rain) = 0.49$$

$$P(Game = cancelled \mid Disruption, Heavy\,Rain) = 0.02$$

$$P(Game = delayed \mid Disruption, Heavy\,Rain) = 0.49$$

The dilemma occurs between the two decision options *On* and *Delayed*. Based on the above probability distribution, the decision-maker may decide that the game will not be cancelled.

Though the propagation of evidence in the above aggregation process is always forward, the network in Figure 9-12 can also be used for speculative computations via diagnostic reasoning. For example, one can set the state of the variable *Game* as *cancelled* and then examine the likely state of the other variables if the game was to be cancelled.

9.7   Further Readings

More details of the Domino model, the knowledge representation language, and the underlying theoretical foundation via $\mathcal{LAB}$ can be found in (Fox and Das, 2000) and also in (Das, 2006). See (Das, 2007) for an extension of the Domino model.

References

AGENTS (1997-2001). *Proceedings of the International Conferences on Autonomous Agents* (1st – Marina del Rey, California; 2nd – Minneapolis, Minnesota; 3rd – Seattle, Washington; 4th – Barcelona, Spain; 5th – Montreal, Canada), ACM Press.

AAMAS (2002-2006). *Proceedings of the International Joint Conferences on Autonomous Agents and Multi-Agent Systems* (1st – Bologna, Italy; 2nd – Melbourne, Australia; 3rd – New York, NY; 4th – The Netherlands; 5th – Hakodate, Japan), ACM Press.

Alechina, N. (1994). "Logics with probabilistic operators." *ILLC Research Report CT-94-11*, University of Amsterdam.

Apt, K. R. and Van Emden, M. H. (1982). "Contributions to the theory of logic programming." *Journal of the Association for Computing Machinery*, **29**, pp. 841–862.

Bacchus, F. (1990). *Representing and Reasoning with Probabilistic Knowledge*. MIT Press.

Barbuti, R. and Martelli, M. (1986). "Completeness of the SLDNF-resolution for a class of logic programs." In *Proceedings of the 3rd International Conference on Logic Programming* (London, July), pp. 600–614.

Becker, O. (1930). "Zur Logik der Modalitaten." *Jahrbuch fur Philosophie und Phanomenologische Forschung*, Vol. 11, pp. 497-548.

Blackburn, P., De Rijke, M., and Venema, Y. (2001). *Modal Logic*. Cambridge Tracts in Theoretical Computer Science, No 53. Cambridge, England: Cambridge University Press.

Blackburn, P., Wolter, F., and van Benthem, J. (eds.) (2006). *Handbook of Modal Logic*, Elsevier Science & Technology Books.

Bradley, R. and Swartz N. (1979). *Possible Worlds: An Introduction to Logic and Its Philosophy*, Hackett, Indianapolis, IN.

Bratko, I. (2000). *Prolog Programming for Artificial Intelligence*, 3rd Edition. Wokingham, England: Addison-Wesley.

Cavedon, L. and Lloyd, J. W. (1989). "A Completeness theorem for SLDNF-resolution." *Journal of Logic Programming*, **7**, pp. 177–191.

Chang, C. L. and Lee, R. C. T. (1973). *Symbolic Logic and Mechanical Theorem Proving*. New York: Academic Press.

Chatalic, P. and Froidevaux, C. (1992). "Lattice-based graded logic: a multimodal approach." *Proceedings of the Conference on Uncertainty in Artificial Intelligence*, pp. 33-40.

Chellas, B. F. (1980). *Modal Logic: An Introduction. Cambridge.* Cambridge, England: Cambridge University Press.

Chung, K. L. (2000). *A Course in Probability Theory*, Academic Press.

Clark, K. L. (1978). "Negation as failure." In *Logic and Databases*, H. Gallaire and J. Minker (eds.), New York: Plenum Press, pp. 293–322.

Clocksin, W. F. and Mellish, C. S. (2003). *Programming in Prolog*, 5th edition. Berlin, Germany: Springer-Verlag.

Cohen, P. R. and Levesque, H. J. (1990). "Intention is choice with commitment." *Artificial Intelligence*, **42**, 13–61.

Cook, S. A. (1971). "The complexity of theorem proving procedures." *Proceedings of the Third Annual ACM Symposium on the Theory of Computing*, ACM, New York, 151–158

Cooper, G. (1990), "The computational complexity of probabilistic inference using Bayesian belief networks." *Artificial Intelligence*, **42**.

Copi, I. M. (1979). *Symbolic Logic*. New York: Macmillan.

Dantsin, E., Eiter, T., Gottlob, G., and Voronkov, A. (2001). "Complexity and Expressive Power of Logic Programming." *ACM Computing Surveys*, **33**(3), pp. 374–425.

Das, S. K. (1992). *Deductive Databases and Logic Programming*. Wokingham, England: Addison-Wesley.

Das, S., Fox, J., Elsdon, D., and Hammond, P. (1997). "A flexible architecture for autonomous agents", *Journal of Experimental and Theoretical Artificial Intelligence*, **9**(4): 407–440

Das, S. and Grecu, D. (2000). "COGENT: Cognitive agent to amplify human perception and cognition." *Proceedings of the 4th International Conference On Autonomous Agents*, Barcelona, June.

Das, S., Shuster, K., and Wu, C. (2002). "ACQUIRE: Agent-based Complex QUery and Information Retrieval Engine," Proceedings of the 1st International Joint Conference on Autonomous Agents and Multi-Agent Systems, Bologna, Italy.

Das, S. (2005). "Symbolic Argumentation for Decision Making under Uncertainty," *Proceedings of the 8th International Conference on Information Fusion*, Philadelphia, PA.

Das, S. (2007). "Envelope of Human Cognition for Battlefield Information Processing Agents," *Proceedings of the 10th International Conference on Information Fusion*, Quebec, Canada.

Davis, M. and Putnam, H. (1960). "A computing procedure for quantification theory." *Journal of the Association for Computing Machinery*, **7**, 201–215.

Deo, N. (1974). Graph Theory with Applications to Engineering and Computer Science, Prentice-Hall.

Dubois, D. and Prade, H. (1988). *Possibility Theory*. Plenum Press, New York.

Fagin, R. 1988. "Belief, Awareness, and Limited Reasoning." *Artificial Intelligence*, **34**(1):39–76.

Fagin, R., Halpern, J. Y., and Megiddo, N. (1990). "A logic for reasoning about probabilities." *Information and Computation*, **87**, Nos. 1 & 2, pp. 78–128.

Fagin, R., Halpern, J. Y., Moses Y. and Vardi, M. Y. (1995). *Reasoning about Knowledge*. Cambridge: MIT Press.

Fattorosi-Barnaba, M. and Amati, G. (1987). "Modal operators with probabilistic interpretations I." *Studia Logica*, **46**, pp. 383–393.

Feller, W. (1968). *An Introduction to Probability Theory and Its Applications*, Vol. 1, John Wiley & Sons.

Fitting, M. and Mendelsohn, R. L. (1998). *First-Order Modal Logic*, Kluwer Academic Publishers.

Fox, J. and Das, S. K. (2000). *Safe and Sound: Artificial Intelligence in Hazardous Applications*. AAAI-MIT Press, June.

Fox, J., Krause, P, and Ambler, S. (1992). "Arguments, contradictions, and practical reasoning." *Proceedings of the 10th European Conference on Artificial Intelligence*, Vienna, August, pp. 623–626.

Gallaire, H., Minker, J. and Nicolas, J. M. (1984). "Logic and databases: a deductive approach," *ACM Computing Surveys*, **16**(2), 153–185.

Gallier, J. H. (2003). Logic for Computer Science, New York: John Wiley & Sons. Revised On-Line Version (2003): http://www.cis.upenn.edu/~jean/gbooks/logic.html.

Garson, J. W. (1984). Quantification in modal logic. In *Handbook of Philosophical Logic*, D. Gabbay and F. Guenthner (eds.), **2**, 249–307, Reidel Publication Co.

Ginsberg, M. L. (ed.) (1987). *Readings in Nonmonotonic Reasoning*, Los Altos, CA: Morgan Kaufmann.

Goldblatt, R. (2001). "Mathematical modal logic: a view of its evolution." In *A History of Mathematical Logic*, van Dalen et al. (Eds).

Halpern, J. Y. and Moses, Y. O. (1985). "A guide to the modal logics of knowledge and belief." *Proc. of the 9th International Joint Conference on Artificial Intelligence*, 480–490.

Halpern, J. and Rabin, M. (1987). "A logic to reason about likelihood." *Artificial Intelligence*, **32**, 379–405.

Halpern, J. Y. (1995a). "Reasoning about knowledge: a survey." *Handbook of Logic in Artificial Intelligence and Logic Programming*, **4**, D. Gabbay, et al., eds., Oxford University Press, 1–34.

Halpern, J. Y. (1995b). "The effect of bounding the number of primitive propositions and the depth of nesting on the complexity of modal logic." *Artificial Intelligence*, **75**.

Heckerman, D. E. and Shortliffe. E. H. (1991). "From Certainty Factors to Belief Networks." *Artificial Intelligence in Medicine*, **4**, pp. 35–52.

Hintikka, J. (1962). *Knowledge and Belief: An Introduction to the Logic of the Two Notions*. N.Y.: Cornell University Press.

Howard, R. A. and Matheson, J. E. (1984). "Influence diagrams." In *Readings on the Principles and Applications of Decision Analysis*, 721–762. Strategic Decisions Group.

Huang, C. and Darwiche, A. (1996). "Inference in belief networks: A procedural guide." *International Journal of Approximate Reasoning volume*,**15** (3), 225-263.

Hughes, G. E., and M. J. Cresswell. (1996). *A New Introduction to Modal Logic*. London: Routledge.

Jackson, P. (1998). *Introduction to Expert Systems*. Addison-Wesley.

Jensen, F. V., Lauritzen, S. L., and Olesen. K. G. (1990). ""Bayesian updating in causal probabilistic networks by Local Computations." Computational Statistics, Q. 4, 269-282.

Jensen, F., Jensen, F. V. and Dittmer, S. L. (1994). "From influence diagrams to junction trees." *Proceedings of the 10th Conference on Uncertainty in Artificial Intelligence (UAI)*.

Jensen, F.V. (1996). *An Introduction to Bayesian Networks*. Springer-Verlag.

Jensen, F. V. (2002). *Bayesian Networks and Decision Graphs*, Springer.

Karp, R. (1972) "Reducibility among Combinatorial Problems." "Complexity of Computer Computations", ed. R. Miller and J. Thatcher (eds), 85–103, in Plenum Press, New York.

Knight, K. (1989). "Unification: A multidisciplinary survey." *ACM Computing Surveys*, **21**(1), pp. 93–124.

Kowalski, R. A. (1979a). *Logic for Problem Solving*, New York: North-Holland.

Kowalski, R. A. (1979b). "Algorithm = Logic + Control." *Communications of the ACM*, **22**(7), 424–435.

Kowalski, R. A. and Kuehner, D. (1971). "Linear resolution with selection function." *Artificial Intelligence*, **2**, 227–260.

Kripke, S. A. (1963). Semantical Analysis of Modal Logic I: Normal Modal Propositional Calculi. *Zeitschrift fur Mathematische Logik und Grundladen der Mathematik (ZMLGM)*, **9**, 67–96.

Ladner, R. (1977). "The computational complexity of provability in systems of modal propositional logic." *SIAM Journal of Computing*, **6**, 467–480.

Lauritzen, S. L. and D. J. Spiegelhalter (1988). "Local computations with probabilities on graphical structures and their applications to expert systems." *Journal of the Royal Statistical Society*, B **50**(2), pp.154–227.

Lemmon, E. J. (1960). Quantified S4 and the Barcan formula. *Journal of Symbolic Logic*, **24**, 391–392.

Lemmon, E. J. (1977). *An Introduction to Modal Logic*. American Philosophical Quarterly, Monograph No. 11.

Lewis, C. I. (1918). *A Survey of Symbolic Logic*. University of California Press.

Lewis, C. I. and Langford, C. H. (1932). *Symbolic Logic*. Dover Publications, Inc., New York.

Lloyd, J. W. (1987). *Foundations of Logic Programming*, 2nd Edition, Berlin, Germany: Springer Verlag.

Lukasiewicz, J. (1951). *Aristotle's Syllogistic*, Oxford University Press.

McCarthy, J. (1980). "Circumscription: A form of non-monotonic reasoning." *Artificial Intelligence*, 13, 27–39.

McKinsey, J. C. C. (1934). "A reduction in the number of postulates for C. I. Lewis' system of strict implication." *Bulletin of American Mathematical Society*, **40**, pp. 425-427.

McKinsey, J. C. C. (1940). "Proof that there are infinitely many modalities in Lewis's system S2." *Journal of Symbolic Logic*, **5**(3), pp. 110-112.

McKinsey, J. C. C. (1941). "A solution of the decision problem for the Lewis systems S2 and S4, with an application to topology." *Journal of Symbolic Logic*, **6**, pp. 117-134.

Mendelson, E. (1987). Introduction to Mathematical Logic. California, USA: Wadsworth & Brooks/Cole Advanced Books and Software.

Mitchell, T. (1997). *Machine Learning*. McGraw-Hill, 1997.

Meyer, J.-J.Ch and Hoek, W. van der (1995). *Epistemic Logic for AI and Computer Science*. Cambridge Tracts in Theoretical Computer Science 41. Cambridge: Cambridge University Press.

Meyer, J.-J.Ch and Hoek, W. van der (1995). *Epistemic Logic for AI and Computer Science*. Cambridge Tracts in Theoretical Computer Science 41. Cambridge: Cambridge University Press.

O'Keefe, R. A. (1990). *Craft of Prolog*. Cambridge, MA: MIT Press.

Ohlbach, H. J. (1988). "A resolution calculus for modal logics." Proceedings of the 9th International Conference on Automated Deduction, Argonne, USA.

Papadimitriou, C. H. (1993). *Computational Complexity*. Addison Wesley, Reading, MA, USA.

Parry, W. T. (1939). "Modalities in the *survey* system of strict implication." *Journal of Symbolic Logic*, **4**(4), pp. 137-154.

Pearl, J. (1986). "A constraint-propagation approach to probabilistic reasoning." *Proceedings of the Conference on Uncertainty in Artificial Intelligence*, pp. 357-369, Amsterdam: North-Holland.

Pearl, J. (1988). *Probabilistic Reasoning in Intelligent Systems: Networks of Plausible Inference*. San Mateo, CA, Morgan Kaufmann.

Quine, W. V. O. (1980). *Elementary Logic*. Cambridge, MA: Harvard University Press.

Rao, A. S. and Georgeff, M. P. (1991). "Modeling rational agents within a BDI-architecture." *Proceedings of Knowledge Representation and Reasoning*, pp. 473-484. Morgan Kaufmann Publishers: San Mateo, CA.

Rao, A. S. and Georgeff, M. P. (1991). "Modelling rational agents within a BDI-architecture." *Proceedings of the Knowledge Representation and Reasoning*, 473–484.

Rasmussen, J. (1983). "Skills, Rules and Knowledge: Signals, Signs and Symbolism, and Other Distinctions in Human Performance Models." *IEEE Transactions on Systems, Man, and Cybernetics*, **12**, 257–266.

Reiter, R. (1978b). "On closed world databases." *Logic and Databases*, H. Gallaire and J. Minker (eds.), New York: Plenum Press, 55–76.

Reike, R. D. and Sillars, M. O. (2001). *Argumentation and Critical Decision Making*. Longman, New York.

Robinson, J. A. (1965). "A machine-oriented logic based on the resolution principle." *Journal of the Association for Computing Machinery*, **12**, 23–41.

Russell, S. J. and Norvig, P. (2002). *Artificial Intelligence: A Modern Approach*. Prentice Hall.

Shachter, R. D. (1986). "Evaluating influence diagrams." *Operations Research*, **34**, 871–882.

Shafer, G. (1976). *A Mathematical Theory of Evidence*. Princeton, NJ, Princeton University Press.

Shenoy, P. P. (1992). "Valuation-based systems for Bayesian decision analysis." *Operations Research*, **40**(3), 463-484.

Smets, P. (1991). "Varieties of ignorance and the need for well-founded theories," *Information Sciences*, 57-58: 135-144.

Sterling, L. and Shapiro, E. (1994). *The Art of Prolog*. Cambridge, MA: MIT Press.

Stockmeyer, L. J. and Meyer, A.R. (1973). "Word problems requiring exponential time." *Proceedings of the 5th ACM Symposium on Theory of Computing*, 1–9.

Stephanou, H. E. and Lu, S.-Y. (1988). "Measuring consensus effectiveness by a generalized entropy criterion." *IEEE Transactions on Pattern Analysis and Machine Intelligence*, **10**(4), pp. 544-554.

Stoll, R. R. (1963). *Set Theory and Logic*, New York: W. H. Freeman and Company. Reprint edition (1979) by Dover Publications.

Toulmin, S. 1956. *The Uses of Argument*. Cambridge, U.K.: Cambridge University Press.

Ullman, J. D. (1984). *Principles of Database Systems*, 2nd edition. Maryland, USA: Computer Science Press.

Van Emden, M. H. and Kowalski, R. A. (1976). "The semantics of predicate logic as a programming language." *Journal of the Association for Computing Machinery*, **23**(4), pp. 733–742.

Vardi, M. Y. (1982). "The complexity of relational query languages." *Proceedings of the 14th Annual ACM Symposium on Theory of Computing, San Francisco*, California, pp. 137–146.

Von Wright, G., H. (1951). *An Essay on Modal Logic*. Amsterdam: North-Holland Publishing Company.

Voorbraak, F. (1996). Epistemic Logic and Uncertainty. Workshop on Quantitative and Symbolic Approaches to Uncertainty, *European Summer School on Logic, Language, and Information*, August 12–23, Prague.

Wainer, J. (1994). "Yet Another Semantics of Goals and Goal Priorities." *Proceedings of the Eleventh European Conference on Artificial Intelligence, Amsterdam*, The Netherlands, pp. 269-273.

Wen W. X. (1990). "Optimal decomposition of belief networks." *Proceedings of the 6th Conference on Uncertainty in Artificial Intelligence*, 245–256. Morgan Kaufmann.

Whitehead, A. N. and Russell, B. (1925–1927). *Principia Mathematica*, 1, 2 & 3, 2nd Edition. Cambridge, England: Cambridge University Press.

Wilson, R. (1996). *Introduction to Graph Theory*, Addison-Wesley.

Yager, R. R., Kacprzyk, J., and Fedrizzi, M. (Editors) (1994). *Advances in the Dempster-Shafer Theory of Evidence*, John Wiley & Sons.

Yager, R. R. (2004). "On the determination of strength of belief for decision support under uncertainty – Part II: fusing strengths of belief." *Fuzzy Sets and Systems*, **142**, pp. 129–142.

Zadeh, L. A. (1965). "Fuzzy Sets." *Information and Control*, 8: 338-353.

Zadeh, L. A. (1978). "Fuzzy Sets as the Basis for a Theory of Possibility." *Fuzzy Sets and Systems*, **1**:3-28.

Index